Studies in Computational Intelligence

Volume 1235

The series "Studies in Computational Intelligence" (SCI) publishes new developments and advances in the various areas of computational intelligence—quickly and with a high quality. The intent is to cover the theory, applications, and design methods of computational intelligence, as embedded in the fields of engineering, computer science, physics and life sciences, as well as the methodologies behind them. The series contains monographs, lecture notes and edited volumes in computational intelligence spanning the areas of neural networks, connectionist systems, genetic algorithms, evolutionary computation, artificial intelligence, cellular automata, self-organizing systems, soft computing, fuzzy systems, and hybrid intelligent systems. Of particular value to both the contributors and the readership are the short publication timeframe and the world-wide distribution, which enable both wide and rapid dissemination of research output.

Indexed by SCOPUS, DBLP, WTI AG (Switzerland), zbMATH, SCImago.

All books published in the series are submitted for consideration in Web of Science.

Laura Sparacino

Spectral Information Dynamics in Network Neuroscience and Physiology

Springer

Laura Sparacino
Department of Engineering
University of Palermo
Palermo, Italy

ISSN 1860-949X ISSN 1860-9503 (electronic)
Studies in Computational Intelligence
ISBN 978-3-032-05415-9 ISBN 978-3-032-05416-6 (eBook)
https://doi.org/10.1007/978-3-032-05416-6

This Springer imprint is published by the registered company Springer Nature Switzerland AG
The registered company address is: Gewerbestrasse 11, 6330 Cham, Switzerland

Preface

Recent advances in signal processing and information theory are boosting the development of new approaches for the data-driven modeling of complex network systems. In the fields of Network Physiology and Network Neuroscience, where the signals of interest are rich of oscillatory content, the spectral representation of network systems is essential to ascribe interactions to specific oscillations with physiological meaning. This book introduces a coherent framework integrating several information dynamics approaches to quantify node-specific, pairwise, and high-order interactions in network systems. A hierarchical organization of interactions of different order is established using measures of information rate to quantify the dynamics of each individual node of the network, the links between pairs of nodes, and the redundant/synergistic hyperlinks in groups of nodes. All measures are formulated in the time domain and then expanded to the spectral domain to obtain frequency-specific information in the context of Gaussian data characterized by linear parametric models. The framework is first illustrated using simulation examples where the properties of the measures are displayed in benchmark simulated network systems. Then, it is applied to several representative datasets of multivariate time series in the context of Network Neuroscience and Network Physiology. The utilization of high-order measures of information rate with spectral meaning has been proven successful to highlight the respiratory-driven redundant nature of cardiovascular, cardiorespiratory, and cerebrovascular interactions, as well as the overall prevalence of redundancy for high-order brain interactions together with the emergence of synergistic circuits not retrievable from a pairwise analysis.

Palermo, Italy　　　　Laura Sparacino

Acknowledgements The research outlined in this work was carried out in the Biosignals and Information Theory Laboratory (BITLAB) headed by Prof. Luca Faes, in the *Department of Engineering, University of Palermo, Palermo (Italy).* Various previous established and newborn research collaborations were fundamental for carrying out the research activity, since they either provided me access to historical datasets, or contributed with extensive feedback on the physiological interpretation of the results. Moreover, the time I spent in foreign academic institutions really opened my mind and provided new insights to think about my commitment to research.

Part of the research outlined here was conducted in strong collaboration with the *Department of Physiology, Comenius University, Jessenius Faculty of Medicine, Martin (Slovakia)*, headed by Prof. Michal Javorka, and the *Laboratorio di Modellistica di Sistemi Complessi, IRCCS Policlinico San Donato, San Donato Milanese, Milano (Italy)*, headed by Prof. Alberto Porta. Professor Javorka dispensed his unique contribution with the physiological interpretation of the results and gave important feedback during the brainstorming sessions aimed at finding the best way to analyze physiological data. Collaboration with the bioengineers of IRCCS Policlinico San Donato provided interesting research insights and a large amount of clinical data acquired in outpatient controlled conditions, which was then exploited for applying our measures of pairwise and high-order connectivity in complex networks of multiple physiological variables. Moreover, it is worth mentioning the role of clinicians and researchers from *IRCCS ISMETT—Istituto Mediterraneo per i Trapianti, Palermo (Italy)*, who gave us the possibility to analyze and interpret clinical data of significant relevance in the context of brain functional connectivity in diseased patients assessed through fMRI.

Both the exploited signal processing pipelines and the practical implementation tools of the proposed measures were thoroughly reviewed by numerous colleagues of the BITLAB group, whom I thank for their remarkable contribution.

Last but not least, I dedicate this milestone to all those people who have influenced the few past intense years and my whole life with their presence, their precious words, and their love. You are all part of my growth as a person and a researcher.

Competing Interests The author has no competing interests to declare that are relevant to the content of this manuscript.

Contents

1 Introduction 1
1.1 Problem Statement 1
1.2 The Network Representation of Complex Physiological Systems 2
1.3 Aims of the Book 3
1.4 Outline of the Book 6
References 7

2 Linear Modelling of Stochastic Interactions 11
2.1 Introduction to Static and Dynamic Stochastic Interactions 11
2.1.1 Basic Concepts of Probability 11
2.1.2 Network Analysis Definitions 13
2.2 Static Models of Random Variables 17
2.3 Dynamic Models of Random Processes in the Time Domain 17
2.3.1 The Single Unit: AR Models 18
2.3.2 Interactions Between Two Nodes: ARX Models 19
2.3.3 The Whole Network: VAR Models 20
2.3.4 Restricted Models in the Time Domain 21
2.3.5 Model Identification 22
2.4 Dynamic Models of Random Processes in the Frequency Domain 26
2.4.1 Restricted Models in the Frequency Domain 27
2.5 Summary 29
References 30

3 Static Networks of Random Variables 33
3.1 Single-Node and Pairwise Connectivity 33
3.1.1 Linear Parametric Formulation 35
3.2 High-Order Interactions 36
3.2.1 Linear Parametric Formulation 45

3.3 Simulation Examples ... 45
3.3.1 Pairwise and High-Order Functional Dependencies Assessed via Mutual Information and O-Information ... 45
3.3.2 High-Order Links in a Simulated Static System: The B-Index Reveals Redundant and Synergistic Patterns of Link-Specific Interactions ... 49
3.4 Summary ... 51
References ... 52

4 Dynamic Networks of Random Processes ... 55
4.1 Behavior of a Single Network Unit: Self-predictable Dynamics ... 56
4.1.1 Time Domain ... 56
4.1.2 Frequency Domain ... 58
4.1.3 Simulation Example: Spectral Decomposition of the Linear Self-predictability of a Simulated AR Process ... 62
4.2 Beyond the Single Network Unit: The Interactions Between Two Nodes ... 64
4.2.1 Time Domain ... 64
4.2.2 Frequency Domain ... 69
4.2.3 Simulation Examples: Pairwise Measures of Causality, Isolation and Autonomy ... 76
4.3 Beyond Pairwise Interactions: High-Order Interactions of Multiple Network Units ... 85
4.3.1 Time Domain ... 85
4.3.2 Frequency Domain ... 93
4.3.3 Simulation Examples ... 98
4.4 Summary ... 112
References ... 114

5 Applications to Physiological Networks ... 119
5.1 Arterial Compliance Short-Term Variability Analysis ... 121
5.2 The Rate of Mutual Information Exchanged in Networks of Cardiovascular, Respiratory and Cerebrovascular Variables ... 127
5.3 A Portable Multisensor System to Assess Cardiorespiratory Interactions Through Photoplethysmography ... 129
5.4 Linear Parametric Assessment of Cardiorespiratory Interactions During Spontaneous and Controlled Breathing ... 133
5.5 Spectral Decomposition of Cerebrovascular and Cardiovascular Interactions in Patho-Physiological States ... 136

5.6 Granger Causality, Isolation and Autonomy in Closed-Loop Systems of Cerebrovascular Variables 145
5.7 Cardiovascular, Cardiorespiratory and Cerebrovascular High-Order Interactions Assessed Through the O-Information Rate 151
5.8 Gradients of O-Information in Multi-organ Networks 153
5.9 Cardiovascular and Respiratory Interactions During Paced Breathing: An Approach to Decompose the O-Information Rate 156
5.10 Partial Information Rate Decomposition in Physiological Networks 159
5.11 High-Order Links in Cardiovascular and Respiratory Networks 162
5.12 Summary 167
References 168

6 Applications to Brain Networks 175
6.1 Static Networks of Random Variables: Applications to fMRI Data 176
6.1.1 Pairwise and High-Order Brain Functional Connectivity Signatures on a Single-Subject Basis 178
6.1.2 Gradients of O-Information Highlight Synergistic and Redundant Informational Circuits 186
6.2 Dynamic Networks of Random Processes: Applications to EEG Data 189
6.2.1 Human Brain Activity During Motor Execution Assessed via the OIR Framework 190
6.2.2 Neural Interactions From ECoG Signals in the Anesthetized Macaque Monkey 192
6.2.3 Rehabilitation Modulates High-Order Interactions Among Large-Scale Brain Networks in Subacute Stroke 195
6.2.4 Brain Interactions at Different Orders: Exemplary Single-Subject Analysis 203
6.2.5 High-Order Behaviours Uncover the Hierarchical Organization of Interactions in the Motor Network of the Human Brain 205
6.3 Summary 214
References 215

7 Conclusions and Future Directions 225
7.1 A Picture of the Complex Interactions in Physiological and Brain Networks of Multiple Nodes 225
7.2 Simultaneous Evaluation of Information-Theoretic Measures at Different Hierarchical Levels 227

7.3 Future Perspectives 228
References 229

Appendix A 231

Appendix B 241

Abbreviations

ABP	Arterial Blood Pressure
AC	Arterial Compliance
AIC	Akaike Information Criterion
ANES	Anesthesia
ANS	Autonomic Nervous System
AP	Arterial Pressure
AR	Auto-Regressive
ARX	Auto- and Cross-Regressive
ASR	Artifacts Subspace Reconstruction
B	B-Index
BA	Bivariate Analysis
BIC	Bayesian Information Criterion
BOLD	Blood Oxygen Level-Dependent
BPV	Blood Pressure Variability
CA	Cerebral Autoregulation
CB	Cerebellar
CBF	Cerebral Blood Flow
CBFV	Cerebral Blood Flow Velocity
CE	Conditional Entropy
CI	Confidence Interval
ciS	conditional information Shared
CMI	Conditional Mutual Information
CMIR	Conditional Mutual Information Rate
CO	Cardiac Output
Coh	Coherence
DA	Dorsal Attention
DAP	Diastolic Arterial Pressure
DARE	Discrete Algebraic Riccati Equation
DC	Directed Coherence
DMN	Default Mode Network
DTC	Dual Total Correlation

ECG	Electrocardiography
ECoG	Electrocorticography
ECN	Executive Control Network
EEG	Electroencephalography
eLORETA	exact LOw REsolution Tomography Approach
EMD	Empirical Mode Decomposition
EN	Entropy
EPI	Echo-Planar Imaging
ER	Entropy Rate
FAC	Functional Ambulation Category
FMA	Fugl-Meyer Assessment
fMRI	functional Magnetic Resonance Imaging
FP	Frontoparietal
FT	Fourier Transform
GA	Granger Autonomy
GC	Granger Causality
GI	Granger Isolation
HE	Hepatic Encephalopathy
HF	High Frequency
HG	Handgrip
hHON	hierarchical High-Order Network
hHONA	hierarchical High-Order Network Analysis
HOI	High-Order Interaction
HONA	High-Order Network Analysis
HP	Heart Period
HRV	Heart Rate Variability
iAAFT	iterative Amplitude Adjusted Fourier Transform
IAF	Individual Alpha Frequency
ICA	Independent Component Analysis
ICG	Impedance Cardiography
II	Interaction Information
IIR	Interaction Information Rate
i.i.d	independent identically distributed
IID	Interaction Information Decomposition
iS	information Shared
IS	Information Storage
IC	Instantaneous Causality
IT	Instantaneous Transfer
LF	Low Frequency
LSP	Linear Self-Predictability
LVET	Left Ventricular Ejection Time
M1	primary motor cortex
MAP	Mean Arterial Pressure
MCBFV	Mean Cerebral Blood Flow Velocity
MI	Mutual Information

MIR	Mutual Information Rate
MMI	Minimum Mutual Information
MN	Motor Network
MNI	Montreal Neurological Institute
MR	Magnetic Resonance
MRI	Magnetic Resonance Imaging
niRS	net information Rate Shared
niS	net information Shared
OI	O-Information
OIR	O-Information Rate
OLS	Ordinary Least-Squares
PCC	Posterior Cingulate Cortex
PFC	Prefrontal Cortex
PI	Partial Information
PID	Partial Information Decomposition
PIRD	Partial Information Rate Decomposition
PMC	Premotor Cortex
PPG	Photoplethysmography
PPI	P-P Interval
PSD	Power Spectral Density
RESP	Respiration
REST	Resting State
rest-fMRI	Resting-state functional Magnetic Resonance Imaging
RIGHT	Right fist
ROI	Region Of Interest
RRI	R-R Interval
RS	Random Shuffling
RSA	Respiratory Sinus Arrhythmia
RSN	Resting State Network
SAL	Salience
SAP	Systolic Arterial Pressure
SM	Sensory Motor
SMA	Supplementary Motor Area
SS	State-Space
SV	Stroke Volume
TC	Total Correlation
TD	Total Dependence
TE	Transfer Entropy
TF	Transfer Function
TILT	Head-up Tilt
TPR	Total Peripheral Resistance

UA	Univariate Analysis
VAR	Vector Auto-regressive
VLF	Very Low Frequency
VS	Visual network
WSS	Wide-Sense Stationarity
YW	Yule-Walker

Chapter 1
Introduction

1.1 Problem Statement

In the emerging research fields of Network Physiology [1, 2] and Network Neuroscience [3], the human organism is modeled as a network in which nodes correspond to organ systems or brain units, and edges map the functional dependencies between pairs of physiological systems or brain areas [4, 5]. The possibility to assess node activity and interactions between nodes through measures defined in different domains has already been proven useful to investigate physiological states in healthy and pathological conditions (see, e.g., [6–12]). However, since physiological systems operate across different hierarchical levels and time scales, focusing only on individual dynamics and pairwise interactions and limiting the analysis to a time-domain description is often insufficient to provide a complete description of the human organism as a complex network [13]. A main ongoing research trend is indeed the investigation of how multi-organ regulation emerges from *high-order interactions*, i.e., interactions involving more than two network nodes. Moreover, the need of performing the analysis of individual, pairwise and high-order interactions with specific focus on the distinct oscillations that characterize physiological and brain activities is also emerging. In this context, this book focuses on describing new information-theoretic and spectral measures for the assessment of complex high-order interactions between multiple signals originating from different body district. A comprehensive picture on the topic will be given below.

L. Sparacino, *Spectral Information Dynamics in Network Neuroscience and Physiology*, Studies in Computational Intelligence 1235,
https://doi.org/10.1007/978-3-032-05416-6_1

1.2 The Network Representation of Complex Physiological Systems

The increasing availability of large-scale and fine-grained recordings of biomedical signals is opening the way to the network representation of complex physiological systems. For instance, in neuroscience the organizational principles of functional segregation and integration in the human brain are typically studied through the theoretical and empirical tools of *Network Neuroscience* [3], while in integrative physiology the reductionist approach of studying in isolation the function of an organ system is nowadays complemented by the holistic investigation of collective interactions among diverse organ systems performed in the field of *Network Physiology* [1]. Network Neuroscience and Network Physiology are sub-fields of *Network Science*, a large interdisciplinary area that develops theoretical and practical techniques to improve the understanding of natural and man-made networks with hierarchical structures [14]. Data-driven methods for network inference play a key role in Network Science, being devised to build a network model out of a set of observed multivariate time series. Such a model is typically encoded by a graph where the observed dynamic system (e.g., the brain or the human organism) is represented by distinct nodes (e.g., neural units or organ systems) connected by edges mapping functional dependencies (e.g., brain connectivity or cardiovascular interactions) [4, 5]. Beyond this basic description, the need to deepen the exploration of real-world systems has led network scientists to enrich the way to represent the system properties captured by a network model [15]. Several augmented network descriptions have been proposed exploiting, for instance, active nodes encoding self-dependencies within an individual process, and directed and/or weighted edges depicting cause-effect relations and quantifying the intensity of pairwise interactions. These representations are well accommodated in functional brain and physiological networks through the definition of measures to assess complexity or regularity of individual time series [16, 17], or coupling and causality between pairs of time series [18, 19].

In spite of the usefulness of network models encoded by graphs, the representation with self-effects and pairwise (*dyadic*) interactions is often insufficient to provide a complete description of a complex system. It is now firmly acknowledged that many real-world systems exhibit high-order (*polyadic*) interactions (HOIs), i.e., interactions involving more than two network nodes [13]; in these systems the network behavior is integrated at different hierarchical levels and time scales. This occurs in Network Neuroscience and Network Physiology, where it is important to distinguish between brain regions or organ systems that interact as a pair, or as a part of a more complex structure, to produce the observed dynamics. For instance, brain dynamics display mesoscopic or macroscopic behaviors requiring multiple-unit interactions to be predicted accurately (see, e.g., [20]), and cardiovascular interactions may arise autonomously from self-sustained mechanisms or as a result of the effects of respiration on the measured dynamics (see, e.g., [21, 22]). The generalized network structure which allows to go beyond the framework of pairwise interactions is the so-called *hierarchical high-order network* (hHON), described by mathematical constructs such as simplicial complexes and hypergraphs [23]. This novel representation is impacting strongly on the ability to describe the real-world systems studied in the context of Network Science [13].

While the implementation of hHONs is straightforward for networks inherently defined as sets of interactions, it is much less striking in systems where interactions are not already identified but need to be inferred from data, as in the case of brain and physiological networks. The main reason for this difficulty is that measures to quantify polyadic interactions from time series data have not been defined unequivocally [11, 21, 24, 25]. Different information-theoretic frameworks performing entropy decomposition of the multiple time series mapping the activity of network systems provide tools, such as the measures of redundant and synergistic information shared by groups of source time series about a target series, akin to the detection of high-order effects [11, 21]. Thus, these frameworks could be exploited for the detection and estimation of high-order interactions in practical settings. In this context, the unification and the extension to high-order interactions of emerging approaches to treat diverse types of physiological activity would open new perspectives for the use of hHON structures in Network Neuroscience and Network Physiology. Moreover, it is known that physiological systems and the brain very often present oscillatory behavior deployed across several distinct time scales [21, 26]; since these different oscillatory activities are collapsed into the recorded time series and cannot be distinguished by time-domain measures, the definition of the framework of information dynamics in the spectral domain would open the possibility to focus each measure in specific bands with physiological meaning (e.g., identifying different interaction mechanisms in different frequency bands).

1.3 Aims of the Book

Due to the intricate and non-deterministic nature of physiological and brain oscillations, Network Physiology and Network Neuroscience necessitate the development of integrated tools for the multivariate analysis of biomedical signals that can account for the multifaceted interactions possibly occurring in network systems. In this context, the present work of book introduces an integrated framework for the assessment of hierarchically organized interactions in brain and physiological networks, based on information-theoretic concepts and focused on connecting the time-domain and spectral representations of such interactions.

To reach this overarching goal, the book performs a thorough integration, extension, unification and illustration of several classic and recent approaches proposed for the analysis of network systems which are gaining wide interest in the fields of Network Neuroscience and Physiology (see, e.g., [26–39]). These approaches are based on the information-theoretic and spectral representations of *multi-order interactions* in network systems, and are here presented along three different lines of development (Fig. 1.1). First, we move from the standard *static analysis* of physiological processes, which draws a parallel between physiological networks and vectors of

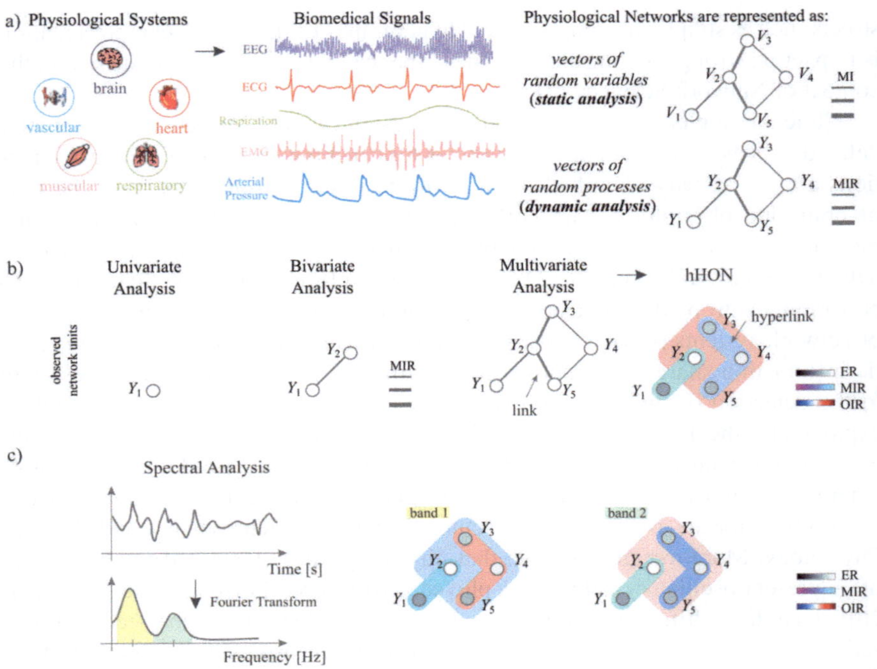

Fig. 1.1 **a** Collective interactions among organ systems are investigated recording biosignals whose activity is represented by realizations of (vector) random variables or processes, in turn investigated through static or dynamic **b** univariate, bivariate or hierarchical high-order network analysis. **c** The frequency domain representation captures the wide range of time scales characterizing the dynamic activity at the nodes. The figure is adapted from [27]

random variables (V_i), to a *dynamic analysis* which models the observed network system in terms of vector *random processes* (Y_i) (Fig. 1.1a); this is achieved moving from the use of measures of entropy computed for random variables (such as the mutual information—MI) to the use of measures of entropy rate which explicitly consider the temporal correlations within and between random processes (such as the mutual information rate—MIR). Second, when we consider the interactions in a network, we should first be aware of the number of observed network units. Thus, *univariate* and *bivariate* analyses, i.e., analyses considering only one process/node or only a pair of processes/nodes, respectively, will be characterized straightforwardly using classical measures of self-predictability (e.g., the entropy rate—ER), coupling or causality (e.g., the mutual information rate or the transfer entropy) (Fig. 1.1b, left). Then, when more than two nodes are observed, we shift from the paradigm of *multivariate analysis* to the paradigm of *high-order interactions*. While multivariate analysis is still anchored to the concept of interaction between two nodes even though the other nodes are taken into account, thus yielding the standard network representation where the link is the building block and is assessed by pairwise measures such as the MIR or conditional causality measures such as conditional transfer entropies, the analysis of *high-order interactions* encodes such interactions by the hyperlinks of a hHON, where interactions of different order are represented using different dynamic information measures (e.g., order 1: ER; order 2: MIR; order >2: O-information rate (OIR)); the latter allow to provide an overall quantification of the collective interaction through the concepts of redundancy and synergy (see Fig. 1.1b). Third, when we study networks where the activity at the nodes is rich of oscillatory content, we shift from the *time domain* to the *frequency domain* representation of the network, where information measures are expanded in the spectral domain to obtain frequency-specific information on the hHON interactions occurring within distinct frequency band (Fig. 1.1c).

In this work, different approaches for the analysis of multi-order interactions in network systems are unified in a coherent framework where spectral information measures are linked with their time-domain analogous measures, and are categorized hierarchically on the basis of the *number of observed network units*, and the number of network nodes involved in the computation of each measure (i.e., the *order of interactions*). Specifically, as depicted in Fig. 1.1b, we will characterize networks focusing on individual nodes, where the relevant interaction measures are those quantifying the static/dynamic activity at the node intended as amount of information content through measures of complexity or self-predictability. Then, we will describe the *pairwise interactions* taking place between two nodes, where the links are representative of coupling and/or causality effects. The last step is straightforward, and consists in enlarging our view of the possible arising interactions between multiple network nodes. Here, the notion of *high-order structures* emerges and brings the new concept of *hyperlink* quantifying interactions of multiple orders through a hierarchical high-order network analysis (hHONA). The different types of high-order behaviors in network systems mapped by multivariate time series, i.e., HOIs operating simultaneously across different levels of resolution including the entire network, individual links, and specific nodes, will be assessed through the recently developed O-Information (OI)-based framework [27, 34, 37, 40]. Specifically, we will define *(i) network-specific* measures capturing HOIs among all the analyzed time series, which however cannot consider high-order effects that are specific to certain parts of the network (e.g., individual nodes or links); then, we will resort to *(ii)* local OI-based measures, quantifying the net information shared between two time series and the rest of the network [40], here taken as *link-specific* measures of HOIs, and to *(iii)* OI gradients [41], quantifying the information shared between one time series and the rest of the network, here taken as *node-specific* measures of HOIs. Further, the majority of dynamic measures (e.g., entropy rate, mutual information rate, O-information rate) will be expanded into the frequency domain by defining the spectral counterparts of the time domain measures; the two representations are linked, in the case of Gaussian processes, via the *spectral integration property*, which guarantees that the observation of a given spectral measure extended over the entire frequency range leads it back to the corresponding time domain measure.

A comprehensive description of the structure of this book is provided below.

1.4 Outline of the Book

This book is split up in seven chapters. The first introductory chapter focuses on the problem of studying complex networks through information-theoretic approaches expanded in the spectral domain, and describes the structure of this book (the present chapter). The second chapter discusses the theoretical background of strictly causal linear parametric modelling. The third and fourth chapters thoroughly describe the time domain and spectral measures of self-effect, pairwise dependencies and high-order activity used to investigate static and dynamic networks of multiple nodes. The fifth and sixth chapters present exemplary applications to Network Physiology and Network Neuroscience, respectively. The final chapter summarizes the main research conclusions outlining future directions for the methods introduced in this book.

A more detailed description of each chapter is summarized herein.

This chapter, briefly introduces the problem statement and research objectives of this book. Then, it initiates the reader to the concept of *Network Science*, which requires novel approaches for the data-driven modelling of complex network systems. *What is already known?*: the background of the addressed topic is described with reference to existing approaches to the representation of network systems, e.g., augmented network descriptions which make use, for instance, of active nodes encoding self-dependencies within an individual process, and of directed and/or weighted edges depicting cause-effect relations and quantifying the intensity of interactions. *What is lacking in the field?*: the limitations of the representation with self-effects and pairwise interactions are described, with reference to the need of using measures capturing high-order interactions, i.e., interactions involving more than two network nodes [13]. *What do we aim to do?*: finally, the chapter briefly states the main objectives of the book and its major contribution, i.e., the thorough integration, extension and unification of several classic and recent approaches proposed for the analysis of network systems which are nowadays gaining wide interest in the field of Network Science.

Chapter 2 provides an overview of the theoretical background related to the linear parametric representation of network systems, which allows the expansion of the proposed information-theoretic measures in the frequency domain. The latter is essential to investigate the oscillatory content of individual physiological signals and to retrieve amounts of information shared by the observed processes within specific frequency bands representative of well-known physiological mechanisms.

Chapter 3 reviews the entropy measures exploited to describe static networks mapped by random variables, and relevant to the information-theoretic analysis of random processes.

Chapter 4 introduces a framework to measure dynamic interactions of different orders in networks of multiple interconnected systems, expanded in the frequency domain to explore their spectral patterns. The framework is defined for dynamic systems, whose activity is described by dynamic information measures which extend to random processes the measures defined in Chap. 3 for random variables.

In both chapters, validation of the proposed measures has been performed through theoretical simulations in controlled conditions of multivariate coupling.

Chapters 5 and 6 present applications to cardiovascular, respiratory and cerebrovascular data, and to neural networks probed by functional magnetic resonance imaging (fMRI) and electroencephalography (EEG), respectively, with the main aim to assess the effectiveness of multi-order information measures in providing non-invasive indicators of physiological states and clinical markers of pathological states in different contexts.

Chapter 7 summarizes the main findings of the research presented in this book. The fundamental results will be summarized in terms of the innovative potential of our information-theoretic framework as well as of the most relevant physiological insights in Network Physiology and Network Neuroscience. In addition, we will have a look at future directions for the research outlined in this book.

Appendices

Approaches for the statistical validation of the proposed measures including well-known and more sophisticated techniques, based on the concepts of surrogate and bootstrap data analyses, will be discussed with fine detail in Appendix A. In Appendix B, details on the theoretical background of extended linear parametric modelling will be given.

References

1. Amir Bashan et al. "Network physiology reveals relations between network topology and physiological function". In: *Nature communications* 3.1 (2012), p. 702.
2. Plamen Ch Ivanov and Ronny P Bartsch. "Network physiology: mapping interactions between networks of physiologic networks". In: *Networks of Networks: the last Frontier of Complexity*. Springer, 2014, pp. 203–222.
3. Danielle S Bassett and Olaf Sporns. "Network neuroscience". In: *Nature neuroscience* 20.3 (2017), pp. 353–364.
4. Mikail Rubinov and Olaf Sporns. "Complex network measures of brain connectivity: uses and interpretations". In: *Neuroimage* 52.3 (2010), pp. 1059–1069.
5. Klaus Lehnertz, Timo Bröhl, and Thorsten Rings. "The human organism as an integrated interaction network: Recent conceptual and methodological challenges". In: *Frontiers in Physiology* 11 (2020), p. 598694.
6. Koichi Sameshima and Luiz Antonio Baccalá. "Using partial directed coherence to describe neuronal ensemble interactions". In: *Journal of neuroscience methods* 94.1 (1999), pp. 93–103.
7. Luca Faes, Giandomenico Nollo, et al. "Multivariate frequency domain analysis of causal interactions in physiological time series". In: *Biomedical Engineering, Trends in Electronics, Communications and Software* 8 (2011), pp. 403–428.

8. André M. Bastos and Jan Mathijs Schoffelen. “A tutorial review of functional connectivity analysis methods and their interpretational pitfalls”. In: *Frontiers in Systems Neuroscience* 9.JAN2016 (2016). Publisher: Frontiers Research Foundation. ISSN: 16625137. DOI: https://doi.org/10.3389/fnsys.2015.00175.
9. Vlasta Bari et al. “Cerebrovascular and cardiovascular variability interactions investigated through conditional joint transfer entropy in subjects prone to postural syncope”. In: *Physiological Measurement* 38.5 (2017), p. 976.
10. Michal Javorka et al. “Basic cardiovascular variability signals: mutual directed interactions explored in the information domain”. In: *Physiological Measurement* 38.5 (2017), p. 877.
11. Joseph T Lizier et al. *Information decomposition of target effects from multi-source interactions: Perspectives on previous, current and future work.* 2018.
12. Maja Elstad et al. “Cardiorespiratory interactions in humans and animals: rhythms for life”. In: *American Journal of Physiology-Heart and Circulatory Physiology* 315.1 (2018), H6–H17.
13. Federico Battiston et al. “Networks beyond pairwise interactions: Structure and dynamics”. In: *Physics reports* 874 (2020), pp. 1–92.
14. Albert-László Barabási. “Network science”. In: *Philosophical Transactions of the Royal Society A: Mathematical, Physical and Engineering Sciences* 371.1987 (2013), p. 20120375.
15. Carter T Butts. “Revisiting the foundations of network analysis”. In: *science* 325.5939 (2009), pp. 414–416.
16. Steven M Pincus and Ary L Goldberger. “Physiological time-series analysis: what does regularity quantify?” In: *American Journal of Physiology-Heart and Circulatory Physiology* 266.4 (1994), H1643–H1656.
17. Alberto Porta et al. “Measuring regularity by means of a corrected conditional entropy in sympathetic outflow”. In: *Biological cybernetics* 78 (1998), pp. 71–78.
18. Ernesto Pereda, Rodrigo Quian Quiroga, and Joydeep Bhattacharya. “Nonlinear multivariate analysis of neurophysiological signals”. In: *Progress in neurobiology* 77.1-2 (2005), pp. 1–37.
19. Alberto Porta and Luca Faes. “Wiener–Granger causality in network physiology with applications to cardiovascular control and neuroscience”. In: *Proceedings of the IEEE* 104.2 (2015), pp. 282–309.
20. Sebastiano Stramaglia, Jesus M Cortes, and Daniele Marinazzo. “Synergy and redundancy in the Granger causal analysis of dynamical networks”. In: *New Journal of Physics* 16.10 (2014), p. 105003.
21. Luca Faes et al. “Information decomposition in multivariate systems: definitions, implementation and application to cardiovascular networks”. In: *Entropy* 19.1 (2016), p. 5.
22. Jana Krohova et al. “Multiscale information decomposition dissects control mechanisms of heart rate variability at rest and during physiological stress”. In: *Entropy* 21.5 (2019), p. 526.
23. Owen T Courtney and Ginestra Bianconi. “Generalized network structures: The configuration model and the canonical ensemble of simplicial complexes”. In: *Physical Review E* 93.6 (2016), p. 062311.
24. Michael Wibral, Raul Vicente, and Joseph T Lizier. *Directed information measures in neuroscience.* Vol. 724. Springer, 2014.
25. Ryan G James, Nix Barnett, and James P Crutchfield. “Information flows? A critique of transfer entropies”. In: *Physical review letters* 116.23 (2016), p. 238701.
26. Luca Faes et al. “Information decomposition in the frequency domain: A new framework to study cardiovascular and cardiorespiratory oscillations”. In: *Philosophical Transactions of the Royal Society A* 379.2212 (2021), p. 20200250.
27. Laura Sparacino et al. “Measuring hierarchically-organized interactions in dynamic networks through spectral entropy rates: theory, estimation, and illustrative application to physiological networks”. In: *Neurocomputing* 630 (2025), p. 129675.
28. William McGill. “Multivariate information transmission”. In: *Transactions of the IRE Professional Group on Information Theory* 4.4 (1954), pp. 93–111.
29. Izrail Moiseevich Gelfand and AM IAglom. *Calculation of the amount of information about a random function contained in another such function.* American Mathematical Society Providence, 1959.

30. Andrei Nikolaevitch Kolmogorov. "Entropy per unit time as a metric invariant of automorphisms". In: *Dokl. Akad. Nauk SSSR*. Vol. 124. 4. 1959, pp. 754–755.
31. Tyrone E Duncan. "On the calculation of mutual information". In: *SIAM Journal on Applied Mathematics* 19.1 (1970), pp. 215–220.
32. Thomas M Cover. *Elements of information theory*. John Wiley & Sons, 1999.
33. Daniel Chicharro. "On the spectral formulation of Granger causality". In: *Biological cybernetics* 105 (2011), pp. 331–347.
34. Fernando E Rosas et al. "Quantifying high-order interdependencies via multivariate extensions of the mutual information". In: *Physical Review E* 100.3 (2019), p. 032305.
35. Yuri Antonacci et al. "Measuring high-order interactions in rhythmic processes through multivariate spectral information decomposition". In: *IEEE Access* 9 (2021), pp. 149486–149505.
36. Riccardo Pernice et al. "Spectral decomposition of cerebrovascular and cardiovascular interactions in patients prone to postural syncope and healthy controls". In: *Autonomic Neuroscience* (2022), p. 103021.
37. Luca Faes et al. "A new framework for the time-and frequency-domain assessment of high-order interactions in networks of random processes". In: *IEEE Transactions on Signal Processing* 70 (2022), pp. 5766–5777.
38. Laura Sparacino et al. "A method to assess granger causality, isolation and autonomy in the time and frequency domains: theory and application to cerebrovascular variability". In: *IEEE Transactions on Biomedical Engineering* (2023).
39. Laura Sparacino et al. "A method to assess linear self-predictability of physiologic processes in the frequency domain: application to beat-to-beat variability of arterial compliance". In: *Frontiers in Network Physiology* 4 (2024), p. 1346424.
40. Gorana Mijatovic et al. "Assessing High-Order Links in Cardiovascular and Respiratory Networks via Static and Dynamic Information Measures". In: *IEEE Open Journal of Engineering in Medicine and Biology* (2024).
41. Tomas Scagliarini et al. "Gradients of O-information: Low-order descriptors of high-order dependencies". In: *Physical Review Research* 5.1 (2023), p. 013025.

Chapter 2
Linear Modelling of Stochastic Interactions

This chapter first introduces basic concepts of probability and provides key network analysis definitions (Sect. 2.1). Then, it illustrates the theory of linear vector modelling applied to random variables (Sect. 2.2) and random processes in both time (Sect. 2.3) and frequency domains (Sect. 2.4), which will be exploited respectively in Chaps. 3 and 4 to provide a mathematical framework used to describe network interactions in physiological and neural systems, and to define the measures of information dynamics descriptive of the structure of these networks.

2.1 Introduction to Static and Dynamic Stochastic Interactions

2.1.1 Basic Concepts of Probability

Static networks of random variables. A *random variable* is a mathematical variable whose value is subject to variations due to chance. Specifically, continuous random variables can take values inside an infinite-dimensional set usually denoted as the domain. The generic scalar random variable V with domain D_V is characterized by its distribution function, which assigns a probability to each measurable subset of D_V. Formally, the probability for the variable V of taking values within the interval $[a, b] \subseteq D_V$ is determined by the integral $Pr\{a \leq V \leq b\} = \int_a^b p_V(v)\mathrm{d}v = F_V(b) - F_V(a)$, where p_V is the *marginal probability density function* of the variable and F_V is its cumulative distribution function. The cumulative distribution quantifies the probability that the variable V has v as its upper bound, $F_V(v) = Pr\{V \leq v\}$, while the probability density is mathematically defined as the derivative of the cumulative distribution, in a way such that $F_V(v) = \int_{-\infty}^{v} p_V(u)\mathrm{d}u$. These definitions extend in a straightforward way to the generic k-dimensional variable $\mathbf{V} = \{V_1, \ldots, V_k\}$ by defining the joint probability density $p_\mathbf{V}(v_1, \ldots, v_k)$ and

L. Sparacino, *Spectral Information Dynamics in Network Neuroscience and Physiology*, Studies in Computational Intelligence 1235,
https://doi.org/10.1007/978-3-032-05416-6_2

performing multiple integration over the domain of each scalar component to get the cumulative distribution. Moreover, the *conditional probability density function* of, e.g., V_1 given V_2 expresses the probability of observing the value v_1 for V_1 given that the value v_2 has been observed for V_2: $p_{V_1|V_2}(v_1|v_2) = \frac{p_{V_1,V_2}(v_1,v_2)}{p_{V_2}(v_2)}$, where $p_{V_1,V_2}(v_1, v_2)$ is the *joint probability density function*.

In this book, we will consider the static network system $\mathcal{V}$ composed of M nodes $\mathcal{V}_1, \ldots, \mathcal{V}_M$, where the activity at each node is described by the (possibly vector) random variable $\mathbf{V}_i$, $i = 1, \ldots M$ (Fig. 2.1a). The multiple interactions between the M variables $\mathbf{V}_1, \ldots, \mathbf{V}_M$ can be investigated by means of a *static analysis* of multiple realizations of these variables available in the form of multiple data sequences. Static analysis implicitly disregards temporal correlations, taking into account only zero-lag effects between these sequences. This type of analysis is typically carried out in the field of brain functional connectivity [1, 2], which, indeed, quantifies the zero-lag dependency of neuronal activation patterns of anatomically separated brain regions [3]. Here, the different static systems analyzed may be M (subsets of) resting state networks, where each variable $\mathbf{V}_i$, $i = 1, \ldots, M$, may represent the neural activity of a group of voxels within that network and thus map the time course of the ith neural signal extracted through a seed-based correlation approach (e.g., the blood oxygen level-dependent functional magnetic resonance imaging (fMRI) signal describing the activity of a given brain region inside a resting state network) [4].

Dynamic networks of random processes. Contrarily to static systems, dynamic systems take over diverse states at different instants of time and are assumed to be stochastic, meaning that their current state does not depend only on the inputs to the system and on its initial state but also on the outcome of a random experiment. Therefore, the evolution over time of these systems can be only described in probabilistic terms using *stochastic* (or *random*) *processes*, which can be thought of as sequences of random variables ordered according to time. The states visited by the generic dynamic system $\mathcal{X}$ over time are described as a stochastic process $X = \{X_n\}$, $n = 1, 2, \ldots$, where the random variable X_n describes the nth state assumed by $\mathcal{X}$ at the nth time step. Then, a realization of the stochastic process X is the time series $\mathbf{x} = \{x(1), \ldots, x(L)\}$, containing the values of X collected over L time points. When considered individually, each stochastic process describing the dynamic activity of a node of the observed network is characterized by a family of probability density functions, or equivalently by the corresponding family of cumulative distribution functions. Accordingly, the temporal statistical structure of the process X is described by the probability density functions $p_{X_n}(x_n)$ or by the cumulative distribution functions $F_{X_n}(x_n)$, $n = 1, 2, \ldots, L$. The probability density is then defined in a straightforward way for joint variables taken from the same process, $p_{X_{n_1},\ldots,X_{n_k}}(x_{n_1}, \ldots, x_{n_k})$, $\forall n_1, \ldots, n_k$, $\forall k \geq 1$. Setting a temporal reference frame in which n represents the present time, we denote as X_n the random variable describing the present state of X, and as $X_n^- = [X_{n-1}, X_{n-2}, \ldots]$ the random variable that sample the process over the whole past history. In general, the operation of separating the present from the past allows to consider the flow of time and to study the causal interactions within and between processes by looking at the statistical dependencies among these variables. In fact, the dynamic properties of a system are studied in the information domain introducing the concept of *transition probability*, which is the probability associated with the transition of the system from its past states to its present state. Thus, the state transition of the history of X relevant to the present state X_n is described by the conditional probability density $p_{X_n|X_n^-}(x_n|x_n^-)$. A useful property of stochastic processes is *wide-sense stationarity* (WSS), which defines the time-invariance of any joint probability density taken from the process. When the process is stationary, the fact that the probability density is the same at all times allows to pool together the observations measured across time order to estimate the densities, thus enabling the estimation of probabilities from individual realizations, i.e., an individual time series [5, 6]. For a stationary stochastic process, also the transition probabilities are time-independent, i.e., $p_{X_n|X_n^-}(x_n|x_n^-) = p_X(x_n|x_n^-)$. An important class of dynamic processes is represented by *Markov processes*, for which the present depends on the past only through a finite number of time steps. Specifically, the process X is a Markov process of order k if its transition probability function satisfies the condition $p_X(x_n|x_n^-) = p_X(x_n|x_{n-1}, x_{n-2}, \ldots, x_{n-k})$. With this notation, we define as $X_n^k = [X_{n-1}, \ldots, X_{n-k}]$ the random variable that samples the process over the past k lags, with $X_n^- = \lim_{k\to\infty} X_n^k$.

In this work, we will deal with the generic dynamic network system $\mathcal{X}$ composed of M nodes $\mathcal{X}_1, \ldots, \mathcal{X}_M$, where the activity at each node is described in terms of random processes. Specifically, Q stationary stochastic processes $\mathbf{Y} = \{Y_1, \ldots, Y_Q\}$, possibly grouped in M blocks $\mathbf{X} = \{\mathbf{X}_1, \ldots, \mathbf{X}_M\}$, are considered, where each block process $\mathbf{X}_i$ describes the dynamic activity of the network node $\mathcal{X}_i$, $i = 1, \ldots, M$; the ith block has dimension M_i, so that $Q = \sum_{i=1}^{M} M_i$ (Fig. 2.1b). The different dynamic systems analyzed may be M brain regions or M organ systems, where each group process $\mathbf{X}_i$, $i = 1, \ldots, M$, represents the neural activity of a given brain region or organ system, and each scalar process $Y_j \in \mathbf{X}_i$, $j = 1, \ldots, M_i$, maps the time course of the jth neural signal recorded inside the ith region (e.g., the EEG signal at one frontal electrode) or the jth physiological time series belonging to the ith organ system (e.g., systolic or diastolic pressure for the circulatory system). Without loss of generality, we assume that the processes are real-valued, defined at discrete time ($Y_q = \{Y_{q,n}\}$, $q = 1, \ldots, Q$; e.g., are sampled versions of the continuous time processes $Y_{q,t}$, taken at the times $t_n = nT$, with T the sampling period) and have zero mean ($\mathbb{E}[Y_{q,n}] = 0$, where $\mathbb{E}[\cdot]$ is the statistical expectation operator).

2.1.2 Network Analysis Definitions

In this section, we provide some key definitions regarding network analysis, since distinct linear models can be identified depending on *(i)* the *order of interactions*, as well as the need to describe linear interactions relevant to a given subset of units, and *(ii)* the *number of observed units*.

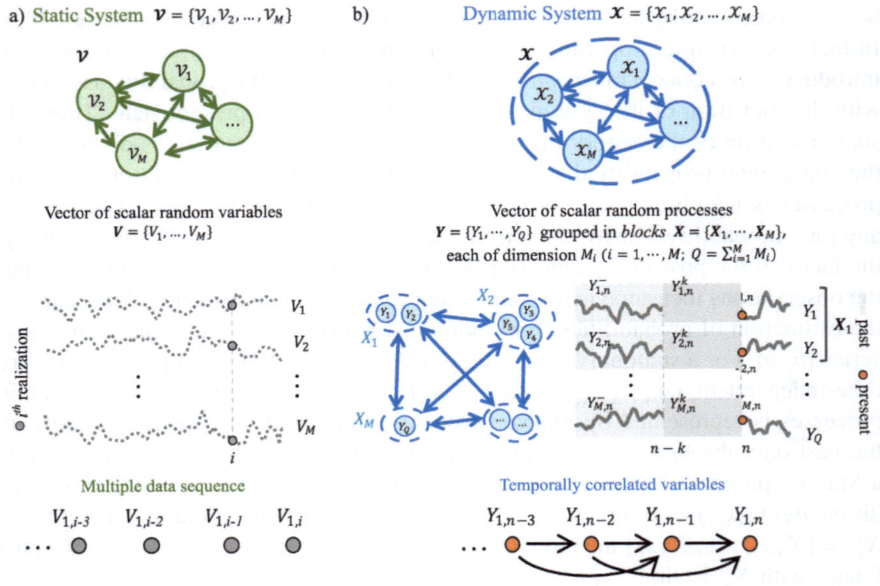

Fig. 2.1 **a** Static and **b** dynamic network systems

Full and restricted models. The static $\mathcal{V}$ and dynamic $\mathcal{X}$ network systems are always described as a whole with the aim of gathering information on the overall behaviour of the network or subparts of it. In this context, a distinction between ***full*** and ***restricted*** models is needed: while a full model provides a global representation of the overall activity of the observed network, there is often the need to describe interactions relevant to a given subset of variables or processes, e.g., whenever explicit physiological regulatory mechanisms or neural patterns require to be investigated. To this end, restricted models involving only those relevant variables or processes are defined. In the case of linear dynamic models, the restricted model parameters can be directly retrieved from the full model parameters through different procedures (see Sect. 2.3.5.1).

Networks with different number of units. Based on the number of observed units of the analyzed static or dynamic systems, we distinguish between the following three types of network analysis. In Fig. 2.2, we schematize these definitions with regard to dynamic networks.

- A ***univariate analysis*** (UA) is performed if only one scalar process Y is observed ($Q = M = 1$) within a dynamic system; the corresponding linear model will be referred to as auto-regressive (AR) model on Y, and allows to investigate the *single-node activity* of Y ($\mathbb{I}_1$). Univariate analysis of the short-term beat-to-beat variability of cardiovascular parameters such as arterial compliance is a common example (Sect. 5.1). The same rationale can be applied whether the observed unit

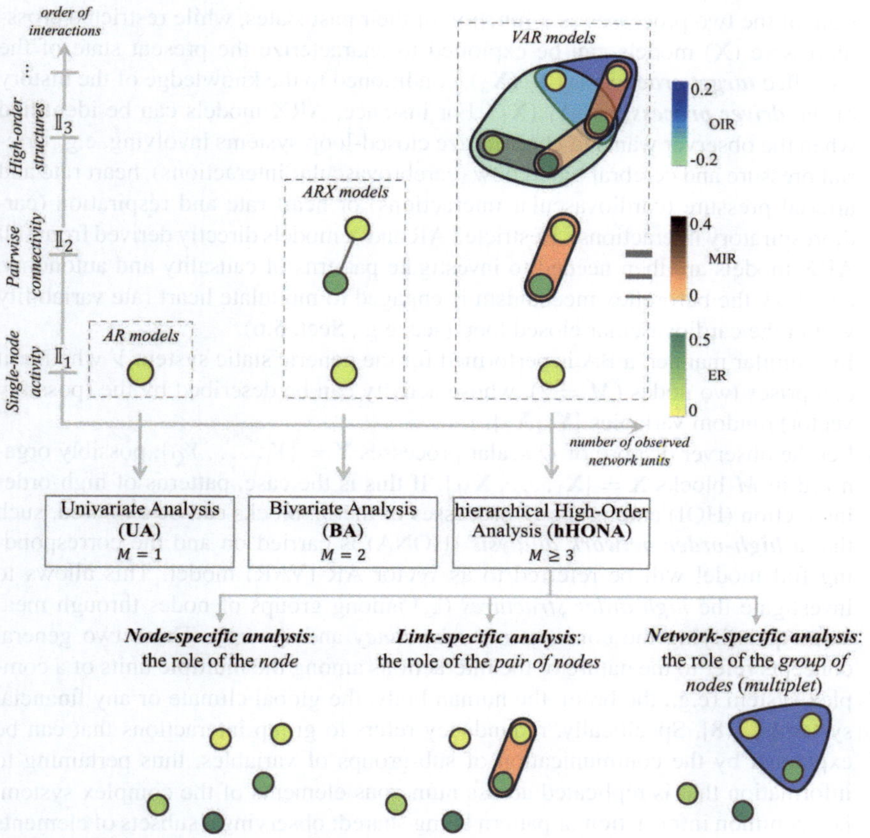

Fig. 2.2 Multi-order interactions in dynamic network systems (ER: entropy rate; MIR: mutual information rate; OIR: O-Information rate)

is the block process $\mathbf{X} = \{Y_1, \ldots, Y_{M_X}\}$ of dimension M_X ($Q = M_X$, $M = 1$), e.g., in the case of a group of M_X EEG electrodes located over the contralateral motor area.
Similarly, a UA is performed for the generic static system $\mathcal{V}$ whether it comprises only one node ($M = 1$), whose activity is described by the (possibly vector) random variable $\mathbf{V}$.

- If two scalar (vector) processes $\{Y_1, Y_2\}$ ($\{\mathbf{X}_1, \mathbf{X}_2\}$) are observed, a ***bivariate analysis*** (BA) is carried on and the corresponding full model will be referred to as auto- and cross-regressive (ARX) model on $\{Y_1, Y_2\}$ ($\{\mathbf{X}_1, \mathbf{X}_2\}$). This allows to investigate the *pairwise connectivity* ($\mathbb{I}_2$) between the two nodes through bivariate measures of coupling (information shared) and causality (information transfer). Restricted models can be formulated to either explore the *single-node activities* in terms of self- or cross-dependencies: bivariate AR models describe the present

state of the two processes as a function of their past states, while restricted cross-regressive (X) models can be exploited to characterize the present state of the so-called ***target process***, say Y_2 ($\mathbf{X}_2$), conditioned to the knowledge of the history of the ***driver process***, say Y_1 ($\mathbf{X}_1$). For instance, ARX models can be identified when the observer wants to characterize closed-loop systems involving, e.g., arterial pressure and cerebral blood flow (cerebrovascular interactions), heart rate and arterial pressure (cardiovascular interactions) or heart rate and respiration (cardiorespiratory interactions). Restricted AR and X models directly derived from full ARX models are then needed to investigate patterns of causality and autonomy, e.g., how the baroreflex mechanism is engaged to modulate heart rate variability within the cardiovascular closed loop (see, e.g., Sect. 5.6).
In a similar manner, a BA is performed for the generic static system $\mathcal{V}$ whether it comprises two nodes ($M = 2$), whose activity can be described by the (possibly vector) random variables $\{\mathbf{V}_1, \mathbf{V}_2\}$.

- Let the observer dispose of Q scalar processes $\mathbf{Y} = \{Y_1, \ldots, Y_Q\}$, possibly organized in M blocks $\mathbf{X} = \{\mathbf{X}_1, \ldots, \mathbf{X}_M\}$. If this is the case, patterns of high-order interaction (HOI) among the Q processes or the M blocks can be explored, such that a ***high-order network analysis*** (HONA) is carried on and the corresponding full model will be referred to as vector AR (VAR) model. This allows to investigate the *high-order structures* ($\mathbb{I}_3$) among groups of nodes through measures quantifying the concepts of *redundancy* and *synergy*. These two general concepts refer to the nature of the interactions among the multiple units of a complex system (e.g., the brain, the human body, the global climate or any financial system) [7, 8]. Specifically, redundancy refers to group interactions that can be explained by the communication of sub-groups of variables, thus pertaining to information that is replicated across numerous elements of the complex system, i.e., common information or pattern being shared: observing a subsets of elements can resolve uncertainty across all the other elements of that system. Conversely, synergistic information sharing takes place when the joint state of three or more variables is necessary to resolve uncertainty, arising from statistical interactions that can be found collectively in a network but not in parts of it considered separately. Moreover, restricted models can be formulated to investigate the linear interactions involving subgroups of processes selected from $\mathbf{Y}$ or $\mathbf{X}$, i.e., the *single-node* and *pairwise activities* through a so-called ***hierarchical HONA*** (hHONA). In addition, when more than two units are observed, the HOIs can be examined at different levels of resolution, i.e., characterizing

  (i) the role of the single node within the selected subset (***node-specific analysis***),
  (ii) the nature of the link between two nodes with respect to the remaining nodes (***link-specific analysis***),
  (iii) the overall activity of groups of nodes (*multiplets*) or the whole network (***network-specific analysis***).

This type of analysis can be carried on whenever the observer wants to describe complex physiological or neural networks involving parameters such as heart rate, respiration, arterial pressure and cerebral blood flow (see, e.g., Sects. 5.7–5.11), or groups of electroencephalographic signals extracted from electrodes located over different areas of the cerebral cortex (see Sects. 6.2.4–6.2.5).

A static HONA is performed for the generic static system $\mathcal{V}$ when it comprises M nodes, whose activity can be described by the (possibly vector) random variables $\{\mathbf{V}_1, \ldots, \mathbf{V}_M\}$.

2.2 Static Models of Random Variables

Well-established measures defined in the framework of information theory can be exploited to study the interactions between pairs and/or groups of variables taken from the static network $\mathcal{V}$. Assuming that the observed variables have a joint Gaussian distribution, the analysis can be performed by exploiting linear parametric regression models. Specifically, two generic zero-mean vector variables $\mathbf{V}_i$ and $\mathbf{V}_j$, containing respectively v_i and v_j scalar variables, are related by the following linear regression model:

$$\mathbf{V}_i = \mathbf{A}_j \mathbf{V}_j + \mathbf{U}_j, \tag{2.1}$$

where $\mathbf{V}_i$ is predicted using a $v_i \times v_j$ coefficient matrix $\mathbf{A}_j$ which weights the regressors $\mathbf{V}_j$, and $\mathbf{U}_j = [U_{j,1} \ldots U_{j,v_i}]^\intercal$ is a vector of v_i independent Gaussian variables with zero mean. Now, let us collect the remaining $M-2$ variables in the vector $\mathbf{V}_z = \mathcal{V} \setminus [\mathbf{V}_i, \mathbf{V}_j]$. The zero-mean vector variables $\mathbf{V}_i$ and $\mathbf{V}_z$ are related by a linear regression model similar to (2.1), where $\mathbf{V}_i$ is predicted using a coefficient matrix $\mathbf{A}_z$ which weights the regressors $\mathbf{V}_z$ and $\mathbf{U}_z = [U_{z,1} \ldots U_{z,v_i}]^\intercal$ is a vector of v_i independent Gaussian variables, i.e., $\mathbf{V}_i = \mathbf{A}_z \mathbf{V}_z + \mathbf{U}_z$. Further, the vector variable $\mathbf{V}_i$ can be predicted using the coefficient matrices $\mathbf{A}_{jz}^{(j)}$ and $\mathbf{A}_{jz}^{(z)}$ which weight the regressors $\mathbf{V}_j$ and $\mathbf{V}_z$ used simultaneously rather than individually, i.e., $\mathbf{V}_i = \mathbf{A}_{jz}^{(j)} \mathbf{V}_j + \mathbf{A}_{jz}^{(z)} \mathbf{V}_z + \mathbf{U}_{jz}$, where $\mathbf{U}_{jz} = [U_{jz,1} \ldots U_{jz,v_i}]^\intercal$ is a vector of v_i independent Gaussian variables.

2.3 Dynamic Models of Random Processes in the Time Domain

The time-, frequency- and information-domain measures which can be derived from the static linear models described in Sect. 2.2 suffer from the limitation that they only allow a static analysis of random variables where the temporal information is disregarded. To perform a dynamic analysis, one needs to consider random processes, intended as collections of random variables sorted in temporal order. The processes are assumed to be jointly Gaussian distributed, to exploit the formalism linking information-theoretic measures with linear regression models [6, 9] and spectral quantities [10–12]. The parametric implementation exploits the knowledge that linear regression models capture all of the entropy differences relevant to the various information measures when the observed processes have a joint Gaussian distribution [6, 9]. We remark that our definitions of bivariate and multivariate models, i.e., when two or more than two (groups of) processes are observed, limit to past values only the possible influences of one process to another, excluding instantaneous effects (i.e., effects occurring within the same lag). The absence of instantaneous effects is denoted as *strict causality* of the process [13, 14] and will be assumed henceforth $\forall Q, M$ in this book. However, since in the past years we also worked with extended models including zero-lag effects (see, e.g., [15]), brief technical details about their formulation are given in Appendix B, while two practical applications are shown in Sect. 5.3 and 5.5.

In the following subsections, we will first describe linear AR (Sect. 2.3.1), ARX (Sect. 2.3.2) and VAR (Sect. 2.3.3) models in the time domain, representing the self-dynamics of the single node, the interactions between two nodes, and the multiple interactions among several nodes, respectively; then, the time domain formulations of restricted parametric models obtained from the full models will be described in Sect. 2.3.4, and the problem of their identification starting from the full model parameters will be detailed in Sect. 2.3.5.

2.3.1 The Single Unit: AR Models

Autoregressive models can be identified on single (vector) processes to characterize their own dynamics. In case of a generic zero-mean scalar random process Y with variance σ_Y^2 ($Q = M = 1$), the following AR model is identified [16]:

$$Y_n = \sum_{k=1}^{p} a_{Y,k}^{(ar)} Y_{n-k} + U_{Y,n}^{(ar)}, \tag{2.2}$$

where p is the model order, defining the maximum lag used to quantify interactions, Y_n is the present state of Y, $a_{Y,k}^{(ar)}$ is the AR model coefficient relating the present with the past of the process at lag k and $U_Y^{(ar)}$ is a zero-mean white process, i.e., a process composed by independent identically distributed (i.i.d.) variables, with variance $\sigma_{U_Y}^{2(ar)}$. Further, a block AR model can be identified when the overall activity at the node is represented by the generic zero-mean vector process $\mathbf{X} = \{Y_1, \ldots, Y_{M_X}\}$ of dimension M_X ($Q = M_X$, $M = 1$):

$$\mathbf{X}_n = \sum_{k=1}^{p} \mathbf{A}_{\mathbf{X},k}^{(ar)} \mathbf{X}_{n-k} + \mathbf{U}_{\mathbf{X},n}^{(ar)}, \tag{2.3}$$

where $\mathbf{X}_n = [Y_{1,n}, \ldots, Y_{M_X,n}]^\intercal$ is a M_X-dimensional vector collecting the present state of all processes, $\mathbf{A}_{\mathbf{X},k}^{(ar)}$ the $M_X \times M_X$ matrix of the model coefficients relating the present with the past of the processes at lag k, and $\mathbf{U}_{\mathbf{X},n}^{(ar)} = [U_{Y_1,n}^{(ar)}, \ldots, U_{Y_{M_X},n}^{(ar)}]^\intercal$ is a vector of M_X zero-mean white noises with $M_X \times M_X$ positive definite covariance matrix $\Sigma_{\mathbf{U_X}}^{(ar)}$. The block process $\mathbf{X}_n$ has a covariance matrix $\Sigma_{\mathbf{X}} = \mathbb{E}[\mathbf{X}_n \mathbf{X}_n^\intercal]$, where the diagonal elements represent the variances of the scalar processes in $\mathbf{X}$, i.e., $\sigma_{Y_j}^2$, $j = 1, \ldots, M_X$.

Note that the model (2.3) is, as a matter of fact, a VAR model formally equivalent to that defined in Sect. 2.3.3, but here is presented as *block AR* model to highlight that the relevant dynamics are analyzed collectively as representative of the activity of a single node of the analyzed vector.

2.3.2 *Interactions Between Two Nodes: ARX Models*

Auto- and cross-regressive models are identified whether two (groups of) network units are observed. Generally, the two random processes representing the dynamic activity of the units interact in a closed-loop manner, i.e., through bidirectional causal relationships which allow to identify driver-response patterns. Different information-theoretic measures of (non-)directed coupling and autonomy can be defined, whose linear Gaussian formulation requires the identification of full ARX models from which restricted AR and X models can be derived through a nontrivial mathematical procedure detailed in Sect. 2.3.5.1. Note that ARX models have been generally referred to as bivariate AR models [9, 17].

These models feature two model equations, where the present states of the two processes are written as linear combinations of the past states of both processes weighted by a set of model coefficients plus the residuals. Assuming that $\mathbf{Y}$ is the generic vector process comprising the two scalar processes $\{Y_1, Y_2\}$ ($Q = 2$, $M = 2$), the following ARX model can be identified [16]:

$$Y_{1,n} = \sum_{k=1}^{p} a_{Y_1 Y_1,k}^{(arx)} Y_{1,n-k} + a_{Y_1 Y_2,k}^{(arx)} Y_{2,n-k} + U_{Y_1,n}^{(arx)}, \tag{2.4a}$$

$$Y_{2,n} = \sum_{k=1}^{p} a_{Y_2 Y_1,k}^{(arx)} Y_{1,n-k} + a_{Y_2 Y_2,k}^{(arx)} Y_{2,n-k} + U_{Y_2,n}^{(arx)}, \tag{2.4b}$$

where $\mathbf{Y}_n = [Y_{1,n}, Y_{2,n}]^\intercal$ is the 2-dimensional vector collecting the present state of the two processes, $\mathbf{A}_{\mathbf{Y},k}^{(arx)} = \begin{bmatrix} a_{Y_1 Y_1,k}^{(arx)} & a_{Y_1 Y_2,k}^{(arx)} \\ a_{Y_2 Y_1,k}^{(arx)} & a_{Y_2 Y_2,k}^{(arx)} \end{bmatrix}$ is the 2×2 matrix of the model coefficients relating the present with the past of the two processes at lag k, and $\mathbf{U}_{\mathbf{Y},n}^{(arx)} = [U_{Y_1,n}^{(arx)}, U_{Y_2,n}^{(arx)}]^\intercal$ a vector of 2 zero-mean white noises with 2×2 positive definite covariance matrix $\Sigma_{\mathbf{U_Y}}^{(arx)}$. The process $\mathbf{Y}_n$ has a 2×2 covariance matrix $\Sigma_{\mathbf{Y}} = \mathbb{E}[\mathbf{Y}_n \mathbf{Y}_n^\intercal]$, where the diagonal elements represent the variances of the scalar processes $\{Y_1, Y_2\}$, i.e., $\sigma_{Y_1}^2$, $\sigma_{Y_2}^2$. If the network is composed of two nodes whose activity is mapped by the two blocks $\{\mathbf{X}_1, \mathbf{X}_2\}$ of dimensions M_1, M_2 ($Q = M_1 + M_2$, $M = 2$), the following block ARX model can be identified:

$$\mathbf{X}_{1,n} = \sum_{k=1}^{p} \mathbf{A}^{(arx)}_{\mathbf{X}_1\mathbf{X}_1,k}\mathbf{X}_{1,n-k} + \mathbf{A}^{(arx)}_{\mathbf{X}_1\mathbf{X}_2,k}\mathbf{X}_{2,n-k} + \mathbf{U}^{(arx)}_{\mathbf{X}_1,n}, \tag{2.5a}$$

$$\mathbf{X}_{2,n} = \sum_{k=1}^{p} \mathbf{A}^{(arx)}_{\mathbf{X}_2\mathbf{X}_1,k}\mathbf{X}_{1,n-k} + \mathbf{A}^{(arx)}_{\mathbf{X}_2\mathbf{X}_2,k}\mathbf{X}_{2,n-k} + \mathbf{U}^{(arx)}_{\mathbf{X}_2,n}, \tag{2.5b}$$

where $\mathbf{X}_n = [\mathbf{X}^\intercal_{1,n}, \mathbf{X}^\intercal_{2,n}]^\intercal$ is the $(M_1 + M_2)$-dimensional vector collecting the present state of the two processes, $\mathbf{A}^{(arx)}_{\mathbf{X},k} = \begin{bmatrix} \mathbf{A}^{(arx)}_{\mathbf{X}_1\mathbf{X}_1,k} & \mathbf{A}^{(arx)}_{\mathbf{X}_1\mathbf{X}_2,k} \\ \mathbf{A}^{(arx)}_{\mathbf{X}_2\mathbf{X}_1,k} & \mathbf{A}^{(arx)}_{\mathbf{X}_2\mathbf{X}_2,k} \end{bmatrix}$ is the $(M_1 + M_2) \times (M_1 + M_2)$ matrix of the model coefficients relating the present with the past of the processes at lag k, with $\mathbf{A}^{(arx)}_{\mathbf{X}_1\mathbf{X}_1,k}, \mathbf{A}^{(arx)}_{\mathbf{X}_1\mathbf{X}_2,k}, \mathbf{A}^{(arx)}_{\mathbf{X}_2\mathbf{X}_1,k}, \mathbf{A}^{(arx)}_{\mathbf{X}_2\mathbf{X}_2,k}$ block matrices of dimensions $(M_1 \times M_1)$, $(M_1 \times M_2)$, $(M_2 \times M_1)$, $(M_2 \times M_2)$, respectively, and $\mathbf{U}^{(arx)}_{\mathbf{X},n} = [\mathbf{U}^{(arx)\,\intercal}_{\mathbf{X}_1,n}, \mathbf{U}^{(arx)\,\intercal}_{\mathbf{X}_2,n}]^\intercal$ is a vector of $(M_1 + M_2)$ zero-mean white noises with $(M_1 + M_2) \times (M_1 + M_2)$ positive definite covariance matrix $\Sigma^{(arx)}_{\mathbf{U}_\mathbf{X}}$, with $\mathbf{U}^{(arx)}_{\mathbf{X}_1,n} = [U^{(arx)}_{Y_1,n}, \ldots, U^{(arx)}_{Y_{M_1},n}]^\intercal$, $\mathbf{U}^{(arx)}_{\mathbf{X}_2,n} = [U^{(arx)}_{Y_{M_1+1},n}, \ldots, U^{(arx)}_{Y_Q,n}]^\intercal$. The process $\mathbf{X}_n$ has a $(M_1 + M_2) \times (M_1 + M_2)$ covariance matrix $\Sigma_\mathbf{X} = \mathbb{E}[\mathbf{X}_n\mathbf{X}^\intercal_n]$, which can be partitioned into a 2×2 matrix where the diagonal blocks are the covariance matrices of the two block processes in $\mathbf{X}_n$, i.e., $\Sigma_{\mathbf{X}_1}$ and $\Sigma_{\mathbf{X}_2}$.

2.3.3 *The Whole Network: VAR Models*

The generic Q-dimensional stationary vector random process $\mathbf{Y}$ can be represented by the VAR linear model [16]:

$$\mathbf{Y}_n = \sum_{k=1}^{p} \mathbf{A}^{(var)}_{\mathbf{Y},k}\mathbf{Y}_{n-k} + \mathbf{U}^{(var)}_{\mathbf{Y},n}, \tag{2.6}$$

where $\mathbf{Y}_n = [Y_{1,n}, \ldots, Y_{Q,n}]^\intercal$ is a Q-dimensional vector collecting the present state of all processes, $\mathbf{A}^{(var)}_{\mathbf{Y},k}$ is the $Q \times Q$ matrix of the model coefficients relating the present with the past of the processes at lag k, and $\mathbf{U}^{(var)}_{\mathbf{Y},n} = [U^{(var)}_{Y_1,n}, \ldots, U^{(var)}_{Y_Q,n}]^\intercal$ is a vector of Q zero-mean white noises with $Q \times Q$ positive definite covariance matrix $\Sigma^{(var)}_{\mathbf{U}_\mathbf{Y}} = \mathbb{E}[\mathbf{U}^{(var)}_{\mathbf{Y},n}\mathbf{U}^{(var)\,\intercal}_{\mathbf{Y},n}]$. The main diagonal of $\Sigma^{(var)}_{\mathbf{U}_\mathbf{Y}}$ contains the variances of the processes in $\mathbf{U}^{(var)}_{\mathbf{Y},n}$, i.e., $\sigma^{2(var)}_{U_{Y_i}}$, $i = 1, \ldots, Q$. The process $\mathbf{Y}_n$ has a $Q \times Q$ covariance matrix $\Sigma_\mathbf{Y} = \mathbb{E}[\mathbf{Y}_n\mathbf{Y}^\intercal_n]$, where the diagonal elements represent the variances of the scalar processes $\{Y_1, \ldots, Y_Q\}$, i.e., $\sigma^2_{Y_1}, \ldots, \sigma^2_{Y_Q}$. We remark that Eqs. (2.2), (2.3), (2.4), (2.5) are special cases of (2.6) at varying the number of observed nodes (M), as well as the nature of the nodes (i.e., scalar or block processes).

2.3.4 Restricted Models in the Time Domain

While the VAR model (2.6) provides a global representation of the overall multivariate process, to describe the linear interactions relevant to a given subset of processes we need to define a restricted VAR model involving only those processes. The same reasoning can be applied to the ARX model (2.4), from which restricted AR and X models describing causality and autonomy patterns can be derived.

Restricted VAR models. From the *full VAR model* (2.6), a restricted VAR model involving only the selected processes, say $\mathbf{Z}$, can be formulated as [18]

$$\mathbf{Z}_n = \sum_{k=1}^{\infty} \mathbf{B}_{\mathbf{Z},k}\mathbf{Z}_{n-k} + \mathbf{W}_{\mathbf{Z},n}, \tag{2.7}$$

where $\mathbf{Z}_n$ and $\mathbf{W}_{\mathbf{Z},n}$ are column vectors of dimension R, and $\mathbf{B}_{\mathbf{Z},k}$ is an $R \times R$ coefficient matrix. We denote as $\Sigma_{\mathbf{W_Z}} = \mathbb{E}[\mathbf{W}_{\mathbf{Z},n}\mathbf{W}_{\mathbf{Z},n}^{\intercal}]$ the $R \times R$ covariance matrix of the residuals in $\mathbf{W_Z}$. Note that the order of the restricted VAR model is theoretically infinite, and see Sect. 2.3.5.1 for the identification of its parameters starting from the parameters of the full model (2.6).

Restricted AR and X models. Similarly, from the *full ARX model* (2.4), a restricted AR model involving only the target process, say Y_2, and an X model involving only the target as the predicted variable and both processes as the regressors can be defined [19]. To implement this concept, the present state of the target, $Y_{2,n}$, is described first from the past of Y_2 only through the *restricted AR model* [19]

$$Y_{2,n} = \sum_{k=1}^{\infty} b_{Y_2Y_2,k}Y_{2,n-k} + W_{Y_2|Y_2,n}, \tag{2.8}$$

where $b_{Y_2Y_2,k}$ are AR coefficients and $W_{Y_2|Y_2}$ is a white noise process with variance $\lambda^2_{W_{Y_2|Y_2}}$. Then, a *restricted X model* is derived whereby $Y_{2,n}$ is described only from the past of the driver Y_1 [19]:

$$Y_{2,n} = \sum_{k=1}^{\infty} b_{Y_2Y_1,k}Y_{1,n-k} + W_{Y_2|Y_1,n} \tag{2.9}$$

where $b_{Y_2Y_1,k}$ are cross-regression coefficients and $W_{Y_2|Y_1}$ is an innovation process with variance $\lambda^2_{W_{Y_2|Y_1}}$. Note that the orders of the restricted AR and X models are theoretically infinite, and see Sect. 2.3.5.1 for the identification of their parameters starting from the parameters of the full ARX model (2.4). Straightforwardly, equations (2.8) and (2.9) can be written taking Y_1 as the target process.

2.3.5 Model Identification

The identification procedure of the VAR model (2.6) is typically performed by means of estimation methods based on minimizing the prediction error, i.e., the difference between actual and predicted data [16, 20]. While several approaches have been proposed throughout the years [21, 22], the most common estimator is the multivariate version of the *ordinary least-squares* (OLS) method [16]. Briefly, defining the past history of $\mathbf{Y}$ truncated at p lags as the pQ-dimensional vector $\mathbf{Y}_n^p = [\mathbf{Y}_{n-1}^\intercal, \ldots, \mathbf{Y}_{n-p}^\intercal]^\intercal$ and considering L consecutive time steps, a compact representation of (2.6) can be defined as $\mathbf{y} = \mathbf{A}_\mathbf{Y}^{(var)}\mathbf{y}^p + \mathbf{U}_\mathbf{Y}^{(var)}$, where $\mathbf{A}_\mathbf{Y}^{(var)} = [\mathbf{A}_{\mathbf{Y},1}^{(var)}, \ldots, \mathbf{A}_{\mathbf{Y},p}^{(var)}]$ is the $Q \times pQ$ matrix of unknown coefficients, $\mathbf{y} = [\mathbf{Y}_{p+1}, \ldots, \mathbf{Y}_L]$ and $\mathbf{U}_\mathbf{Y}^{(var)} = [\mathbf{U}_{\mathbf{Y},p+1}^{(var)}, \ldots, \mathbf{U}_{\mathbf{Y},L}^{(var)}]$ are $Q \times (L-p)$ matrices, and $\mathbf{y}^p = [\mathbf{Y}_{p+1}^p, \ldots, \mathbf{Y}_L^p]$ is a $pQ \times (L-p)$ matrix collecting the regressors. The method estimates the coefficient matrices through the OLS formula, $\widehat{\mathbf{A}}_\mathbf{Y}^{(var)} = \mathbf{y}(\mathbf{y}^p)^\intercal[\mathbf{y}^p(\mathbf{y}^p)^\intercal]^{-1}$. The innovation process is estimated as the residual time-series $\widehat{\mathbf{U}}_\mathbf{Y}^{(var)} = \mathbf{y} - \widehat{\mathbf{A}}_\mathbf{Y}^{(var)}\mathbf{y}^p$, whose covariance matrix $\widehat{\mathbf{\Sigma}}_{\mathbf{U}_\mathbf{Y}}^{(var)}$ is an estimate of $\mathbf{\Sigma}_{\mathbf{U}_\mathbf{Y}}^{(var)}$. After identification, the model (2.6) can be analyzed in the frequency domain (see Sect. 2.4).

As regards the selection of the model order p, several criteria exist for its determination (see, e.g., [16, 23]). One commonly used approach is to set the order according to the *Akaike Information Criterion* (AIC) [24], or the *Bayesian Information Criterion* (BIC) [25]. The primary difference between AIC and BIC lies in how they penalize model complexity and their underlying theoretical foundations. AIC is based on information theory and aims to minimize the information lost when using a model to approximate the true process. It focuses on predictive accuracy and is more likely to select models that perform well for future data. The penalty for the number of parameters is more moderate with AIC, which indeed balances goodness of fit with model complexity but places less emphasis on the number of parameters. On the other hand, BIC approximates the posterior probability of a model given the data, and seeks the model that is most likely to be the true one, based on the given data, and is more concerned with identifying the correct model. Penalty grows with the sample size using BIC, which heavily penalizes models with more parameters, especially in larger datasets. Practically, it is crucial to find the right balance between excessively low orders, which might lead to an inadequate description of crucial oscillatory information in the vector process, and overly high orders, which could result in overfitting, with the outcome that the model captures not only the desired information but also includes noise.

2.3.5.1 Identification of Restricted Models

An issue with great practical relevance is that the order of the restricted models in Eqs. (2.7), (2.8), (2.9) is typically infinite and thus very difficult to identify from finite-length time series. The approach followed to face this issue in the context of causality analysis is essentially based on truncating the order of the restricted model to p, and estimating its parameters from the relevant subset of the original data. Though simple, this approach exposes to a trade-off between bias and variance of the estimates that prevents reliable model identification in most cases [26]. To solve this issue, methods which essentially extract the parameters of the restricted model from those of the full model have been proposed, i.e., methods based on *state-space (SS) models* [27, 28] and on the resolution of the *Yule-Walker (YW) equations* [6, 17, 19, 29]. In the following, the two methods will be thoroughly described in the context of VAR and ARX models.

(1) State-space models. The issue related to the formation of the restricted VAR models (2.7) can be overcome working in the frame of SS models [30]. This class of models is the most appropriate to use because it is closed under the formation of restricted models: in fact, any restricted process obtained from the VAR process (2.6) is actually a VAR process with a moving average component, or equivalently a finite-order SS process [28]. Therefore, using SS models allows to identify restricted models from the parameters of the original VAR model estimated with a single regression, thus guaranteeing high computational reliability. We exploit the SS modeling to describe the original process $\mathbf{Y}$ obeying the VAR representation (2.6) using the SS model

$$\mathbf{S}_{n+1} = \mathbf{A}\mathbf{S}_n + \mathbf{K}\mathbf{U}^{(var)}_{\mathbf{Y},n}, \tag{2.10a}$$
$$\mathbf{Y}_n = \mathbf{C}\mathbf{S}_n + \mathbf{U}^{(var)}_{\mathbf{Y},n}, \tag{2.10b}$$

where $\mathbf{S}_n = [\mathbf{Y}^\intercal_{n-1}, \ldots, \mathbf{Y}^\intercal_{n-p}]^\intercal$ is the pQ-dimensional state process and the SS parameters $(\mathbf{A}, \mathbf{C}, \mathbf{K}, \mathbf{V})$ are given by the matrices $\mathbf{C} = [\mathbf{A}^{(var)}_{\mathbf{Y},1}, \ldots, \mathbf{A}^{(var)}_{\mathbf{Y},p}]$, $\mathbf{K} = [\mathbf{I}_Q \mathbf{0}_{Q\times Q(p-1)}]^\intercal$, $\mathbf{A} = [\mathbf{C}; \mathbf{I}_{Q(p-1)}\mathbf{0}_{Q(p-1)\times Q}]$, and $\mathbf{V} = \mathbb{E}[\mathbf{U}^{(var)}_{\mathbf{Y},n}\mathbf{U}^{(var)\intercal}_{\mathbf{Y},n}] = \Sigma^{(var)}_{\mathbf{U}_\mathbf{Y}}$ ($\mathbf{I}$ and $\mathbf{0}$ are the identity and null matrices, respectively). Then, to represent the R-dimensional process $\mathbf{Z}$ (see (2.7)) formed by taking from $\mathbf{Y}$ the subset of processes indexed by the elements of $\mathbf{r} \subset \{1, \ldots, Q\}$, we replace (2.7) with a restricted SS model with state equation (2.10a) and observation equation $\mathbf{Z}_n = \mathbf{C}^{(\mathbf{r},:)}\mathbf{S}_n + \mathbf{W}_{\mathbf{Z},n}$. The parameters of the model are $(\mathbf{A}, \mathbf{C}^{(\mathbf{r},:)}, \mathbf{K}\mathbf{V}\mathbf{K}^\intercal, \mathbf{V}^{(\mathbf{r},\mathbf{r})}, \mathbf{K}\mathbf{V}^{(:,\mathbf{r})})$, where the superscripts denote selection of the rows and/or columns with indices $\mathbf{r}$ in a matrix. To exploit the restricted SS model for the linear causality analysis of $\mathbf{Z}$ it is necessary to lead its form back to that of (2.10), which reads [30]

$$\mathbf{S}_{n+1} = \tilde{\mathbf{A}}\mathbf{S}_n + \tilde{\mathbf{K}}\mathbf{W}_{\mathbf{Z},n}, \tag{2.11a}$$
$$\mathbf{Z}_n = \tilde{\mathbf{C}}\mathbf{S}_n + \mathbf{W}_{\mathbf{Z},n}. \tag{2.11b}$$

The parameters of the restricted model (2.11) are $(\tilde{\mathbf{A}}, \tilde{\mathbf{C}}, \tilde{\mathbf{K}}, \tilde{\mathbf{V}})$, of dimension $pQ \times pQ$, $R \times pQ$, $pQ \times R$, $R \times R$, and can be derived directly from the parameters $\mathbf{A}^{(var)}_{\mathbf{Y},k}$ and $\Sigma^{(var)}_{\mathbf{U}_\mathbf{Y}}$ of the original full VAR model (2.6) [30]: while the state and observation matrices are easily determined as $\tilde{\mathbf{A}} = \mathbf{A}$ and $\tilde{\mathbf{C}} = \mathbf{C}^{(\mathbf{r},:)}$, the gain $\tilde{\mathbf{K}}$ and the restricted innovation covariance $\tilde{\mathbf{V}} = \Sigma_{\mathbf{W}_\mathbf{Z}}$ must be obtained by solving a discrete algebraic Riccati equation (DARE) (see [30, 31] for detailed derivations). After identification, the model (2.11) can be analyzed in the frequency domain to study spectral interactions relevant to the subset $\mathbf{Z}$ (see Sect. 2.4).

The method based on SS models can also be applied to the ARX model (2.4) to derive the parameters of the two corresponding restricted AR models of Y_1, Y_2 in the form of (2.8), i.e., $\{b_{Y_1Y_1}, \lambda^2_{W_{Y_1|Y_1}}\}$ and $\{b_{Y_2Y_2}, \lambda^2_{W_{Y_2|Y_2}}\}$, respectively. However, since the (groups of) predicted variables and regressors must be the same in SS models, this approach is not feasible when dealing with restricted X models derived from ARX models, where the predictors are represented by the past states of the driver process while the predicted variable is the present state of the target. The X model parameter identification thus requires a different approach to the problem.

(2) Resolution of the Yule-Walker equations. The issue related to the formation of X restricted models from the ARX model (2.4) can be overcome by solving the YW equations. The restricted model coefficients, $b_{Y_2Y_2,k}$ and $b_{Y_2Y_1,k}$, and the variance of the residuals, $\lambda^2_{W_{Y_2|Y_2}}$ and $\lambda^2_{W_{Y_2|Y_1}}$, appearing in (2.8) and (2.9) respectively, can be identified starting from the covariance and cross-covariance matrices between the present and past variables of the two scalar processes Y_1 and Y_2 [19]. Using these matrices allows to identify restricted models from the parameters of the original ARX model estimated with a single regression up to an arbitrarily large order q, thus guaranteeing high computational reliability. For jointly Gaussian processes, these matrices contain as scalar elements the covariance between two time-lagged variables taken from the processes Y_1 and Y_2, which in turn appear as elements of the 2×2 autocovariance of the whole observed 2-dimensional process $\mathbf{Y}_n = [Y_{1,n} Y_{2,n}]^\intercal$, defined at each lag $k \geq 0$ as $\Gamma_k = \mathbb{E}[\mathbf{Y}_n \mathbf{Y}_{n-k}^\intercal]$. The procedure described in [6, 17] exploits the possibility to compute Γ_k from the parameters of the ARX formulation of the process $\mathbf{Y}_n$ via the well-known YW equations:

$$\Gamma_k = \sum_{l=1}^{p} \mathbf{A}_{\mathbf{Y},l}^{(arx)} \Gamma_{k-l} + \delta_{k0} \Sigma_{\mathbf{U_Y}}^{(arx)}, \tag{2.12}$$

where δ_{k0} is the Kronecher product. In order to solve this equation for Γ_k, with $k = 0, \ldots, p-1$, we first express the ARX model (2.4) in compact form as $\psi_n = \mathbf{A}\psi_{n-1} + \mathbf{E}_n$, where:

$$\begin{aligned}
\psi_n &= [\mathbf{Y}_n^\intercal \mathbf{Y}_{n-1}^\intercal, \ldots, \mathbf{Y}_{n-p+1}^\intercal]^\intercal; \\
\mathbf{A} &= \begin{bmatrix} \mathbf{A}_{\mathbf{Y},1}^{(arx)} & \ldots & \mathbf{A}_{\mathbf{Y},p-1}^{(arx)} & \mathbf{A}_{\mathbf{Y},p}^{(arx)} \\ \mathbf{I} & \ldots & \mathbf{0} & \mathbf{0} \\ \vdots & \ddots & \vdots & \vdots \\ \mathbf{0} & \ldots & \mathbf{I} & \mathbf{0} \end{bmatrix}; \\
\mathbf{E}_n &= [\mathbf{U}_{\mathbf{Y},n}^{(arx)\intercal} \mathbf{0}_{1\times 2(p-1)}]^\intercal.
\end{aligned} \tag{2.13}$$

Then, the $2p \times 2p$ covariance matrix of ψ_n, which is defined as $\Psi = \mathbb{E}[\psi_n \psi_n^\intercal]$ and has the form

$$\Psi = \begin{bmatrix} \Gamma_0 & \Gamma_1 & \dots & \Gamma_{p-1} \\ \Gamma_1^\intercal & \Gamma_0 & \dots & \Gamma_{p-2} \\ \vdots & \vdots & \ddots & \vdots \\ \Gamma_{p-1}^\intercal & \Gamma_{p-2}^\intercal & \dots & \Gamma_0 \end{bmatrix}, \tag{2.14}$$

can be expressed as $\Psi = \mathbf{A}\Psi\mathbf{A}^\intercal + \Xi$ where $\Xi = \mathbb{E}[\mathbf{E}_n\mathbf{E}_n^\intercal]$ is the $2p \times 2p$ covariance of $\mathbf{E}_n$. This last equation is a discrete-time Lyapunov equation, which can be solved for Ψ yielding the autocovariance matrices $\Gamma_0, \dots, \Gamma_{p-1}$ [17]. Note that $\Gamma_0 \equiv \Sigma_{\mathbf{Y}}$. Finally, the autocovariance can be calculated recursively for any lag $k \geq p$ by repeatedly applying YW equations (2.12) up to the desired lag q, starting from the parameters of the ARX representation (2.4) of the observed Gaussian vector process $\mathbf{Y}$.

To summarize, the above-described procedure is based first on computing the autocovariance sequence of the bivariate process $\mathbf{Y}$ from its parameters ($\mathbf{A}_{\mathbf{Y},l}^{(arx)}$, with $l = 1, \dots, p$, and $\Sigma_{\mathbf{U_Y}}^{(arx)}$), which are previously identified through the vector OLS approach, and then on rearranging the elements of the autocovariance matrices for building the auto- and cross-covariances to be used in the computation of the AR parameters $\{b_{Y_2Y_2,k}, \lambda^2_{W_{Y_2|Y_2}}\}$, and of the X parameters $\{b_{Y_2Y_1,k}, \lambda^2_{W_{Y_2|Y_1}}\}$, appearing in (2.8) and (2.9) respectively. The identification of the restricted models (2.8) and (2.9) can be represented in the frequency domain to study spectral patterns of causality and autonomy (see Sect. 2.4).

The parameter determining the accuracy of the procedure is the number of lags used to truncate the past history of the process: considering the past up to lag q corresponds to calculating the autocovariance of the process (2.4) up to the matrix Γ_q. Given that the autocovariance of a stable vector AR process decays exponentially with the lag, with a rate of decay depending on the modulus of the largest eigenvalue of $\mathbf{A}$, $\rho(A)$, it has been suggested to compute the autocovariance up to a lag q such that $\rho(A)$ is smaller than a predefined numerical tolerance [29]. It has been found that computation of very long autocovariance sequences is not necessary for the purpose of evaluating information dynamics, because all measures stabilize to constant values already for small lags (typically $q = 10$) even for reasonably high values of $\rho(A)$ [17, 32, 33]. Remarkably, the procedure described above yields results similar to the method of SS models with q sufficiently large.

Restricted AR model. The AR model (2.8) can be written in compact form as $Y_{2,n} = \mathbf{B}_{Y_2Y_2}\mathbf{Y}_{2,n}^q + W_{Y_2|Y_2,n}$, where $\mathbf{B}_{Y_2Y_2} = [b_{Y_2Y_2,1}, \dots, b_{Y_2Y_2,q}]$ is the vector collecting all coefficients up to lag q. From this representation, taking the expectation $\mathbb{E}[Y_{2,n}\mathbf{Y}_{2,n}^{q\intercal}]$ and solving for $\mathbf{B}_{Y_2Y_2}$ yields:

$$\mathbf{B}_{Y_2Y_2} = \Sigma_{Y_{2,n},Y_{2,n}^q} \cdot \Sigma_{Y_{2,n}^q}^{-1}, \tag{2.15}$$

where $\Sigma_{Y_{2,n}^q}$ is the $q \times q$ autocovariance matrix of $\mathbf{Y}_{2,n}^q$ defined as $\Sigma_{Y_{2,n}^q} = \mathbb{E}[\mathbf{Y}_{2,n}^q\mathbf{Y}_{2,n}^{q\intercal}]$, while $\Sigma_{Y_{2,n},Y_{2,n}^q}$ is the $1 \times q$ cross-covariance matrix of $Y_{2,n}$ and $\mathbf{Y}_{2,n}^q$, defined as

$\Sigma_{Y_{2,n}Y_{2,n}^q} = \mathbb{E}[Y_{2,n}\mathbf{Y}_{2,n}^{q\intercal}]$. The matrices $\Sigma_{Y_{2,n}^q}$ and $\Sigma_{Y_{2,n}Y_{2,n}^q}$ are extracted from Γ_k. Then, the variance of the AR residuals $\lambda^2_{W_{Y_2|Y_2}}$ in (2.8) is computed as [19, 34]:

$$\lambda^2_{W_{Y_2|Y_2}} = \sigma^2_{Y_2} - \Sigma_{Y_{2,n},Y_{2,n}^q} \cdot \Sigma^{-\intercal}_{Y_{2,n}^p} \cdot \Sigma^{\intercal}_{Y_{2,n},Y_{2,n}^q}, \tag{2.16}$$

where $\sigma^2_{Y_2}$ is the variance of Y_2.

Restricted X model. Analogously to (2.15), the X model coefficients $b_{Y_2Y_1,k}$ in (2.9) are computed as:

$$\mathbf{B}_{Y_2Y_1} = \Sigma_{Y_{2,n},Y_{1,n}^q} \cdot \Sigma^{-1}_{Y_{1,n}^q}, \tag{2.17}$$

where $\mathbf{B}_{Y_2Y_1} = [b_{Y_2Y_1,1}, \ldots, b_{Y_2Y_1,q}]$, $\Sigma_{Y_{1,n}^q}$ is the $q \times q$ autocovariance matrix of $\mathbf{Y}_{1,n}^q$ defined as $\Sigma_{Y_{1,n}^q} = \mathbb{E}[\mathbf{Y}_{1,n}^q \mathbf{Y}_{1,n}^{q\intercal}]$, while $\Sigma_{Y_{2,n},Y_{1,n}^q}$ is the cross-covariance matrix of $Y_{2,n}$ and $\mathbf{Y}_{1,n}^q$, defined as $\Sigma_{Y_{2,n},Y_{1,n}^q} = \mathbb{E}[Y_{2,n}\mathbf{Y}_{1,n}^{q\intercal}]$. The matrices $\Sigma_{Y_{1,n}^q}$ and $\Sigma_{Y_{2,n},Y_{1,n}^q}$ are extracted from Γ_k. Then, the variance of the X residuals $\lambda^2_{W_{Y_2|Y_1}}$ in (2.9) is computed as [19]:

$$\lambda^2_{W_{Y_2|Y_1}} = \sigma^2_{Y_2} - \Sigma_{Y_{2,n},Y_{1,n}^q} \cdot \Sigma^{-\intercal}_{Y_{1,n}^q} \cdot \Sigma^{\intercal}_{Y_{2,n},Y_{1,n}^q}. \tag{2.18}$$

2.4 Dynamic Models of Random Processes in the Frequency Domain

In the linear signal processing framework, the VAR model (2.6) can be suitably represented in the frequency domain. To this end, the Fourier Transform (FT) of (2.6) is computed to derive

$$\mathbf{Y}(\omega) = \left[\mathbf{I} - \sum_{k=1}^{p} \mathbf{A}^{(var)}_{\mathbf{Y},k} e^{-\mathbf{j}\omega k}\right]^{-1} \mathbf{U}^{(var)}_{\mathbf{Y}}(\omega) = \mathbf{H}^{\mathbf{Y}(var)}(\omega)\mathbf{U}^{(var)}_{\mathbf{Y}}(\omega), \tag{2.19}$$

where $\mathbf{Y}(\omega)$ and $\mathbf{U}^{(var)}_{\mathbf{Y}}(\omega)$ are the FTs of $\mathbf{Y}_n$ and $\mathbf{U}^{(var)}_{\mathbf{Y},n}$, $\omega \in [-\pi, \pi]$ is the normalized angular frequency ($\omega = 2\pi\frac{f}{f_s} = 2\pi\bar{f}$, with $\bar{f} = f/f_s$, $f \in [-\frac{f_s}{2}, \frac{f_s}{2}]$, being f_s the sampling frequency of the processes), $\mathbf{j} = \sqrt{-1}$ and $\mathbf{I}$ is the Q-dimensional identity matrix. The $Q \times Q$ matrix $\mathbf{H}^{\mathbf{Y}(var)}(\omega)$ contains the transfer functions (TF) relating the FTs of the innovation processes in $\mathbf{U}^{(var)}_{\mathbf{Y}}$ to the FTs of the processes in $\mathbf{Y}$. Computing $\mathbf{H}^{\mathbf{Y}(var)}(\omega)$ on the unit circle in the complex plane ($\mathbf{H}^{\mathbf{Y}(var)}(\bar{f}) = \mathbf{H}^{\mathbf{Y}(var)}(\omega)|_{z=e^{\mathbf{j}\omega}}$) and exploiting spectral factorization, it is possible to derive the $Q \times Q$ power spectral density (PSD) matrix of the stationary vector random process $\mathbf{Y} = \{Y_1, \ldots, Y_Q\}$ in the frequency domain as

$$\mathbf{P}^{(var)}_{\mathbf{Y}}(\bar{f}) = \mathbf{H}^{\mathbf{Y}(var)}(\bar{f})\Sigma^{(var)}_{\mathbf{U}_{\mathbf{Y}}}\mathbf{H}^{\mathbf{Y}(var)*}(\bar{f}), \tag{2.20}$$

where $*$ stands for the Hermitian transpose. The PSD matrix contains the individual PSD of the process Y_i, $P^{(var)}_{Y_i}(\bar{f})$, as ith diagonal element and the cross-PSD between the processes Y_i and Y_j, $P^{(var)}_{Y_iY_j}(\bar{f})$, as off-diagonal elements in the position ij $(i, j = 1, \ldots, Q)$. When the subdivision of $\mathbf{Y}$ in M blocks is considered to yield $\mathbf{X} = \{\mathbf{X}_1, \ldots, \mathbf{X}_M\}$, the TF matrix matrix can be partitioned in $M \times M$ blocks to evidence the spectral properties related to the internal dynamics, through the $M_i \times M_i$ diagonal blocks $\mathbf{H}^{\mathbf{Y}(var)}_{\mathbf{X}_i\mathbf{X}_i}(\omega)$, or to the causal interactions between $\mathbf{X}_i$ and $\mathbf{X}_j$, through the $M_i \times M_j$ off-diagonal blocks $\mathbf{H}^{\mathbf{Y}(var)}_{\mathbf{X}_i\mathbf{X}_j}(\omega)$, $i, j \in \{1, \ldots, M\}, i \neq j$. Similarly, the PSD matrix $\mathbf{P}^{(var)}_{\mathbf{Y}}(\bar{f})$ can be partitioned in $M \times M$ blocks as follows:

$$\mathbf{P}^{(var)}_{\mathbf{X}}(\bar{f}) = \begin{bmatrix} \mathbf{P}^{(var)}_{\mathbf{X}_1}(\bar{f}) & \cdots & \mathbf{P}^{(var)}_{\mathbf{X}_1\mathbf{X}_M}(\bar{f}) \\ \vdots & \ddots & \vdots \\ \mathbf{P}^{(var)}_{\mathbf{X}_M\mathbf{X}_1}(\bar{f}) & \cdots & \mathbf{P}^{(var)}_{\mathbf{X}_M}(\bar{f}) \end{bmatrix}, \tag{2.21}$$

where the ijth block has dimension $M_i \times M_j$.

The PSD matrix (2.20) can be estimated as $\widehat{\mathbf{P}}^{(var)}_{\mathbf{Y}}(\bar{f}) = \widehat{\mathbf{H}}^{\mathbf{Y}(var)}(\bar{f})\widehat{\mathbf{\Sigma}}^{(var)}_{\mathbf{U}_\mathbf{Y}}\widehat{\mathbf{H}}^{\mathbf{Y}(var)*}(\bar{f})$, where the transfer matrix is $\widehat{\mathbf{H}}^{\mathbf{Y}(var)}(\bar{f}) = [\mathbf{I} - \widehat{\mathbf{A}}^{(var)}_{\mathbf{Y}}(\bar{f})]^{-1}$, with $\widehat{\mathbf{A}}^{(var)}_{\mathbf{Y}}(\bar{f}) = \sum_{k=1}^{p} \widehat{\mathbf{A}}^{(var)}_{\mathbf{Y},k} e^{-\mathrm{j}2\pi\bar{f}k}$ (see Sect. 2.3.5 for the VAR identification procedure).

The frequency domain representation of the AR (2.2) and ARX (2.4) models, though not shown here for brevity, follows the footprints of the representation given here for VAR models, with suitable mathematical adaptations related to the dimensions of the involved processes.

2.4.1 Restricted Models in the Frequency Domain

(1) Frequency-domain expansion of SS VAR models. The linear parametric representation of the dynamic interactions among the selected subset of processes can be translated in the frequency domain. Starting from the subset $\mathbf{Z}$ of the observed multivariate process, described by the SS model (2.11), taking the FT of the state equation (2.11a) yields

$$\mathbf{S}(\bar{f}) = \tilde{\mathbf{A}}\mathbf{S}(\bar{f})e^{-\mathrm{j}2\pi\bar{f}} + \tilde{\mathbf{K}}\mathbf{W}_{\mathbf{Z}}(\bar{f})e^{-\mathrm{j}2\pi\bar{f}}, \tag{2.22}$$

where $\mathbf{S}(\bar{f})$ and $\mathbf{W}_{\mathbf{Z}}(\bar{f})$ are the FTs of $\mathbf{Z}_n$ and $\mathbf{W}_{\mathbf{Z},n}$. From (2.22) it is easy to derive the PSD of the state process, $\mathbf{S}(\bar{f})$, to be substituted in the FT of (2.11b) to obtain $\mathbf{Z}(\bar{f}) = \tilde{\mathbf{H}}^{\mathbf{Z}}(\bar{f})\mathbf{W}_{\mathbf{Z}}(\bar{f})$, which evidences the TF matrix

$$\tilde{\mathbf{H}}^{\mathbf{Z}}(\bar{f}) = \left(\mathbf{I}_R + \tilde{\mathbf{C}}[\mathbf{I}_{pQ} - \tilde{\mathbf{A}}e^{-\mathrm{j}2\pi\bar{f}}]^{-1}\tilde{\mathbf{K}}e^{-\mathrm{j}2\pi\bar{f}}\right)\mathbf{W}_{\mathbf{Z}}(\bar{f}). \tag{2.23}$$

The $R \times R$ matrix $\tilde{\mathbf{H}}^{\mathbf{Z}}(\bar{f})$ contains the TFs relating the FTs of the innovation processes in $\mathbf{W_Z}$ to the FTs of the processes in $\mathbf{Z}$, and can be used together with the innovation covariance matrix to derive the $R \times R$ PSD matrix of the process $\mathbf{Z}$ using spectral factorization:

$$\mathbf{S_Z}(\bar{f}) = \tilde{\mathbf{H}}^{\mathbf{Z}}(\bar{f})\Sigma_{\mathbf{W_Z}}\tilde{\mathbf{H}}^{\mathbf{Z}*}(\bar{f}). \tag{2.24}$$

Assuming that, e.g., $\mathbf{Z} = \{\mathbf{Z}_1, \mathbf{Z}_2\}$, with the two blocks having dimensions R_1 and R_2, the matrix $\mathbf{S_Z}(\bar{f})$ can be then factorized in blocks to make explicit the PSDs of $\mathbf{Z}_1$ and $\mathbf{Z}_2$, $\mathbf{S}_{\mathbf{Z}_1}(\bar{f})$ and $\mathbf{S}_{\mathbf{Z}_2}(\bar{f})$, as diagonal blocks, and the cross-spectral densities between $\mathbf{Z}_1$ and $\mathbf{Z}_2$, $\mathbf{S}_{\mathbf{Z}_1\mathbf{Z}_2}(\bar{f})$ and $\mathbf{S}_{\mathbf{Z}_2\mathbf{Z}_1}(\bar{f})$, as off-diagonal blocks. This representation allows to define spectral measures of coupling and causality for VAR processes (see Chap. 4).

(2) Frequency-domain expansion of AR and X models. Here, we provide mathematical details regarding the spectral representation of restricted AR and X models derived from the ARX model (2.4).

Specifically, the spectral behaviour of the restricted AR model (2.8) can be straightforwardly derived from the diagonal element $P_{Y_2}^{(arx)}(\bar{f})$ of the 2×2 PSD matrix of the ARX model (2.4), i.e., $\mathbf{P}_{\mathbf{Y}}^{(arx)}(\bar{f}) = \begin{bmatrix} P_{Y_1}^{(arx)}(\bar{f}) & P_{Y_1Y_2}^{(arx)}(\bar{f}) \\ P_{Y_2Y_1}^{(arx)}(\bar{f}) & P_{Y_2}^{(arx)}(\bar{f}) \end{bmatrix}$. This approach is equivalent to identify an AR model on the target as in (2.2), where the PSD can be written as $P_Y^{(ar)}(\bar{f}) = |H^{Y(ar)}(\bar{f})|^2\sigma_{U_Y}^{2\ (ar)}$, with $|\cdot|$ standing for matrix determinant and $H^{Y(ar)}(\bar{f}) = [1 - \sum_{k=1}^{p} a_{Y,k}^{(ar)} e^{-\mathrm{j}2k\pi\bar{f}}]^{-1}$ being the TF relating the FT of the innovation process U_Y to the FT of the process Y. The two equivalent representations allow to define a spectral measure of causality for ARX processes (see Chap. 4) [19].

On the other hand, deriving the spectral representation of restricted X models in the frequency domain is more challenging, though straightforward whether the X model parameters are directly derived from the ARX model parameters through resolution of the YW equations (see Sect. 2.3.5.1). To this end, the following linear model formed by the Eqs. (2.4a) and (2.9) taken together, being representative of the driver and the target dynamics, respectively

$$Y_{1,n} = \sum_{k=1}^{p} a_{Y_1Y_1,k}^{(arx)} Y_{1,n-k} + a_{Y_1Y_2,k}^{(arx)} Y_{2,n-k} + U_{Y_1,n}^{(arx)}, \tag{2.25a}$$

$$Y_{2,n} = \sum_{k=1}^{\infty} b_{Y_2Y_1,k} Y_{1,n-k} + W_{Y_2|Y_1,n}, \tag{2.25b}$$

can be represented in the Z domain as $\mathbf{Y}(z) = \mathbf{G}^{\mathbf{Y}}(z)\mathbf{W_Y}(z)$, where $\mathbf{W_Y}(z)$ is the Z-transform of the noise vector $\mathbf{W}_{\mathbf{Y},n} = [U_{Y_1,n}^{(arx)} W_{Y_2|Y_1,n}]^\intercal$ and the 2×2 transfer matrix is

$$\mathbf{G}^{\mathbf{Y}}(z) = \begin{bmatrix} G_{Y_1Y_1}^{\mathbf{Y}}(z) & G_{Y_1Y_2}^{\mathbf{Y}}(z) \\ G_{Y_2Y_1}^{\mathbf{Y}}(z) & G_{Y_2Y_2}^{\mathbf{Y}}(z) \end{bmatrix} = \begin{bmatrix} 1 - A_{Y_1Y_1}^{(arx)}(z) & -A_{Y_1Y_2}^{(arx)}(z) \\ -B_{Y_2Y_1}(z) & 1 \end{bmatrix}^{-1}, \tag{2.26}$$

with $A^{(arx)}_{Y_1Y_1}(z) = \sum_{k=1}^{p} a^{(arx)}_{Y_1Y_1,k} z^{-k}$, $A^{(arx)}_{Y_1Y_2}(z) = \sum_{k=1}^{p} a^{(arx)}_{Y_1Y_2,k} z^{-k}$, $B_{Y_2Y_1}(z) = \sum_{k=1}^{p} b_{Y_2Y_1,k} z^{-k}$. Computing $\mathbf{G}^{\mathbf{Y}}(z)$ on the unit circle of the complex plane ($z = e^{\mathrm{j}2\pi\bar{f}}$) yields the 2×2 complex transfer function in the frequency domain, $\mathbf{G}^{\mathbf{Y}}(\bar{f})$. This representation allows to define a spectral measure of autonomy for ARX processes (see Chap. 4) [19].

2.5 Summary

This chapter provided an overview of the theoretical background related to the linear parametric representation of static and dynamic network systems. Remarkably, the formalism introduced here allowed the expansion of the dynamic information-theoretic measures quantifying single-node, pairwise and high-order network activity in the frequency domain, which is essential to investigate the oscillatory content of individual time series and to retrieve amounts of information shared by the observed processes within specific frequency bands with plausible physiological meaning.

First, the probabilistic concepts of random (i.e., *stochastic*) variables and processes have been deepened. Stochastic interactions refer to the ways in which random variables and processes influence each other in network systems. These interactions can be broadly categorized into static and dynamic, depending on whether the system's behavior involves time-dependent processes or remains time-invariant. *Static interactions* involve static systems $\mathcal{V}_1, \ldots, \mathcal{V}_M$ where randomness affects variables in a time-independent context, i.e., interactions are contextualized without accounting for the temporal dimension and are characterized by probabilistic distributions, moments (mean, variance) and correlations. The focus here is on *random variables*, i.e., mathematical variables whose values are subject to variations due to chance. The multiple interactions between M random variables $V_1, \ldots, V_M$ can be investigated by analyzing multiple realizations of these variables available in the form of multiple data sequences. Under the assumption of jointly Gaussian variables, the analysis can be performed by exploiting linear parametric regression models. *Dynamic interactions* occur in dynamic systems $\mathcal{X}_1, \ldots, \mathcal{X}_M$ where randomness unfolds over time, influencing processes that evolve dynamically. Such systems are modeled using *random processes*, which can be thought as sequences of random variables ordered according to time. In this context, the multiple interactions between M processes $X_1, \ldots, X_M$ can be investigated by means of a dynamic analysis of multiple realizations of these processes available in the form of time series. Parametric autoregressive models can be exploited for analyzing and predicting time series data, while spectral analysis is essential for characterizing the frequency content of stochastic interactions among these processes. The processes are assumed to be stationary and jointly Gaussian distributed to exploit the formalism linking measures of information rates with linear regression models and spectral quantities.

Distinct linear models can be identified depending on the number of observed units (i.e., related to univariate, bivariate and multivariate analysis, the latter referred to as high-order network analysis) and the order of interactions (i.e., relationships or patterns involving single-node $\mathbb{I}_1$, pairwise $\mathbb{I}_2$ or high-order $\mathbb{I}_3$ activities). *Univariate analysis* (UA) examines a single network node, focusing on the dynamics of the stochastic variable or process describing the activity at the node. In the case of a dynamic system whose activity is described by a stationary Gaussian (possibly vector) process, this type of analysis is generally performed via linear autoregressive (AR) models relating the present state of the process to its past states. *Bivariate analysis* (BA) examines the relationship between two nodes, and identifies correlations or causal links exploring how one variable/process might influence the other one. Auto- and cross-regressive (ARX) models are exploited to describe the linear dependencies among two dynamic systems under the hypotheses of stationarity and gaussianity of the observed data. *High-order network analysis* (HONA) involves examining three or more variables simultaneously. It focuses on understanding the redundant/synergistic relationships among the data, often dealing with complex, interdependent variables/processes whose linear dependencies are described by means of vector AR (VAR) models under the hypotheses of stationarity and gaussianity of the observed data. A hierarchical HONA (hHONA) is performed to investigate the role of the single node, the pair of nodes and the group of nodes (multiplet) in sharing information with the rest of the system.

The identification procedure of a VAR model is typically performed by means of estimation methods based on minimizing the prediction error. The most common estimator is the multivariate version of the ordinary least-squares (OLS) method, where model order is typically selected via the Akaike Information Criterion (AIC) or the Bayesian Information Criterion (BIC). Since working with bivariate or multivariate datasets, as well as dealing with different orders of interactions (i.e., with patterns involving single-node, pairwise or high-order activities), requires the construction of restricted models involving only subsets of data, it is crucial to face the issue related to the order of these models which is typically infinite and thus very difficult to identify from finite-length time series. In this book, we worked with methods which essentially extract the parameters of the restricted model from those of the full model, i.e., methods based on state-space (SS) models and on the resolution of the Yule-Walker (YW) equations, without the need to estimate their coefficients from the observed subsets of data.

References

1. Baxter P Rogers et al. "Assessing functional connectivity in the human brain by fMRI". In: *Magnetic resonance imaging* 25.10 (2007), pp. 1347–1357.
2. Martijn P Van Den Heuvel and Hilleke E Hulshoff Pol. "Exploring the brain network: a review on resting-state fMRI functional connectivity". In: *European neuropsychopharmacology* 20.8 (2010), pp. 519–534.

3. Karl J Friston. "Functional and effective connectivity in neuroimaging: a synthesis". In: *Human brain mapping* 2.1-2 (1994), pp. 56–78.
4. Laura Sparacino et al. "Statistical Approaches to Identify Pairwise and High-Order Brain Functional Connectivity Signatures on a Single-Subject Basis". In: *Life* 13.10 (2023), p. 2075.
5. Michael Wibral, Raul Vicente, and Joseph T Lizier. *Directed information measures in neuroscience*. Vol. 724. Springer, 2014.
6. Luca Faes et al. "Information decomposition in multivariate systems: definitions, implementation and application to cardiovascular networks". In: *Entropy* 19.1 (2016), p. 5.
7. Federico Battiston et al. "Networks beyond pairwise interactions: Structure and dynamics". In: *Physics reports* 874 (2020), pp. 1–92.
8. Plamen Ch Ivanov. "The new field of network physiology: building the human physiolome". In: *Frontiers in Network Physiology* 1 (2021), p. 711778.
9. Adam B Barrett, Lionel Barnett, and Anil K Seth. "Multivariate Granger causality and generalized variance". In: *Physical Review E* 81.4 (2010), p. 041907.
10. Daniel Chicharro. "On the spectral formulation of Granger causality". In: *Biological cybernetics* 105 (2011), pp. 331–347.
11. Luca Faes et al. "Information decomposition in the frequency domain: A new framework to study cardiovascular and cardiorespiratory oscillations". In: *Philosophical Transactions of the Royal Society A* 379.2212 (2021), p. 20200250.
12. Yuri Antonacci et al. "Measuring high-order interactions in rhythmic processes through multivariate spectral information decomposition". In: *IEEE Access* 9 (2021), pp. 149486–149505.
13. Ilkka Korhonen et al. "Linear multivariate models for physiological signal analysis: theory". In: *Computer Methods and Programs in Biomedicine* 51.1-2 (1996), pp. 85–94.
14. Giuseppe Baselli et al. "Spectral decomposition in multichannel recordings based on multivariate parametric identification". In: *IEEE transactions on biomedical engineering* 44.11 (1997), pp. 1092–1101.
15. Riccardo Pernice et al. "Spectral decomposition of cerebrovascular and cardiovascular interactions in patients prone to postural syncope and healthy controls". In: *Autonomic Neuroscience* (2022), p. 103021.
16. Helmut Lütkepohl. *New introduction to multiple time series analysis*. Springer Science & Business Media, 2005.
17. Luca Faes, Alberto Porta, and Giandomenico Nollo. "Information decomposition in bivariate systems: theory and application to cardiorespiratory dynamics". In: *Entropy* 17.1 (2015), pp. 277–303.
18. Luca Faes et al. "A new framework for the time-and frequency-domain assessment of high-order interactions in networks of random processes". In: *IEEE Transactions on Signal Processing* 70 (2022), pp. 5766–5777.
19. Laura Sparacino et al. "A method to assess granger causality, isolation and autonomy in the time and frequency domains: theory and application to cerebrovascular variability". In: *IEEE Transactions on Biomedical Engineering* (2023).
20. Steven M Kay. *Modern spectral estimation: theory and application*. Pearson Education India, 1988.
21. Alois Schlögl. "A comparison of multivariate autoregressive estimators". In: *Signal processing* 86.9 (2006), pp. 2426–2429.
22. Yuri Antonacci et al. "Information transfer in linear multivariate processes assessed through penalized regression techniques: validation and application to physiological networks". In: *Entropy* 22.7 (2020), p. 732.
23. Mahmood Karimi. "Order selection criteria for vector autoregressive models". In: *Signal Processing* 91.4 (2011), pp. 955–969.
24. Hirotugu Akaike. "A new look at the statistical model identification". In: *IEEE transactions on automatic control* 19.6 (1974), pp. 716–723.
25. Gideon Schwarz. "Estimating the dimension of a model". In: *The annals of statistics* (1978), pp. 461–464.

26. Patrick A Stokes and Patrick L Purdon. "A study of problems encountered in Granger causality analysis from a neuroscience perspective". In: *Proceedings of the national academy of sciences* 114.34 (2017), E7063–E7072.
27. Luca Faes, Sebastiano Stramaglia, and Daniele Marinazzo. "On the interpretability and computational reliability of frequency-domain Granger causality". In: *F1000Research* 6 (2017).
28. Lionel Barnett, Adam B Barrett, and Anil K Seth. "Solved problems for Granger causality in neuroscience: A response to Stokes and Purdon". In: *NeuroImage* 178 (2018), pp. 744–748.
29. Lionel Barnett and Anil K Seth. "The MVGC multivariate Granger causality toolbox: a new approach to Granger-causal inference". In: *Journal of neuroscience methods* 223 (2014), pp. 50–68.
30. Lionel Barnett and Anil K Seth. "Granger causality for state-space models". In: *Physical Review E* 91.4 (2015), p. 040101.
31. Luca Faes, Daniele Marinazzo, and Sebastiano Stramaglia. "Multiscale information decomposition: Exact computation for multivariate Gaussian processes". In: *Entropy* 19.8 (2017), p. 408.
32. Luca Faes et al. "Information decomposition of short-term cardiovascular and cardiorespiratory variability". In: *Computing in Cardiology 2013*. IEEE. 2013, pp. 113–116.
33. Luca Faes et al. "Estimating the decomposition of predictive information in multivariate systems". In: *Physical Review E* 91.3 (2015), p. 032904.
34. Lionel Barnett, Adam B Barrett, and Anil K Seth. "Granger causality and transfer entropy are equivalent for Gaussian variables". In: *Physical review letters* 103.23 (2009), p. 238701.

Chapter 3
Static Networks of Random Variables

This chapter reviews the entropy measures adopted to describe static networks mapped by random variables, and relevant to the information-theoretic analysis of random processes that will be discussed in the next chapter (Chap. 4).

Let us consider the static network system $\mathcal{V}$ composed of M nodes. The activities at the single nodes, as well as the interactions between the activities of two nodes, can be assessed straightforwardly using univariate and bivariate entropy-based measures of static interactions, e.g., entropy and mutual information (Sect. 3.1, Fig. 3.1). However, when more than two units are considered, more sophisticated approaches are needed to examine HOIs at different levels of resolution, i.e., characterizing the role of the single node within the selected subset (*node-specific analysis*), the nature of the link between two nodes with respect to the remaining nodes (*link-specific analysis*), as well as the overall activity of groups of nodes or the whole network (*network-specific analysis*), as depicted in Fig. 3.1. To this end, entropy-based measures of static high-order interactions, e.g., Interaction Information, O-Information, and the B-Index (Fig. 3.1), are defined to provide a more comprehensive and detailed description of the analysed network system and will be thoroughly discussed in Sect. 3.2. Simulation examples of the static interaction measures at different orders are instead provided in Sect. 3.3 to show the drawbacks and the strengths of the proposed framework.

3.1 Single-Node and Pairwise Connectivity

By using an information-theoretic perspective, the activity at the ith node of the static network system $\mathcal{V}$ can be assessed through the information carried by the (possibly vector) variable $\mathbf{V}_i$ assuming values v_i in a continuous domain, which is quantified by the well-known Shannon *entropy* (EN) measure [1, 2]:

$$H(\mathbf{V}_i) = \mathbb{E}\left[\log \frac{1}{p_{\mathbf{V}_i}(\mathbf{v}_i)}\right], \tag{3.1}$$

L. Sparacino, *Spectral Information Dynamics in Network Neuroscience and Physiology*, Studies in Computational Intelligence 1235,
https://doi.org/10.1007/978-3-032-05416-6_3

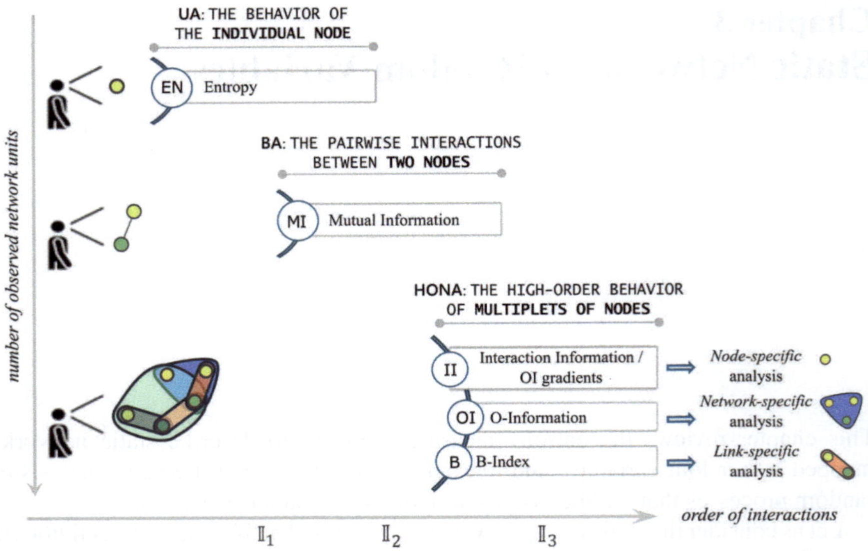

Fig. 3.1 Measures of static interaction in a system of multiple nodes

while the interaction between the activities of the ith and jth nodes can be assessed through the *mutual information* (MI) between $\mathbf{V}_i$ and $\mathbf{V}_j$:

$$I(\mathbf{V}_i; \mathbf{V}_j) = \mathbb{E}\left[\log \frac{p_{\mathbf{V}_i,\mathbf{V}_j}(\mathbf{v}_i, \mathbf{v}_j)}{p_{\mathbf{V}_i}(\mathbf{v}_i)p_{\mathbf{V}_j}(\mathbf{v}_j)}\right]. \quad (3.2)$$

The entropy quantifies the information contained in a random variable intended as the average uncertainty about its outcomes, while the MI quantifies the information shared by two variables intended as the uncertainty about one variable that is resolved by knowing the other variable. Remarkably, the MI is a symmetric measure (i.e., $I(\mathbf{V}_i; \mathbf{V}_j) = I(\mathbf{V}_j; \mathbf{V}_i)$), and is linked to the joint and individual entropies of the two variables by the relation

$$I(\mathbf{V}_i; \mathbf{V}_j) = H(\mathbf{V}_i) + H(\mathbf{V}_j) - H(\mathbf{V}_i, \mathbf{V}_j). \quad (3.3)$$

The latter can be also expressed as $I(\mathbf{V}_i; \mathbf{V}_j) = H(\mathbf{V}_i) - H(\mathbf{V}_i|\mathbf{V}_j)$, where

$$H(\mathbf{V}_i|\mathbf{V}_j) = H(\mathbf{V}_i, \mathbf{V}_j) - H(\mathbf{V}_j) \quad (3.4)$$

is the conditional entropy (CE) of $\mathbf{V}_i$ given $\mathbf{V}_j$, quantifying the information carried by one variable that is not shared with the other, intended as the residual uncertainty which remains in one variable when the other variable is known. The relations between entropy, MI and CE are depicted in the Venn diagram representation in

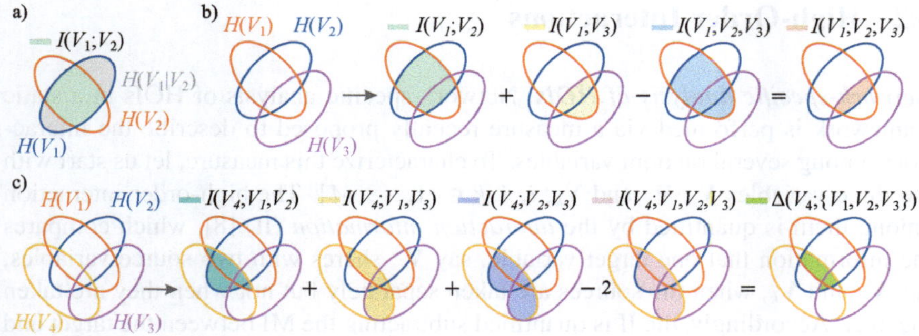

Fig. 3.2 **a** Mutual information between two scalar random variables V_1 and V_2, $I(V_1; V_2)$. **b** Interaction information among the three variables V_1, V_2 and V_3, $I(V_1; V_2; V_3)$. **c** Gradient of the O-information when V_4 is added to the group $\{V_1, V_2, V_3\}$, $\Delta(V_4; \{V_1, V_2, V_3\})$. The figure is adapted from [7]

Fig. 3.2a. The role of the MI as a measure of pairwise connectivity in networks of multiple interacting nodes has been investigated in the frame of functional magnetic resonance imaging data in [3–5]. The application of the measures to fMRI data will be presented and discussed in Sect. 6.1.

3.1.1 *Linear Parametric Formulation*

Let us consider the linear regression model in (2.1). The MI between the two variables $\mathbf{V}_i$ and $\mathbf{V}_j$ can be estimated exploiting the relation between entropy and variance valid for Gaussian variables [6], i.e. expressing the entropy of the predicted variable $\mathbf{V}_i$ as

$$H(\mathbf{V}_i) = \frac{1}{2}\log\left((2\pi e)^{v_i}|\Sigma_{\mathbf{V}_i}|\right), \tag{3.5}$$

and the CE of the predicted variable $\mathbf{V}_i$ given the predictor $\mathbf{V}_j$ as

$$H(\mathbf{V}_i|\mathbf{V}_j) = \frac{1}{2}\log\left((2\pi e)^{v_i}|\Sigma_{\mathbf{U}_j}|\right), \tag{3.6}$$

where $\Sigma_{\mathbf{V}_i}$ is the $v_i \times v_i$ covariance matrix of the predicted variable $\mathbf{V}_i$ and $\Sigma_{\mathbf{U}_j}$ is the $v_i \times v_i$ covariance matrix of the prediction error $\mathbf{U}_j$. Then, combining (3.4), (3.5) and (3.6), the MI (3.3) can be expressed as

$$I(\mathbf{V}_i, \mathbf{V}_j) = \frac{1}{2}\log\left(\frac{|\Sigma_{\mathbf{V}_i}|}{|\Sigma_{\mathbf{U}_j}|}\right). \tag{3.7}$$

3.2 High-Order Interactions

Network-specific analysis of HOIs. Network-specific analysis of HOIs in a static framework is performed via a measure recently proposed to describe the interactions among several random variables. To characterize this measure, let us start with the three variables $\mathbf{V}_i$, $\mathbf{V}_j$ and $\mathbf{V}_k$, $i, j, k \in \{1, \ldots M\}$. The high-order interaction among them is quantified by the *interaction information* (II) [8], which compares the information that one target variable, say $\mathbf{V}_i$, shares with two source variables, say $\mathbf{V}_j$ and $\mathbf{V}_k$, when the sources are taken separately but not when they are taken together. Accordingly, the II is quantified subtracting the MI between the target and the two sources from the sum of the MIs between the target and each source:

$$I(\mathbf{V}_i; \mathbf{V}_j; \mathbf{V}_k) = I(\mathbf{V}_i; \mathbf{V}_j) + I(\mathbf{V}_i; \mathbf{V}_k) - I(\mathbf{V}_i; \mathbf{V}_j, \mathbf{V}_k). \tag{3.8}$$

The computation of the II is illustrated using Venn diagrams in Fig. 3.2b. Importantly, the II is symmetric (i.e., its value does not change modifying the target variable) and can take either positive or negative values. Specifically, the II is positive if the two sources share more information with the target when they are considered individually, denoting *redundancy*; on the contrary, the II is negative if the two sources share more information with the target when they are considered jointly, denoting *synergy*. Redundancy and synergy represent the two basic types of high-order statistical dependencies.

The II has been recently generalized to allow the information-theoretic analysis of HOIs among an arbitrarily large number of random variables through the definition of the so-called *O-information* (OI) [9], which measures the balance between redundancy and synergy in a multiplet of N variables. The two building blocks of OI are the total correlation (TC) [10] and the dual total correlation (DTC) [11], defined as follows for a system described by N stochastic variables $\mathbf{V}^N = \{\mathbf{V}_1, \ldots, \mathbf{V}_N\}$:

$$\mathrm{TC}(\mathbf{V}^N) = \sum_{i=1}^{N} H(\mathbf{V}_i) - H(\mathbf{V}^N), \tag{3.9}$$

$$\mathrm{DTC}(\mathbf{V}^N) = H(\mathbf{V}^N) - \sum_{i=1}^{N} H(\mathbf{V}_i \mid \mathbf{V}^N_{-i}), \tag{3.10}$$

where $\mathbf{V}^N_{-i}$ denotes the set of all the variables in $\mathbf{V}^N$ but $\mathbf{V}_i$. TC quantifies the *collective constraints*, whilst DTC quantifies the *shared randomness* [9]. The OI is defined as the difference TC (3.9) − DTC (3.10) and assumes positive values when the interdependencies among variables can be more efficiently explained as shared randomness, and negative values when collective constraints can be more convenient [9]. Consequently, the OI of the system can be written as [9]

$$\Omega(\mathbf{V}^N) = (N-2)H(\mathbf{V}^N) + \sum_{i=1}^{N} \left[H(\mathbf{V}_i) - H(\mathbf{V}^N_{-i}) \right]. \tag{3.11}$$

If $\Omega > 0$, the system is redundancy-dominated. On the other hand, when $\Omega < 0$ the dependencies are better explained as patterns that can be observed in the joint state of multiple variables but not in subsets of these; in other words, the system is synergy-dominated. It is clear that the main drawback of the OI is the fact it does not put in evidence multiplets of variables which are both redundant and synergistic with equal strength, whilst approaches like partial information decomposition (PID) [12] evaluate both quantities and may, in principle, deal with these cases.

Node-specific analysis of HOIs. While the OI (3.11) is a measure quantifying network-specific HOIs in a static network of multiple variables, node-specific analysis must be based on information-theoretic measures of the additional information brought by adding a new node to a predefined multiplet. Then, in order to measure how much a given variable $\mathbf{V}_i$ plays a role in the informational circuits contained in $\mathbf{V}^N$, its *gradient of OI* is calculated as follows [13]:

$$\begin{aligned}\partial_i \Omega(\mathbf{V}^N) &= \Omega(\mathbf{V}^N) - \Omega(\mathbf{V}^N_{-i}) \\ &= (2-N) I(\mathbf{V}_i; \mathbf{V}^N_{-i}) + \sum_{k=1, k\neq i}^{N} I(\mathbf{V}_k; \mathbf{V}^N_{-ik}),\end{aligned} \tag{3.12}$$

where $\mathbf{V}^N_{-ik}$ denotes all the variables in $\mathbf{V}^N$ except $\mathbf{V}_i$ and $\mathbf{V}_k$. The quantity $\partial_i \Omega(\mathbf{V}^N)$ captures how much the OI changes when $\mathbf{V}_i$ is added to the rest of the system, hence it gives an account of how this variable contributes to the high-order properties of the system. Therefore, $\partial_i \Omega(\mathbf{V}^N) > 0$ means that $\mathbf{V}_i$ introduces mainly redundant information, while $\partial_i \Omega(\mathbf{V}^N) < 0$ indicates that it fosters synergistic interdependencies. Following this rationale, the OI among N variables can be defined as the sum of the OI for a subset including $N-1$ variables plus the OI gradient (3.12) quantifying the increment obtained when a new variable is added, i.e., $\Omega(\mathbf{V}^N) = \Omega(\mathbf{V}^N_{-i}) + \partial_i \Omega(\mathbf{V}^N)$. This quantity has also been referred to as $\Delta(\mathbf{V}_i; \mathbf{V}^N_{-i})$ (see, e.g., [3, 5]).

In presence of scalar variables $V_1, \ldots, V_N$, it has been shown in [13] that the following bounds hold and are tight:

$$-(N-2)\log|\mathcal{V}_c| \leq \partial_i \Omega(\mathbf{V}^N) \leq \log|\mathcal{V}_c|, \tag{3.13}$$

where $|\mathcal{V}_c|$ is the cardinality of the largest alphabet in $\mathbf{V}^N$. The lower bound is achieved in correspondence of the N-XOR gate, that is $V_1 \ldots V_{N-1}$ as Bernoulli random variables with $p = 1/2$ and $V_N = (\sum_{j=1}^{N-1} V_j) \bmod 2$; the upper bound is achieved by the n-COPY gate, specifically by taking V_1 as a Bernoulli variable with $p = 1/2$ and $V_1 = V_2 = \cdots = V_N$. The asymmetry between these two bounds has the following consequence: while redundancy can be only built step by step, synergy can be established more rapidly. Indeed, adding a variable to a system of size $N-1$ might provide a maximal redundant contribution of $\log|\mathcal{V}_c|$, whilst the maximal synergy that it might lend is $(N-2)\log|\mathcal{V}_c|$, which can be substantial if N is large.

The computation of the OI gradient for the case of $N = 4$ variables is illustrated using Venn diagrams in Fig. 3.2c. Crucially, since the OI for two random variables is null ($\Omega(\mathbf{V}^2) = 0$), the OI for three variables is equal to the gradient, which in turn corresponds to the II (3.8)

$$\begin{aligned}\Omega(\mathbf{V}^3) &= \Delta(\mathbf{V}_3; \mathbf{V}^2) = I(\mathbf{V}_3; \mathbf{V}_2) + I(\mathbf{V}_3; \mathbf{V}_1) - I(\mathbf{V}_3; \mathbf{V}_1, \mathbf{V}_2)\\ &= I(\mathbf{V}_1; \mathbf{V}_2; \mathbf{V}_3);\end{aligned} \tag{3.14}$$

these formulations allow the iterative computation of the OI through progressive inclusion of variables [9]. Importantly, the OI gradient can be positive or negative, reflecting respectively redundant and synergistic high-order interactions in the analyzed set of random variables.

Following a similar rationale to the one that leads to (3.12), one can further introduce a second-order descriptor of high-order interdependencies by considering gradients of gradients. In particular, the *second-order gradient* of a pair of variables $\mathbf{V}_i$ and $\mathbf{V}_j$ can be defined as [14]

$$\partial_i \partial_j \Omega(\mathbf{V}^N) = \partial_i \Omega(\mathbf{V}^N) - \partial_i \Omega(\mathbf{V}^N_{-j}); \tag{3.15}$$

this quantity captures how much the presence of the variable $\mathbf{V}_j$ alters the variation of OI of the system due to the inclusion of $\mathbf{V}_i$. It is direct to verify the symmetry $\partial_i \partial_j \Omega(\mathbf{V}^N) = \partial_j \partial_i \Omega(\mathbf{V}^N)$; therefore, we denoted this quantity as $\partial^2_{ij} \Omega(\mathbf{V}^N)$. An interesting property of $\partial^2_{ij} \Omega(\mathbf{V}^N)$ is that it can be re-written as a *whole-minus-sum* quantity:

$$\begin{aligned}\partial^2_{ij} \Omega(\mathbf{V}^N) &= \left[\Omega(\mathbf{V}^N) - \Omega(\mathbf{V}^N_{-ij})\right]\\ &\quad - \left[\Omega(\mathbf{V}^N_{-i}) - \Omega(\mathbf{V}^N_{-ij})\right] - \left[\Omega(\mathbf{V}^N_{-j}) - \Omega(\mathbf{V}^N_{-ij})\right].\end{aligned} \tag{3.16}$$

In other words, $\partial^2_{ij} \Omega(\mathbf{V}^N)$ measures to what degree the variation to the OI due to the inclusions of both $\mathbf{V}_i$ and $\mathbf{V}_j$ is more than the sum of the variations one obtains when including them separately. It is interesting to evaluate $\partial^2_{ij} \Omega(\mathbf{V}^N)$ on the n-COPY gate and on the n-XOR gate: it is easy to obtain zero and $(2 - N)$ respectively. This means that for the n-COPY gate pairs of variables do not provide further redundancy w.r.t. those provided by single variables; on the other hand, for the n-XOR gate, pairs of variables give an irreducible contribution to the synergy. This is a sign of the sensitivity of gradients to evaluate synergistic informational circuits, which occurs due to the partition of $\mathbf{V}^N_{-ij}$ into parts which is adopted to calculate them.

Successive gradients can be similarly introduced, resulting in a simple chain rule [14]. If γ is a subset of $\{1, \ldots, N\}$ of cardinality $|\gamma|$, then:

$$\partial^{|\gamma|}_{\gamma} \Omega(\mathbf{V}^N) = \sum_{\alpha \subseteq \gamma} (-1)^{|\alpha|} \Omega(\mathbf{V}^N_{-\alpha}), \tag{3.17}$$

the sum being over all the subsets α of γ. For example, for triplets of variables the gradient of the OI reads:

$$\partial^3_{ijk}\Omega(\mathbf{V}^N) = \Omega(\mathbf{V}^N) - \Omega(\mathbf{V}^N_{-i}) - \Omega(\mathbf{V}^N_{-j}) - \Omega(\mathbf{V}^N_{-k}) + \Omega(\mathbf{V}^N_{-ij}) + \Omega(\mathbf{V}^N_{-ik}) + \Omega(\mathbf{V}^N_{-jk}) - \Omega(\mathbf{V}^N_{-ijk}), \quad (3.18)$$

and measures the irreducible contribution to the OI by the triplet $\{i, j, k\}$ which cannot be ascribed to the inclusion of pairs nor single variables of the triplet.

The high-order dependencies emerging from static networks of multiple interconnected nodes have been investigated in [3, 5, 14], with specific applications to cardiovascular, respiratory and cerebral blood flow variability [14], as well as to fMRI data [3, 5, 14], which will be discussed in Sects. 6.1 and 5.8, respectively.

Partial Information Decomposition. The main drawback of the OI is the fact that it does not put in evidence multiplets of variables which are both redundant and synergistic with equal strength. In this context, *partial information decomposition* (PID) has been developed as a comprehensive framework designed to understand how information is distributed in multivariate systems: the foundational work by Williams and Beer [12] introduced PID as a method to decompose multivariate information in non-negative terms, addressing limitations of traditional measures like the II [8] which can yield both positive and negative values often obscuring the interpretation of informational relationships. Considering a *target* random variable and a set of *source* variables, the mathematical redundancy lattice structure defined for the PID [12] identifies a set of atoms whose associated partial information (PI) amounts constitute the building blocks of the analyzed multivariate information shared between the target and the sources, quantified by the MI. Moreover, to overcome the limitation that the number of atoms grows super-exponentially with the number of source variables [15], refinements have been introduced whereby the PI atoms are aggregated meaningfully to highlight how the MI is distributed among the sources [16]; these refined approaches provide a *coarse-grained* decomposition with a small number of atoms that scale gracefully with the system size, highlighting the *unique* information exclusively available from each source, the *redundant* information obtained from at least two different sources, and the *synergistic* information revealed only when multiple sources are considered simultaneously.

Nevertheless, in spite of the universality of the problem posed by PID, the underlying analytical framework presents some inherent limitations that restrict its unambiguous utilization in different contexts. A first issue was recognized since the inception of PID noting that the information atoms of unique, redundant, and synergistic information cannot be defined using classical information theory, rather require the introduction of new axioms whose definition is not yet universally accepted. Consequently, several axiomatic definitions of redundancy have been proposed so far that differ depending on the philosophy followed to satisfy the desired properties (e.g., decision- [17], game- [18], information-theoretic [19]), on the nature (continuous [17, 20, 21] or discrete [12, 18, 22]) of the analyzed variables, and on assumptions made about their distribution (e.g., Gaussian [20]). Popular and simple approaches implement the so-called minimum MI (MMI) PID schemes, whereby redundancy is defined for a given atom as the minimum of the MI (or the specific MI) shared between the target (or a specific state of it) and each source [12, 20]. However, these schemes are limited in the fact that they quantify the minimum amount of information that all variables carry but do not require that such information is the same for all variables [18, 22–24]. To overcome this limitation, less conservative approaches have been proposed which typically define redundancy at the local or pointwise level, rather than at the level of ensemble averages [18, 19, 21]; these approaches allow defining more refined redundancy quantities, but suffer in their turn from the limitation of potentially yielding negative information atoms (essentially because local MI can be negative), which hinders a strightforward interpretation of the results. Thus, the definition of a proper operalization of PID, merging computability and interpretability, is still an open problem in information theory.

Let us consider a generic static network system composed of $M+1$ nodes $\mathcal{V} = \{\mathcal{T}, \mathcal{S}_1, \ldots, \mathcal{S}_M\}$, where the activity at each node is described in terms of random variables $T, S_1, \ldots, S_M$, with T assumed as the *target* variable and $\mathbf{S} = \{S_1, \ldots, S_M\}$ as the vector of *sources*. The PID framework allows to assess and decompose the multivariate information shared in the system, i.e., the MI between T and $\mathbf{S}$, $I(T; \mathbf{S})$, defined in terms of probability distributions as in (3.2), into terms related to the contributions that the individual sources S_i, $i = 1, \ldots, M$, or a collection thereof, share with the target T, in turn related in non-trivial ways to the marginal MI between each source and the target, $I(T; S_i)$. Specifically, the mathematical *redundancy lattice structure* defined for the PID [12] identifies a set of atoms whose associated PI amounts constitute the building blocks of the analyzed multivariate information $I(T; \mathbf{S})$. The lattice is identified by the collection $\mathcal{A}$ of all subsets of sources such that no source is a superset of any other, i.e., the set of anti-chains formed from the indices of the sources in $\mathbf{S}$ under set inclusion [12] (see Fig. 3.3a, b (left) for the

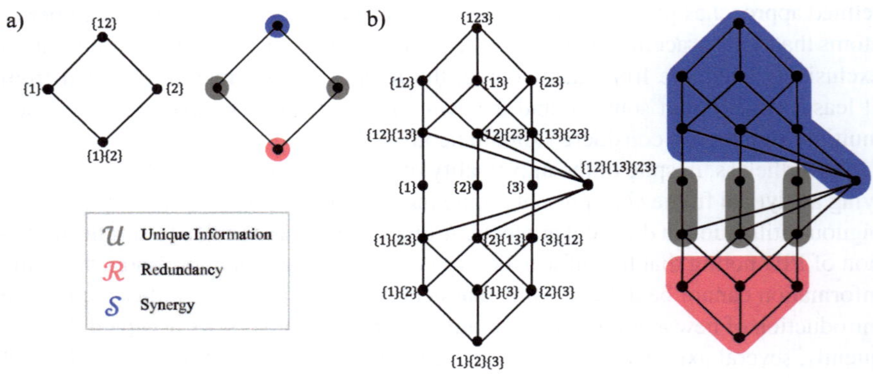

Fig. 3.3 The MI between the target and the sources is decomposed into redundant (magenta), synergistic (blue) and unique (grey) contributions through standard and coarse-grained PID. **a** PID on the redundancy lattice for 2 sources, with alphabet of source combinations $\mathcal{A} = \{\{1\}\{2\}, \{1\}, \{2\}, \{12\}\}$ ($\{1\}$: S_1, $\{2\}$: S_2). **b** Standard and coarse-grained 1st order PID on the redundancy lattice for 3 sources, with alphabet of source combinations $\mathcal{A} = \{\{1\}\{2\}\{3\}, \{1\}\{2\}, \{1\}\{3\}, \{2\}\{3\}, \ldots, \{123\}\}$ ($\{i\}$: S_i, $i = 1, \ldots, M$). The figure is adapted from [25]

cases of $M = 2$ and $M = 3$ sources; e.g., $\mathcal{A} = \{\{1\}\{2\}, \{1\}, \{2\}, \{12\}\}$ if $M = 2$). Formally, the PID expands the MI as

$$I(T; \mathbf{S}) = \sum_{\alpha \in \mathcal{A}} I^{\delta}(T; \mathbf{S}_{\alpha}), \tag{3.19}$$

where $I^{\delta}(\cdot; \cdot)$ is the PI function defined over the atoms of the lattice, $\alpha = \{\alpha_1, \ldots, \alpha_J\} \in \mathcal{A}$, and $\mathbf{S}_{\alpha}$ denotes the set of subsets of source variables indexed by the αth atom, with $\alpha_j \subseteq \{1 \cdots M\}$, $\mathbf{S}_{\alpha_j} \subseteq S$, $j = 1, \ldots, J$. To complete the PID besides the basic statement in (3.19), it is necessary to provide a set of so-called consistency equations which, relating atoms to mutual information, allow to derive the PI terms [15]. The main consistency equations state that the marginal MI terms involving any individual source variable S_i are constructed additively by summing the information of the atoms positioned at the level $\{i\}$ and downwards in the lattice, i.e.,

$$I(T; S_i) = \sum_{\beta \preceq \{i\}} I^{\delta}(T; \mathbf{S}_{\beta}), \tag{3.20}$$

where $\preceq$ identifies precedence based on the partial ordering imposed by the lattice structure [12]. Moreover, as the equations (3.19) and (3.20) do not suffice to solve the PID problem because they provide a number of constraints lower than the number of information atoms to be computed (i.e., $M + 1 < |\mathcal{A}|$, where $|\cdot|$ indicates cardinality), to complete the PID it is necessary to define a so-called *redundancy function*, here denoted as $I^{\cap}(\cdot; \cdot)$, which generalizes the MI over the lattice. The redundancy function extends (3.20) to each atom $\alpha \in \mathcal{A}$, fulfilling

$$I^{\cap}(T; \mathbf{S}_{\alpha}) = \sum_{\beta \preceq \alpha} I^{\delta}(T; \mathbf{S}_{\beta}), \tag{3.21}$$

where β represents the atoms preceding or equal to α and structurally connected to it in the lattice. Finally, the information associated to all atoms can be retrieved, starting from the knowledge of the redundancy function, either iteratively as

$$I^{\delta}(T; \mathbf{S}_{\alpha}) = I^{\cap}(T; \mathbf{S}_{\alpha}) - \sum_{\beta \prec \alpha} I^{\delta}(T; \mathbf{S}_{\beta}), \tag{3.22}$$

or in a compact way via Möbius inversion of (3.21) [12]. Importantly, while the redundancy value of an atom α, $I^{\cap}(T; \mathbf{S}_{\alpha})$, measures the total amount of redundant information shared by all the sources included in that atom, the PI measures the unique information contributed only by that atom. A popular and simple approach to assign a redundancy function is the so-called MMI PID generally applied in case of Gaussian data [20], whereby redundancy is defined for the atom $\alpha = \{\alpha_1, \ldots, \alpha_J\}$ as the minimum of the information shared between the target and each individual information component [20]:

$$I^{\cap}(T; \mathbf{S}_\alpha) = \min_{j=1,\ldots,J} I(T; \mathbf{S}_{\alpha_j}); \tag{3.23}$$

for instance, considering $M = 3$ sources and the atom $\alpha = \{\{1\}, \{23\}\}$, the redundancy becomes the minimum between $I(T; S_1)$ and $I(T; S_2, S_3)$.

An important aspect with practical relevance is that, as an alternative to (3.19), the PID can be formulated in a meaningful way by making explicit the *unique information* that each source S_m holds about the target T ($m = 1, \ldots, M$), the *redundant information* that all source variables in $\mathbf{S}$ hold about T, and the *synergistic information* about T that only arises from knowing all the sources $S_1, \ldots, S_M$. This corresponds to expand multivariate information as

$$I(T; \mathbf{S}) = \sum_{m=1}^{M} \mathcal{U}(T; S_m) + \mathcal{R}(T; \mathbf{S}) + \mathcal{S}(T; \mathbf{S}), \tag{3.24}$$

where

$$\mathcal{U}(T; S_m) = I(T; S_m) - \mathcal{R}(T; \mathbf{S}). \tag{3.25}$$

This approach provides a *coarse-grained* decomposition with a small number of atoms that scale gracefully with the system size: while the full PID (3.19) yields a number of atoms $|\mathcal{A}|$ that grows super-exponentially with M like the Dedekind numbers [15], the coarse-grained PID (3.24) decomposes the multivariate information into exactly $M + 2$ quantities. In particular, the two formulations coincide when $M = 2$ source variables are considered, yielding $\mathcal{R}(T; \mathbf{S}) = I^\delta(T; \mathbf{S}_{\{1\}\{2\}})$, $\mathcal{U}(T; S_1) = I^\delta(T; S_{\{1\}})$, $\mathcal{U}(T; S_2) = I^\delta(T; S_{\{2\}})$, and $\mathcal{S}(T; \mathbf{S}) = I^\delta(T; \mathbf{S}_{\{12\}})$ (Fig. 3.3a, right). On the other hand, when $M \geq 3$ the coarse-graining is implemented by summing the PI of some of the atoms in (3.19). This issue has been addressed in [16], where the construction for $M = 2$ was generalized to $M = 3$ sources through the so-called kth order *coarse-grained* PID, which preserves the intuitive meaning that synergy, redundancy, and unique information have for $M = 2$ sources. Specifically, considering $k = 1$, the 1st-order synergy, $\mathcal{S}(T; \mathbf{S})$, corresponds to the information about the target that is provided by the whole $\mathbf{S}$ but is not contained in any subset of sources when considered separately from the rest (atoms surrounded by the blue shade in Fig. 3.3). Similarly, the 1st-order redundancy, $\mathcal{R}(T; \mathbf{S})$, is the information held by at least two different groups of size 1 (magenta in Fig. 3.3). Finally, the 1st-order unique information provided by the mth source, $\mathcal{U}(T; S_m)$, $m = 1, \ldots, M$, is the information that S_m has access to and no other subset of parts has access to on its own, although bigger groups of other parts may have (grey in Fig. 3.3).

The PID principles have been exploited in [25, 26] to define a dynamic version of the framework by leveraging the use of MI rates instead of MI (see Sects. 4.3.1.1 and 4.3.2.1 for the time and spectral formulations). Benchmark simulations depicting the differences between the two approaches are displayed in Sects. 4.3.3.5 and 4.3.3.6.

Link-specific analysis of HOIs. Link-specific analysis makes use of an information-theoretic measure quantifying the net information shared between the two analyzed nodes and the rest of the network [9, 27, 28]; the measure is refined accounting for the network structure (i.e., its value is not assigned if the dynamics of the two nodes are found to be statistically independent or independent conditioned to the rest of the network [27, 28]). To define this measure, let us consider the variables $\mathbf{V}_i$ and $\mathbf{V}_j$, $i, j = 1, \ldots, M$, $i \neq j$, and the remaining $M-2$ variables collected in the vector $\mathbf{V}_z = \mathcal{V} \setminus [\mathbf{V}_i, \mathbf{V}_j]$. In the framework of information theory applied to static network systems, the relationships between the variables in $\mathcal{V}$ can be analyzed quantifying the *information shared* (iS) between them by using the well-known measure of MI, or the *conditional information shared* (ciS) between them but not with $\mathbf{V}_z$ by using the *conditional MI* (CMI). Given the random variables $\mathbf{V}_i$, $\mathbf{V}_j$ and the random vector $\mathbf{V}_z$ taken from $\mathcal{V}$, the CMI evaluates the information shared between $\mathbf{V}_i$ and $\mathbf{V}_j$ when considering the influence of the third variable $\mathbf{V}_z$, and is defined as [2]

$$I(\mathbf{V}_i; \mathbf{V}_j | \mathbf{V}_z) = H(\mathbf{V}_i | \mathbf{V}_z) - H(\mathbf{V}_i | \mathbf{V}_j, \mathbf{V}_z). \tag{3.26}$$

The iS and ciS measures, for static analysis quantified by the MI and CMI, respectively, assess the link between the two analyzed units from a bivariate or multivariate perspective [27, 28]. Importantly, the comparison between these two quantities highlights the balance between the statistical concepts of redundancy and synergy in the observed network system. In particular, the difference between the iS and the ciS was defined as the *net information shared* (niS) between $\{\mathbf{V}_i, \mathbf{V}_j\}$ and $\mathbf{V}_z$ [27, 28]:

$$I(\mathbf{V}_i; \mathbf{V}_j; \mathbf{V}_z) = I(\mathbf{V}_i; \mathbf{V}_j) - I(\mathbf{V}_i; \mathbf{V}_j | \mathbf{V}_z). \tag{3.27}$$

Given that $I(\mathbf{V}_i; \mathbf{V}_j, \mathbf{V}_z) = H(\mathbf{V}_i) + H(\mathbf{V}_i | \mathbf{V}_j, \mathbf{V}_z)$, it can easily be demonstrated that the quantity in (3.27) is the II between the three variables $\mathbf{V}_i$, $\mathbf{V}_j$ and $\mathbf{V}_z$, defined in (3.8) with $\mathbf{V}_k$ in place of $\mathbf{V}_z$; it can take on both positive and negative values thus reflecting the balance between redundancy and synergy, and has been also referred to as *local OI* [9]. Specifically, when $I(\mathbf{V}_i; \mathbf{V}_j; \mathbf{V}_z) > 0$, the knowledge of $\mathbf{V}_z$ reduces the information shared by $\mathbf{V}_i$ and $\mathbf{V}_j$, thus indicating that (part of) the statistical dependence between $\mathbf{V}_i$ and $\mathbf{V}_j$ is suppressed when $\mathbf{V}_z$ is observed. On the contrary, when $I(\mathbf{V}_i; \mathbf{V}_j; \mathbf{V}_z) < 0$, the knowledge of $\mathbf{V}_z$ increases the information shared by $\mathbf{V}_i$ and $\mathbf{V}_j$, thus indicating that (part of) the statistical dependence between $\mathbf{V}_i$ and $\mathbf{V}_j$ emerges when $\mathbf{V}_z$ is observed.

To emphasize the balance between redundancy and synergy in the interaction among the two observed units $\mathbf{V}_i$ and $\mathbf{V}_j$ and the rest of the system $\mathbf{V}_z$, and to retrieve information about the network topology from such interaction, the so-called *B-index* (B, shorthand for redundancy/synergy balance) is defined by normalizing the niS as follows [27, 28]:

$$B(\mathbf{V}_i; \mathbf{V}_j) = \frac{I(\mathbf{V}_i; \mathbf{V}_j; \mathbf{V}_z)}{\max\{I(\mathbf{V}_i; \mathbf{V}_j), I(\mathbf{V}_i; \mathbf{V}_j | \mathbf{V}_z)\}}. \tag{3.28}$$

In (3.28), the B-index is computed dividing the difference between the MI and the CMI to their maximum, so as to obtain a measure ranging between -1 and

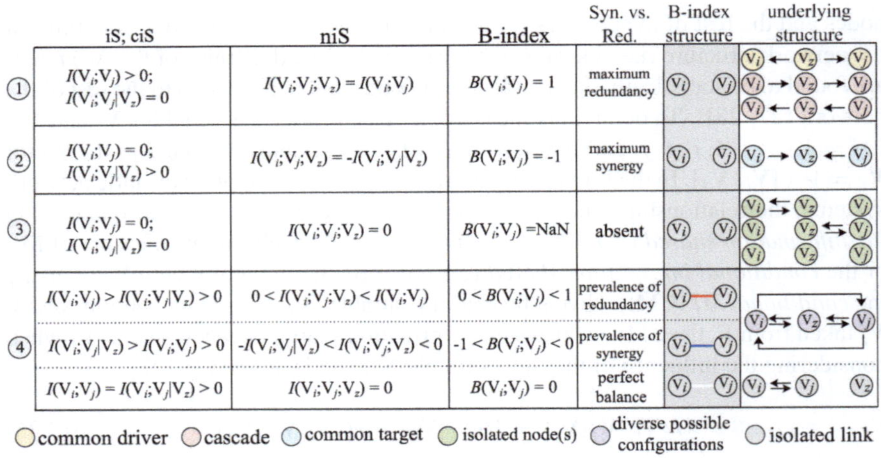

Fig. 3.4 Classification of the redundant/synergistic nature of the high-order link between the two units V_i and V_j and of the types of possible underlying structural mechanisms. The figure is adapted from [28]

1 (Fig. 3.4). The limit values highlight a full imbalance between redundancy and synergy that relates to specific network topologies. In particular, $B(\mathbf{V}_i;\mathbf{V}_j) = 1$ corresponds to maximum redundancy, occurring when the interaction between $\mathbf{V}_i$ and $\mathbf{V}_j$ is fully explained by the rest of the network and indicating the presence of common driver ($\mathbf{V}_i \leftarrow \mathbf{V}_z \rightarrow \mathbf{V}_j$) or cascade ($\mathbf{V}_i \rightarrow \mathbf{V}_z \rightarrow \mathbf{V}_j$ or $\mathbf{V}_j \rightarrow \mathbf{V}_z \rightarrow \mathbf{V}_i$) interactions (Fig. 3.4, case 1). On the other hand, $B(\mathbf{V}_i;\mathbf{V}_j) = -1$ corresponds to maximum synergy, occurring when the interaction between $\mathbf{V}_i$ and $\mathbf{V}_j$ arises fully from their effect on $\mathbf{V}_z$ and indicating the presence of a common cause interaction ($\mathbf{V}_i \rightarrow \mathbf{V}_z \leftarrow \mathbf{V}_j$, Fig. 3.4, case 2). In both cases, the nodes mapped by $\mathbf{V}_i$ and $\mathbf{V}_j$ are topologically disconnected. The two nodes are disconnected also when both the MI and CMI are null, resulting in non-defined B-index and describing a situation in which at least one between $\mathbf{V}_i$ and $\mathbf{V}_j$ is isolated from the rest of the network (Fig. 3.4, case 3). Intermediate values of the B-index ($-1 < B(\mathbf{V}_i;\mathbf{V}_j) < 1$) are obtained when both the MI and CMI are non-null (Fig. 3.4, case 4). In this situation, larger values of the MI denote prevalence of redundancy, while larger values of the CMI denote prevalence of synergy, which topologically corresponds to many possible configurations; the case of identical values denotes perfect balance between synergy and redundancy indicating that the two nodes are linked to each other but disconnected from the rest of the network.

To accurately reconstruct the network structure based on B-index, it is crucial to assess the statistical significance of MI and CMI measures through techniques such as surrogate data analysis (see LINK- SPECIFIC ANALYSIS OF HOIS). This ensures that the B-index accurately reflects fully redundant or synergistic interactions with values of 1 or -1, respectively, (i.e., due to null CMI or MI indicating absence of the link), and identifies the absence of interactions with a return of NaN (obtained when the ratio in (3.28) is $\frac{0}{0}$, i.e., both the MI and the CMI are both null). The procedures used to generate surrogate data is described in SURROGATE DATA ANALYSIS.

The high-order link-specific dependencies emerging from static networks of multiple interconnected nodes have been investigated in [27, 28], with a specific applications to fMRI data [27] that will be discussed in Sect. 6.1.

3.2.1 *Linear Parametric Formulation*

Let us consider the linear regression models defined in Sect. 2.2. The calculation of the information-theoretic measures defined in Sect. 3.2 requires an approach to compute the MI and the CMI between vector random variables. Assuming that the observed variables have a joint Gaussian distribution, the MI between two generic (possibly vector) variables $\mathbf{V}_i, \mathbf{V}_j$ is assessed as in (3.7) [6]. Similarly, the CMI can be assessed exploiting the relation between conditional entropy and partial variance valid for Gaussian variables [6], i.e., $H(\mathbf{V}_i|\mathbf{V}_z) = \frac{1}{2}\log\left((2\pi e)^{v_i}|\Sigma_{\mathbf{U}_z}|\right)$ and $H(\mathbf{V}_i|\mathbf{V}_j, \mathbf{V}_z) = \frac{1}{2}\log\left((2\pi e)^{v_i}|\Sigma_{\mathbf{U}_{jz}}|\right)$, where $\Sigma_{\mathbf{U}_z}$ and $\Sigma_{\mathbf{U}_{jz}}$ are the $v_i \times v_i$ covariance matrices of the prediction errors $\mathbf{U}_z, \mathbf{U}_{jz}$. Then, the CMI is computed straightforwardly from (3.26), yielding

$$I(\mathbf{V}_i; \mathbf{V}_j|\mathbf{V}_z) = \frac{1}{2}\log\left(\frac{|\Sigma_{\mathbf{U}_z}|}{|\Sigma_{\mathbf{U}_{jz}}|}\right). \tag{3.29}$$

The first-order OI gradient, local OI and PID terms are thus computed substituting the relevant MI and CMI linear terms in (3.12), (3.22), (3.23) and (3.27), respectively. The OI can be computed straightforwardly from (3.12), while the OI gradients of higher order are computed from (3.17) exploiting the linear parametric computation of the first-order gradient.

3.3 Simulation Examples

3.3.1 *Pairwise and High-Order Functional Dependencies Assessed via Mutual Information and O-Information*

In [5], the framework for the computation of pairwise and high-order interactions is illustrated making use of a theoretical example of simulated linear regression models for which the MI and OI measures are computed directly from the known model parameters. This simulation is exploited to show that high-order measures can be used to highlight the emergence of patterns of interaction among groups of variables which cannot be traced from pairwise connections alone, as well as to evidence the presence of circuits dominated by synergy or redundancy. Moreover, we show how the methods of surrogate and bootstrap data analysis can help to disregard non-significant interaction pathways among the variables, thus allowing to focus only on specific connectivity links within the network (see SURROGATE DATA ANALYSIS and BOOTSTRAP DATA ANALYSIS for technical details; this specific example exploits the methods described in AN APPROACH TO THE STATISTICAL ASSESSMENT OF HIGH- ORDER INTERACTIONS).

The simulation is focused on the analysis of $M = 7$ scalar random variables, with network structure and interdependencies specified in Fig. 3.5a. The parameters a_i, $i = 1, \ldots, 6$, quantifying the pairwise relationships between the observed variables,

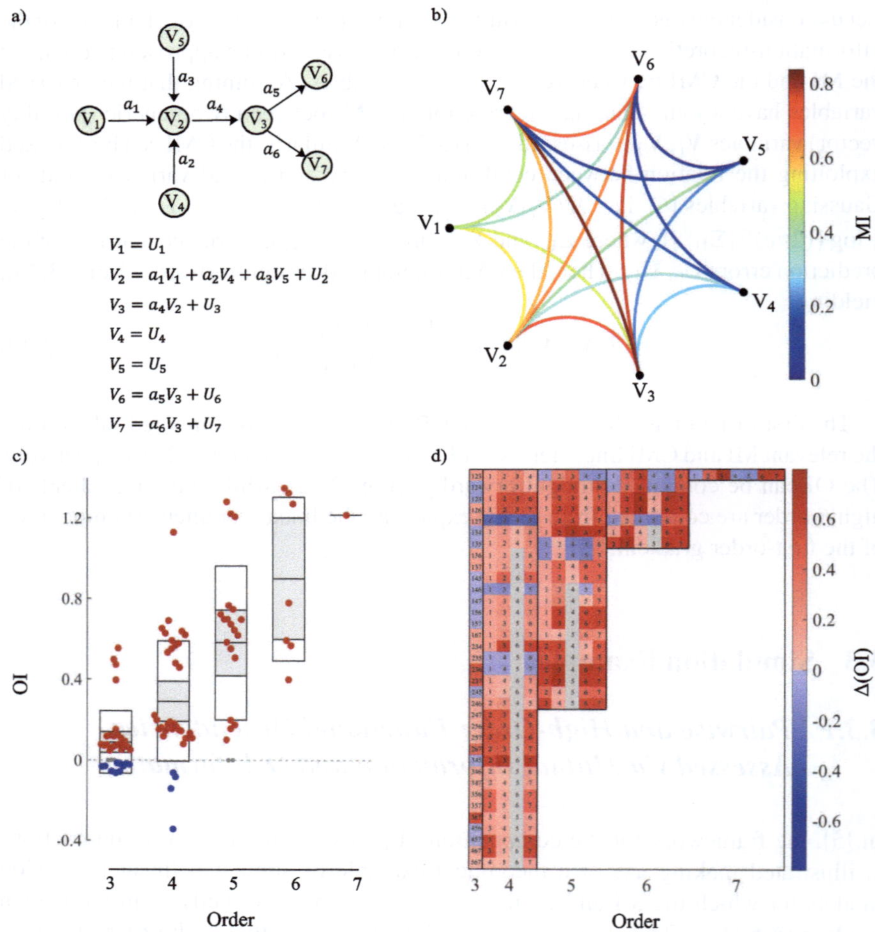

Fig. 3.5 **a** Simulation design and model equations. **b** MI-weighed significant connections among pairs of variables. **c**–**d** OI and $\Delta(OI)$ (red: redundancy; blue: synergy; grey: non-significant). The figure is adapted from [5]

are chosen in the range [0.95–1], arbitrarily setting $a_1 = a_2 = 0.99$, $a_3 = a_6 = 1$ and $a_4 = a_5 = 0.95$. Noteworthy, the parameter range was suitably selected to better highlight and visualize the coupling strength between the observed variables, since lower values of the parameters (i.e., $a_i < 0.95$, $i = 1, \ldots, 6$) would have led to lower and hence non-significant values of the MI measure computed between those variables. The network is designed to simulate three zero-mean random noises V_1, V_4 and V_5, with unit variance, whose sink (commonly called *common child*) is the node V_2. Then, through a *chain* structure, V_2 converges into the node V_3, which in turn acts as a *common driver* for the nodes V_6 and V_7 (Fig. 3.5a). The variable $U_i, i = 1, \ldots, 6$ is zero-mean random noise with unit variance. From the resulting network $\{V_1, \ldots, V_7\}$, implemented via time series realizations of $L = 500$ points, the time-domain MI between pairs of variables was estimated as in (3.7); then, its significance was assessed by applying the method of surrogate data and evaluating the existence of each pairwise link, as described in COUPLED DYNAMICS BETWEEN PAIRS OF NODES (Fig. 3.5b). The OI was computed from (3.12) for all the possible multiplets of orders $N = 3, \ldots, 7$, and deemed as significant when the OI distributions, computed via bootstrap data analysis, did not comprise the zero level (Fig. 3.5c) (we refer to AN APPROACH TO THE STATISTICAL ASSESSMENT OF HIGH- ORDER INTERACTIONS for details on the bootstrap data generation procedure). Moreover, the values of the OI increment were computed as in (3.12), at each order N and for each target V_i within the selected multiplet $\mathbf{V}^N$ of that order, and deemed as significant when the information brought by V_i to $\mathbf{V}^N$ was statistically significant according to the test discussed in AN APPROACH TO THE STATISTICAL ASSESSMENT OF HIGH-ORDER INTERACTIONS (Fig. 3.5d). Values of OI and $\Delta(OI)$ are expressed in *nats*, i.e., natural units.

The MI values shown in the circular graph of Fig. 3.5b reflect the strength of the relationships between pairs of variables; values of MI > 0.5 nats are found for the pairs $\{V_2, V_3\}$, $\{V_2, V_6\}$, $\{V_3, V_6\}$, $\{V_3, V_7\}$, and $\{V_6, V_7\}$. However, not all of these connections are true links of interaction between the investigated variables, as happens, e.g., for the pairs $\{V_2, V_6\}$ and $\{V_6, V_7\}$. Indeed, the nodes V_2 and V_6, as well as V_6 and V_7, are not linked by direct interaction pathways but still show non-zero connectivity (Fig. 3.5a). This finding is related to the existence of common driving and chain effects in these cases, respectively, which determine the appearance of indirect links of interaction between the two investigated variables [29]. Interestingly, this misinterpretation of the network structure does not occur in the case of the common child effect, since truly non-significant MI is found for the pairs $\{V_1, V_4\}$, $\{V_1, V_5\}$ and $\{V_4, V_5\}$, as shown by the absence of links between these variables in Fig. 3.5b.

A high-order representation of the investigated interactions is provided in Fig. 3.5c, d in terms of boxplot distributions (panel c: in each box, the central black mark indicates the mean, and the bottom and top edges of the box indicate the 25th and 75th percentiles, respectively; red, blue and grey circles indicate positive-redundant, negative-synergistic, and non-significant OI values, respectively) and heatmaps (panel d: $\Delta(OI)$ values are computed for each target—numbers in squares—inside the multiplets—sequences of numbers along each row—at all orders—separated by black vertical lines; red, blue and grey squares indicate positive-redundant, negative-synergistic, and non-significant $\Delta(OI)$, respectively, brought by that target to the multiplet for a given order). The OI values in Fig. 3.5c show an expected increase of redundancy as the network size increases (i.e., from order 3 to 7), even though some synergistic multiplets are still found at orders 3 and 4. Specifically, as shown by the values of the OI increment in Fig. 3.5d, synergistic triplets (first column, order 3) are those containing the variables V_1, V_4 and V_5, which indeed are involved in the common child structure (Fig. 3.5a). In addition, the chain structure for which the node V_3 is a sink for V_2, and the same applying for V_6 and V_7 with respect to V_3 (Fig. 3.5a), causes that synergy also involves these variables when combined with V_1, V_4 or V_5. Interestingly, this pattern is maintained at higher orders with most of the multiplets comprising the variables V_1, V_4 and V_5, for which significant synergistic OI increments are found. On the other hand, the triplets $\{V_2, V_3, V_6\}$, $\{V_2, V_3, V_7\}$ and $\{V_3, V_6, V_7\}$, along with others such as $\{V_1, V_2, V_3\}$, $\{V_1, V_2, V_6\}$ and $\{V_1, V_2, V_7\}$, show positive values of the OI increment (Fig. 3.5d, first column, order 3), confirming that the common driver and chain structures are dominated by redundancy (Fig. 3.5a). As happens for the synergistic variables V_1, V_4 and V_5, the addition of the variables V_2, V_3, V_6 and V_7 to form groups of order 4, 5, 6 and 7 is likely to significantly increase the redundancy of the interactions within the network, as shown by red squares containing these variables in Fig. 3.5d (second, third, fourth and fifth column).

The bootstrap data approach, applied on the simulated time series to retrieve confidence intervals for the proposed measures, allowed to statistically validate the OI values and the OI increments (Fig. 3.5c, d). Specifically, in Fig. 3.5c, non-significant OI values are depicted as grey circles around the zero threshold; at order 3, the number of non-significant OI values is the highest (4 over 35 multiplets). Conversely, in Fig. 3.5d, non-significant OI increments are shown as grey squares, where each square corresponds to the target-specific OI increment for that multiplet. Non-significant $\Delta(OI)$ values are found at orders $N = 4, 5$, especially when the targets V_4 and V_5, as well as V_6, are added to form multiplets containing the variables V_2, V_3 and V_7 or V_6.

In conclusion, this simulation example showed that the connectivity maps traced by the MI do not provide a fully-explanatory description of the complex and multiple interactions taking place in the analysed network. Indeed, different network structures, such as common driver, chain and common child ensembles, are not always truly reproduced by these pairwise maps, and the resulting MI values between pairs of observed variables may be biased. The utilization of high-order measures investigating the relationships between more than two variables is then fundamental to provide a more complete description of the connectivity patterns emerging from the network. Noteworthy, the possibility to specify the redundant and/or synergistic character of groups of variables would allow a more faithful representation of the network ground structure. Moreover, the use of surrogate and bootstrap methods is essential in the practical analysis of data sequences to disregard non-significant links between pairs of variables and hyperlinks among groups of variables.

3.3.2 High-Order Links in a Simulated Static System: The B-Index Reveals Redundant and Synergistic Patterns of Link-Specific Interactions

In this section, the simulated static system published in [28] will be discussed in terms of the redundant/synergistic nature of the high-order links within the corresponding network, unveiled through the utilization of the B-index computed as in (3.28). This example showcases how, while it is essential to seek for high-order dependencies led by single nodes or groups of nodes through the OI and its gradients as done in the previous section, understanding the nature of the network structural links and their role in the redundant/synergistic informational circuits can help in reconstructing the ground-truth network topology. Note that this simulation does not refer to Gaussian systems characterized with linear models as in this book, but makes use of discrete (binary) variables connected by specifying joint probabilities. Static measures based on MI can be obtained from observations of discrete random variables collected as sequences of symbols using plug-in entropy estimators. The methods are described in detail in [28]; the inferences apply equally to linear Gaussian systems.

Analyses were performed iterating each simulation 100 times and generating at each iteration datasets of $L \in \{250, 500, 1000\}$ observations, from which the estimator performance was assessed in terms of sensitivity and specificity of the reconstruction of the network topology. The simulation reproduces a network with $M = 10$ nodes connected as depicted in Fig. 3.6a. The node activities are mapped by binary random variables interconnected as follows: V_1, V_3, V_4, V_5 and V_9 are i.i.d. binary variables with equiprobable symbols; V_{10} is a noisy copy of V_9 with coupling strength $\gamma_3 = 0.8$, (i.e., $p(\{v_{10} = v_9\}) = \gamma_3$); V_6 and V_7 are noisy copies of V_5 with coupling strength $\gamma_2 = 0.9$; V_8 is defined via a noisy OR gate from V_6 and V_7, while V_2 is defined via a noisy OR gate from V_3, V_4 and V_5 (noisy gates have coupling strength $\gamma_1 = 0.9$ and are defined according to the conditional probabilities given in Fig. 3.6a).

Figure 3.6b reports, for a simulation of the system collecting $L = 1000$ observations, the matrix of the B-index values computed through the linear parametric estimator, where non-significant values of the estimated MI or CMI leading to $B = 1$, $B = -1$ or $B = \text{NaN}$ were assessed using random shuffling (RS) surrogates (we refer to LINKSPECIFIC ANALYSIS OF HOIS for technical details), and the reconstructed network structure obtained pruning the links with non-significant MI and/or CMI. The B-index correctly identified the presence and nature of the interactions imposed between pairs of variables, recovering for instance the existing isolated link ($B(V_9; V_{10}) = 0$, white) and the expected synergistic (e.g., $B(V_2; V_3) < 0$, $B(V_2; V_4) < 0$, blue) and redundant links (e.g., $B(V_5; V_6) > 0$, $B(V_5; V_7) > 0$, red). Moreover, it allowed to correctly prune out of the reconstructed network several links that would be detected by using only MI due to common drive or cascade effects (e.g., $B(V_2; V_6) = 1$, $B(V_5; V_8) = 1$; dark red), or by using only CMI due to common cause effects (e.g., $B(V_3; V_5) = -1$; dark blue). However, it is important

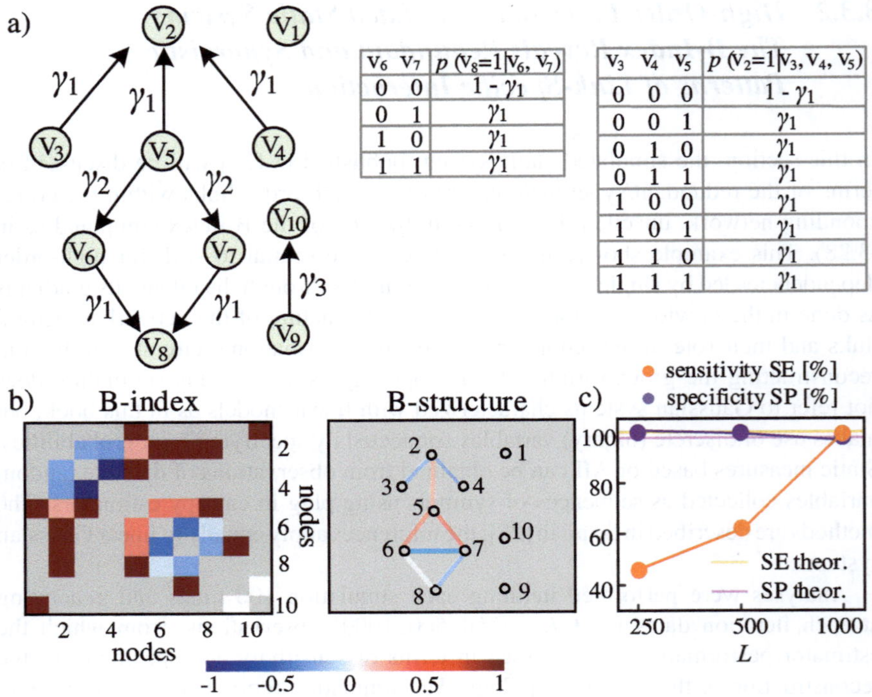

V_6	V_7	$p(V_8=1 \mid V_6, V_7)$
0	0	$1-\gamma_1$
0	1	γ_1
1	0	γ_1
1	1	γ_1

V_3	V_4	V_5	$p(V_2=1 \mid V_3, V_4, V_5)$
0	0	0	$1-\gamma_1$
0	0	1	γ_1
0	1	0	γ_1
0	1	1	γ_1
1	0	0	γ_1
1	0	1	γ_1
1	1	0	γ_1
1	1	1	γ_1

Fig. 3.6 **a** Simulated static system with ten units mapped by binary variables. **b** Matrix of B-index values (left) and the corresponding reconstructed connectivity graph (right). **c** Performance of network reconstruction. The figure is adapted from [28]

to note the false positive detection of a connection between V_6 and V_7: in spite of the lack of a direct link, the two nodes result as connected because they have both a common drive ($V_6 \leftarrow V_5 \rightarrow V_7$) and a common cause ($V_6 \rightarrow V_8 \leftarrow V_7$), inducing significant values for both MI and CMI and thus determining $-1 < B(V_6; V_7) < 1$. This highlights a limitation of the B-index, which cannot guarantee the elimination of all spurious links.

To quantify the ability of the proposed approach to correctly identify the presence and absence of links in the reconstructed network, we assessed the sensitivity and specificity of the B-index at varying the size L of the data collected. Sensitivity and specificity, which measure respectively the impact of false negatives and false positives, were assessed over 100 simulation runs. The results in Fig. 3.6c highlight that the sensitivity increased markedly with the data length, approaching the expected value of 100% when $L = 1000$. On the contrary, the specificity showed less dependence on L, as it remained consistently high exhibiting minimal variations around the expected value of 97%.

3.4 Summary

This chapter reviewed the entropy measures exploited to describe static networks mapped by random variables, which will be then relevant to the information-theoretic analysis of random processes. The linear parametric formulation of such entropy measures is made possible under the assumption of jointly Gaussian variables.

The static network system $\mathcal{V}$ composed of M nodes can be characterized in terms of the activities involving the single nodes, pairs of nodes, and groups of nodes. These activities are assessed straightforwardly using univariate, bivariate and high-order entropy-based measures of static interactions, i.e., the *entropy* $H(V_i)$ of the variable V_i, the *mutual information* (MI) $I(V_i; V_j)$ shared between the variables V_i and V_j, and the *O-information* (OI) $\Omega(\mathbf{V}^N)$ of the multiplet $\mathbf{V}^N = \{V_1, \ldots, V_N\}$ ($N \leq M$). In the case of a high-order network analysis of the static interactions taking place in systems of multiple variables, the OI represents a network-specific measure characterizing the overall informational character (i.e., synergistic or redundant) of groups of nodes (i.e., multiplets). Unlike the OI, the *partial information decomposition* (PID) framework is defined to identify multiplets which are both redundant and synergistic with equal strength, whereby the mutual information shared among target and source variables is decomposed into non-negative redundant, synergistic and unique contributions. Nevertheless, node- and link-specific measures quantifying the additional information brought by adding a new node to a predefined multiplet and the net information shared between a pair of nodes and the rest of the network, respectively, can be used to examine HOIs at different levels of resolution. Then, while the first-order *gradient of OI* $\Delta(V_i; \mathbf{V}^N_{-i})$ is a node-specific measure quantifying how much a given variable V_i plays a role in the informational circuits contained in $\mathbf{V}^N$ (the same rationale holds for higher order gradients), the link-wise measure of the *B-index* $B(V_i; V_j)$ highlights the balance between redundancy and synergy in the interaction among the two observed units V_i and V_j and the rest of the system, and allows to retrieve information about the network topology from such interaction.

Simulated examples emphasized that the connectivity maps traced by the MI do not provide fully-explanatory descriptions of the complex interactions taking place in static networks of multiple nodes, such that the utilization of high-order measures investigating the relationships between more than two variables is then fundamental to provide a more complete description of the connectivity patterns emerging from the network. Furthermore, while it is essential to characterize high-order dependencies led by single nodes or groups of nodes through the OI and its gradients, understanding the role of links between two nodes in contributing to redundant/synergistic informational circuits via the B-index can help reconstructing the ground-truth network topology.

References

1. Claude Elwood Shannon. "A mathematical theory of communication". In: *The Bell system technical journal* 27.3 (1948), pp. 379–423.
2. Thomas M Cover. *Elements of information theory*. John Wiley & Sons, 1999.
3. Simone Valenti et al. "Assessing High-Order Interdependencies Through Static O-Information Measures Computed on Resting State fMRI Intrinsic Component Networks". In: *International Conference on Image Analysis and Processing*. Springer. 2022, pp. 386–397.
4. Laura Sparacino et al. "Statistical Approaches to Characterize Functional Connectivity in Brain and Physiologic Networks on a Single-Subject Basis". In: *2023 45th Annual International Conference of the IEEE Engineering in Medicine & Biology Society (EMBC)*. IEEE. 2023, pp. 1–4.
5. Laura Sparacino et al. "Statistical Approaches to Identify Pairwise and High-Order Brain Functional Connectivity Signatures on a Single-Subject Basis". In: *Life* 13.10 (2023), p. 2075.
6. Adam B Barrett, Lionel Barnett, and Anil K Seth. "Multivariate Granger causality and generalized variance". In: *Physical Review E* 81.4 (2010), p. 041907.
7. Laura Sparacino et al. "Measuring hierarchically-organized interactions in dynamic networks through spectral entropy rates: theory, estimation, and illustrative application to physiological networks". In: *Neurocomputing* 630 (2025), p. 129675.
8. William McGill. "Multivariate information transmission". In: *Transactions of the IRE Professional Group on Information Theory* 4.4 (1954), pp. 93–111.
9. Fernando E Rosas et al. "Quantifying high-order interdependencies via multivariate extensions of the mutual information". In: *Physical Review E* 100.3 (2019), p. 032305.
10. Satosi Watanabe. "Information theoretical analysis of multivariate correlation". In: *IBM Journal of research and development* 4.1 (1960), pp. 66–82.
11. TH Sun. "Linear dependence structure of the entropy space". In: *Inf. Control* 29.4 (1975), pp. 337–368.
12. Paul L Williams and Randall D Beer. "Nonnegative decomposition of multivariate information". In: *arXiv preprint* arXiv:1004.2515 (2010).
13. Tomas Scagliarini et al. "Gradients of O-information: Low-order descriptors of high-order dependencies". In: *Physical Review Research* 5.1 (2023), p. 013025.
14. Tomas Scagliarini et al. "Gradients of O-information highlight synergy and redundancy in physiological applications". In: *Frontiers in Network Physiology* 3 (2024), p. 1335808.
15. Aaron J Gutknecht, Michael Wibral, and Abdullah Makkeh. "Bits and pieces: Understanding information decomposition from part-whole relationships and formal logic". In: *Proc. Roy. Soc. A* 477.2251 (2021), p. 20210110.
16. Fernando E Rosas et al. "Reconciling emergences: An information-theoretic approach to identify causal emergence in multivariate data". In: *PLoS computational biology* 16.12 (2020), e1008289.
17. Ari Pakman et al. "Estimating the unique information of continuous variables". In: *Advances in neural information processing systems* 34 (2021), pp. 20295–20307.
18. Robin AA Ince. "Measuring multivariate redundant information with pointwise common change in surprisal". In: *Entropy* 19.7 (2017), p. 318.
19. Abdullah Makkeh, Aaron J Gutknecht, and Michael Wibral. "Introducing a differentiable measure of pointwise shared information". In: *Physical Review E* 103.3 (2021), p. 032149.
20. Adam B Barrett. "Exploration of synergistic and redundant information sharing in static and dynamical Gaussian systems". In: *Physical Review E* 91.5 (2015), p. 052802.
21. David A Ehrlich et al. "Partial information decomposition for continuous variables based on shared exclusions: Analytical formulation and estimation". In: *Physical Review E* 110.1 (2024), p. 014115.
22. Nils Bertschinger et al. "Quantifying unique information". In: *Entropy* 16.4 (2014), pp. 2161–2183.

23. Malte Harder, Christoph Salge, and Daniel Polani. “Bivariate measure of redundant information”. In: *Physical Review E—Statistical, Nonlinear, and Soft Matter Physics* 87.1 (2013), p. 012130.
24. Virgil Griffith and Tracey Ho. “Quantifying redundant information in predicting a target random variable”. In: *Entropy* 17.7 (2015), pp. 4644–4653.
25. Laura Sparacino et al. “Decomposing Multivariate Information Rates in Networks of Random Processes”. In: *arXiv preprint* arXiv:2502.04555 (2025). URL: https://arxiv.org/abs/2502.04555.
26. Luca Faes et al. “Partial Information Rate Decomposition”. In: *arXiv preprint* arXiv:2502.04550 (2025).
27. Yuri Antonacci et al. “Measuring the Balance Between Synergy and Redundancy in Network Systems by Using Information Theory”. In: *Mediterranean Conference on Medical and Biological Engineering and Computing*. Springer. 2023, pp. 145–154.
28. Gorana Mijatovic et al. “Assessing High-Order Links in Cardiovascular and Respiratory Networks via Static and Dynamic Information Measures”. In: *IEEE Open Journal of Engineering in Medicine and Biology* (2024).
29. Ruben Sanchez-Romero and Michael W Cole. “Combining multiple functional connectivity methods to improve causal inferences”. In: *Journal of cognitive neuroscience* 33.2 (2021), pp. 180–194.

Chapter 4
Dynamic Networks of Random Processes

This chapter introduces a framework to measure dynamic interactions of different orders in networks of multiple interconnected systems, expanded in the frequency domain to explore their spectral patterns. The framework is defined for *dynamic systems*, i.e. systems evolving over time whose activity is modeled appropriately in the context of multivariate random processes.

Let us consider the dynamic network system $\mathcal{X}$ composed of M nodes. The dynamic activities at the single nodes, as well as the interactions between the activities of two nodes, can be assessed straightforwardly using information-theoretic measures which extend to random processes the concepts of entropy and mutual information defined for random variables in Sect. 3.1: these measures are the entropy rate and mutual information rate (Fig. 4.1). Specifically, we will characterize single-node activity in Sect. 4.1, and pairwise interactions involving two network nodes in Sect. 4.2. Nevertheless, when more than two units are considered, more sophisticated approaches are needed to examine *hierarchical* HOIs, i.e., interactions at upraising orders involving the role multiplets of network nodes, at different levels of resolution, i.e., characterizing the role of the single node within the selected subset (*node-specific analysis*), the nature of the link between two nodes with respect to the remaining nodes (*link-specific analysis*), as well as the overall activity of multiplets of nodes or the whole network (*network-specific analysis*). To this end, information-theoretic measures of dynamic HOIs which extend to random processes the concepts of first-order OI gradient, OI, PID and local OI defined for random variables in Sect. 3.2 are required to provide a more comprehensive and detailed description of the analysed dynamic network system, e.g., Interaction Information Rate and O-Information Rate (Fig. 4.1). These approaches will be discussed in Sect. 4.3. Noteworthy, each section of this chapter comprises two formulations of these information-theoretic measures, i.e., a time domain and a spectral formulation, which are linked to each other through the so-called spectral integration property. Furthermore, simulationexamples showing the time and frequency domain behaviors

L. Sparacino, *Spectral Information Dynamics in Network Neuroscience and Physiology*,
Studies in Computational Intelligence 1235,
https://doi.org/10.1007/978-3-032-05416-6_4

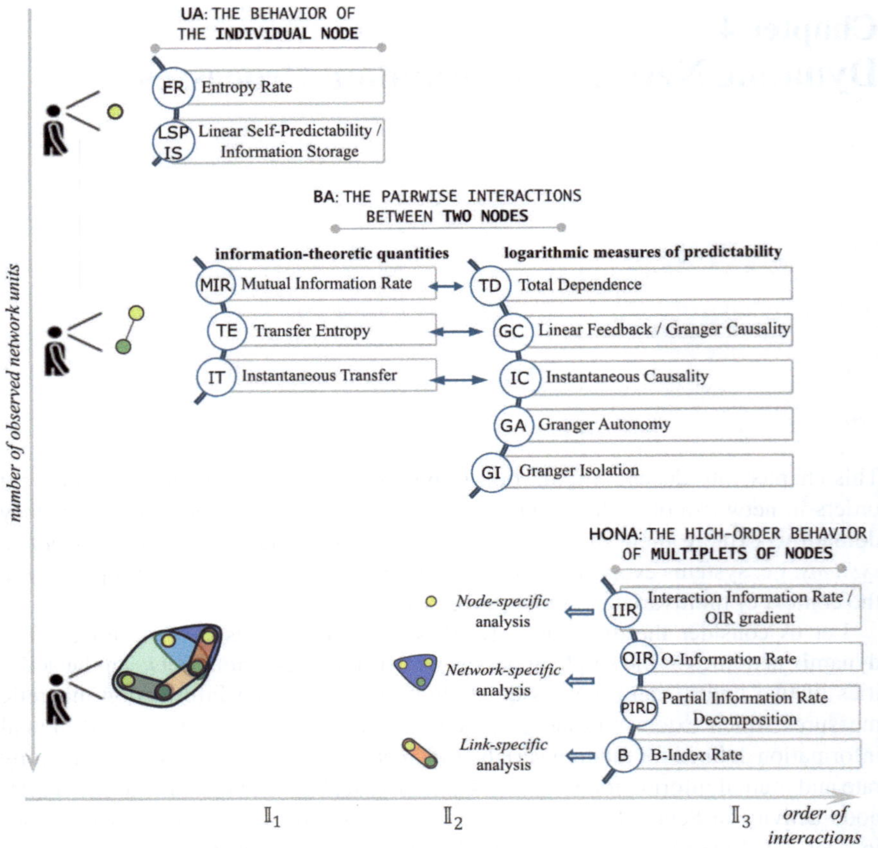

Fig. 4.1 Measures of dynamic interaction in a system of multiple nodes

of the dynamic interaction measures at different orders are provided in Sects. 4.1.3, 4.2.3 and 4.3.3.

4.1 Behavior of a Single Network Unit: Self-predictable Dynamics

4.1.1 Time Domain

In this section, we will characterize the time domain behavior of the generic ith network node taken from the set $\mathcal{X}_1, \ldots, \mathcal{X}_M$, $i = 1, \ldots, M$, in the framework of a UA, i.e., assuming that $M = 1$; the same information-theoretic measures can be defined in the scalar case, i.e., assuming that the block has unitary dimension. To this end, we refer the reader to the mathematical definitions provided in Sect. 2.3.1 related to the time domain representation of (block) AR linear models, from which measures of dynamic *complexity* and *predictability* can be defined and contextualized in the case of linear Gaussian processes.

4.1.1.1 Entropy Rate

The information-theoretic analysis of dynamic information relevant to the single network unit exploits the concept of *entropy rate* (ER), which quantifies the rate of generation of new information in a random process. Specifically, for the vector stationary random process $\mathbf{X}$ of dimension M_X, the ER is defined using these two equivalent definitions [1–3]:

$$H_{\mathbf{X}} = \lim_{m\to\infty} \frac{1}{m} H(\mathbf{X}_{n:n+m}) = H(\mathbf{X}_n|\mathbf{X}_n^-), \tag{4.1}$$

where the second formulation evidences the conditional entropy of the variable representing the present of the process given the variables sampling its past history. Therefore, the ER reflects the complexity of the process intended as the unpredictability of its present state given the past, ranging from $H_{\mathbf{X}} = 0$ for a completely self-predictable process to $H_{\mathbf{X}} = H(\mathbf{X}_n)$ for a fully unpredictable process without temporal statistical structure. The Venn diagram illustration of the ER of a generic vector process is given in Fig. 4.2a. It is worth noting that, when $Q = 1$, (4.1) can be suitably adapted to the generic scalar process Y whose dynamics are described in (2.2), thus resorting to $H_Y = H(Y_n|Y_n^-)$.

Linear parametric formulation. In the linear signal processing framework, the ER (4.1) has a straightforward formulation that involves the $M_X \times M_X$ covariance matrix of the residuals of the block AR representation (2.3) [4]:

$$H_{\mathbf{X}} = \frac{1}{2} \log\left((2\pi e)^{M_X} |\Sigma_{\mathbf{U_X}}^{(ar)}|\right). \tag{4.2}$$

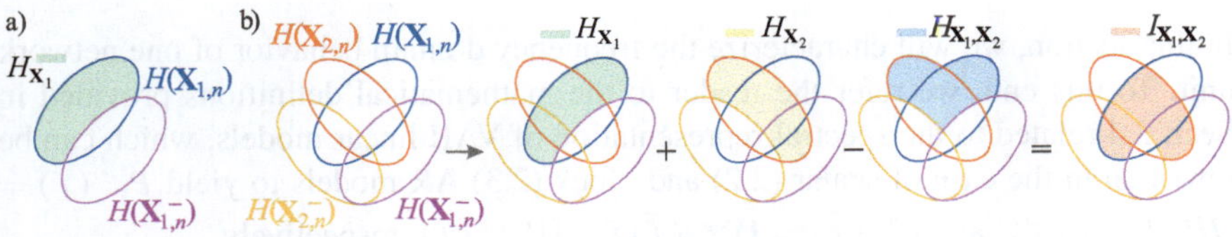

Fig. 4.2 **a** Entropy rate of the process $\mathbf{X}_1$, $H_{\mathbf{X}_1}$. **b** Mutual information rate between two processes $\mathbf{X}_1$ and $\mathbf{X}_2$, $I_{\mathbf{X}_1,\mathbf{X}_2}$. The figure is adapted from [3]

Similarly, (4.2) can be generalized to the scalar case (2.2), thus yielding $H_Y = \frac{1}{2}\log\left(2\pi e \sigma_{U_Y}^{2(ar)}\right)$.

4.1.1.2 Information Storage

While the functional defined in (4.1) is an information-theoretic measure of complexity, the so-called *information storage* (IS) of the vector process $\mathbf{X}$ is a measure of self-predictability quantifying the information stored in $\mathbf{X}$ as the mutual information between the present $\mathbf{X}_n$ and the past $\mathbf{X}_n^-$ [5, 6]. Defined as

$$S_{\mathbf{X}} = I(\mathbf{X}_n; \mathbf{X}_n^-) = H(\mathbf{X}_n) - H(\mathbf{X}_n|\mathbf{X}_n^-), \tag{4.3}$$

it ranges from $S_{\mathbf{X}} = 0$ for a fully unpredictable process without temporal statistical structure to $S_{\mathbf{X}} = H(\mathbf{X})$ for a completely self-predictable process. As in (4.1), (4.3) can be also adapted to the scalar case (2.2), thus obtaining $S_Y = I(Y_n; Y_n^-)$.

Linear parametric formulation. In the linear signal processing framework, the IS has a straightforward formulation that involves the $M_X \times M_X$ covariance matrix of the process and the $M_X \times M_X$ covariance matrix of the prediction error of its block AR representation (2.3) [7, 8]:

$$S_{\mathbf{X}} = \frac{1}{2}\log\left(\frac{|\Sigma_{\mathbf{X}}|}{|\Sigma_{\mathbf{U_X}}^{(ar)}|}\right). \tag{4.4}$$

The quantity defined in (4.4) has been referred to as a measure of *linear self-predictability* (LSP) of the process $\mathbf{X}$ in the time domain, for which we recently proposed a spectral expansion in [3]. Similarly, (4.4) can be adapted to the scalar case, thus yielding $S_Y = \frac{1}{2}\log\left(\frac{\sigma_Y^2}{\sigma_{U_Y}^{2(ar)}}\right)$, where σ_Y^2 is the variance of Y, and $\sigma_{U_Y}^{2(ar)}$ is the variance of the residual in (2.2).

4.1.2 Frequency Domain

In this section, we will characterize the frequency domain behavior of one network unit. To this end, we refer the reader to the mathematical definitions provided in Sect. 2.4 related to the spectral representation of VAR linear models, which can be rewritten in the case of scalar (2.2) and block (2.3) AR models to yield $P_Y^{(ar)}(\bar{f}) = |H^{Y(ar)}(\bar{f})|^2 \sigma_{U_Y}^{2(ar)}$ and $\mathbf{P}_{\mathbf{X}}^{(ar)}(\bar{f}) = \mathbf{H}^{\mathbf{X}(ar)}(\bar{f})\Sigma_{\mathbf{U_X}}^{(ar)}\mathbf{H}^{\mathbf{X}(ar)*}(\bar{f})$, respectively.

4.1.2.1 Spectral Entropy Rate

The spectral densities of AR models can be exploited to provide frequency domain measures of the entropy rate for scalar and block processes, which quantify individual interactions at each specific frequency. Specifically, a spectral measure of the ER of the block process $\mathbf{X}$ of dimension M_X is defined as [2, 3]

$$h_{\mathbf{X}}(\bar{f}) = \frac{1}{2} \log \left((2\pi e)^{M_X} |\mathbf{P}_{\mathbf{X}}^{(ar)}(\bar{f})| \right), \tag{4.5}$$

while a spectral measure of the ER of the scalar process Y described in (2.2) can be written as $h_Y(\bar{f}) = \frac{1}{2} \log \left(2\pi e P_Y^{(ar)}(\bar{f}) \right)$. The spectral ERs satisfy the spectral integration property [9], i.e., these measures are such that their integrals extended over all frequencies return the corresponding time domain ER measures:

$$H_{\mathbf{X}} = \int_0^{\frac{1}{2}} h_{\mathbf{X}}(\bar{f}) \mathrm{d}\bar{f};\ H_Y = \int_0^{\frac{1}{2}} h_Y(\bar{f}) \mathrm{d}\bar{f}. \tag{4.6}$$

One note of remarkable importance is due here. If one wants to compute the ER of the process $Y_j \in \mathbf{X}$, $j = 1, \ldots, M_X$, the diagonal element of the PSD matrix $\mathbf{P}_{\mathbf{X}}^{(ar)}(\bar{f})$ in position (jj)th can be exploited to yield $h_{Y_j}(\bar{f}) = \frac{1}{2} \log \left(2\pi e P_{Y_j}^{(ar)}(\bar{f}) \right)$. This formulation evidences how in a network of multiple interacting nodes possibly organized in one block it is possible to investigate the dynamics of the single nodes starting from the dynamics of the whole block. The same concept can be applied to pairwise connectivity as shown in Sect. 4.2. In this context, the time and frequency domain ER-based first-order dependencies emerging from dynamic networks of multiple interconnected nodes have been investigated in [3], with specific applications to brain dynamics, described and discussed in Sects. 6.2.4 and 6.2.5. Theoretical simulations will be shown in Sect. 4.3.3.

4.1.2.2 Spectral Decomposition of Oscillatory Content

Let us assume $Q = M = 1$, i.e., our network is formed by one unit whose dynamics are described by the scalar process Y. We will see how analysing the spectrum of a single scalar process provides noteworthy information on the frequency-specific location of the oscillations of that process, thus allowing to identify and separate its different spectral components. The approach is particularly useful when the observer wants to characterize the spectral behavior of unexplored physiological short-term beat-to-beat time series, as we did first in [10], and then in [11], by studying the variability of the cardiovascular parameter of arterial compliance (see Sect. 5.1 for the application to real data).

Specifically, we consider the scalar random process Y whose dynamics can be represented through an AR model in the time and frequency domain. Applying the residue theorem, the TF $H^{Y(ar)}(z)$ modelling the relationship between the input

$U_Y^{(ar)}(z)$ and the output $Y(z)$ of the AR model (2.2) represented in the Z-domain, with $z = e^{\mathrm{j}2\pi\bar{f}}$, can be expressed as [12]:

$$H^{Y(ar)}(z) = \frac{z^p}{\prod_{k=1}^{p}(z - p_k)} = \prod_{k=1}^{p} H^{(k)}(z), \tag{4.7}$$

where $p_k, k = 1, \ldots, p$, are the p poles of the AR process, i.e. the roots of $\bar{A}_Y^{(ar)}(z) = [1 - \sum_{k=1}^{p} a_{Y,k}^{(ar)} z^{-k}]$, while the terms $H^{(k)}(z) = \frac{z}{z-p_k} \cdot \frac{1/z^*}{1/z^*-p_k}$ are pole-specific factors associated each to a given pole p_k, with * the Hermitian transpose. Then, the PSD of the process can be written in the Z-domain as $P_Y^{(ar)}(z) = H^{Y(ar)}(z)\sigma_{U_Y}^2 H^{Y(ar)*}(\frac{1}{z^*})$, and expanded exploiting the Heaviside decomposition with simple fractions relevant to all the poles and weighted by the relevant residuals of $P_Y^{(ar)}(z)$, to get [12, 13]:

$$P_Y^{(ar)}(z) = \sum_{\kappa=1}^{K} P_Y^{(\kappa)}(z) = \sum_{\kappa=1}^{K}\left[\frac{r_\kappa p_\kappa}{z - p_\kappa} - \frac{r_\kappa p_\kappa^{-1}}{z - p_\kappa^{-1}}\right], \tag{4.8}$$

where $r_\kappa = \left.\frac{\sigma_{U_Y}^2}{z\prod_{h\neq\kappa}(z-p_h)\cdot\prod(z^{-1}-p_h)}\right|_{z=p_\kappa}$ are the residuals of $P_Y^{(ar)}(z)$, $\kappa = 1, \ldots, K$.

For clarity, we omitted the superscripts $^{Y(ar)}$ and $^{(ar)}$ for the kth components $H^{(k)}(z)$ in (4.7) and the κth components $P_Y^{(\kappa)}(z)$ in (4.8), respectively. Note that the number of components of the PSD is $K \simeq p/2$ depending on the number of real poles; specifically, there is one component for each real pole and for each pair of complex conjugate poles. Then, by computing $P_Y^{(ar)}(z)$ on the unit circle of the complex plane, $P_Y^{(ar)}(\bar{f}) = P_Y^{(ar)}(z)|_{z=e^{\mathrm{j}2\pi\bar{f}}}$, where $f \in [0, f_s/2]$, it is possible to obtain the spectral profile of the process, $P_Y^{(ar)}(\bar{f})$, as well as of the κth component, $P_Y^{(\kappa)}(\bar{f})$. Crucially, each spectral component $P_Y^{(\kappa)}(\bar{f})$ is described by a specific profile that is shaped by the corresponding TF factor $|H^{(\kappa)}(\bar{f})|^2 = H^{(\kappa)}(z)H^{*(\kappa)}(z)|_{z=e^{\mathrm{j}2\pi\bar{f}}}$, and has an associated frequency (related to the pole frequency, $f_\kappa = \frac{arg\{p_\kappa\}}{2\pi}$) and power (related to the pole residual, $\sigma_\kappa^2 = r_\kappa$ for real poles and $\sigma_\kappa^2 = 2 \cdot \mathbb{R}\{r_\kappa\}$ for complex conjugate poles). Note that the sum of the pole variances σ_κ^2, with $\kappa = 1, \ldots, K$, equals the total power of the process, which represents its variance σ_Y^2.

Importantly, in comparison to classical approaches based on integrating the PSD profile within the spectral bands of interest to get band-specific time domain powers [14], the spectral decomposition allows to focus only on the spectral components with frequencies within those bands, thus avoiding spurious contributions due to broadband oscillations in different nearby frequency bands. This peculiarity can be exploited to define a *pole-specific measure of linear self-predictability in the frequency domain*, as we did in [11] and as shown in the next section.

4.1.2.3 Spectral Decomposition of Self-predictability

The characterization of the spectral dynamics of a random scalar process Y can be also carried on by looking at its degree of linear self-predictability, drawing a connection with information theory. The concept of *self-predictability* plays a key role for the analysis of the self-driven dynamics of physiological processes displaying richness of oscillatory rhythms. While time domain measures of self-predictability, as well as time-varying and local extensions, have already been proposed and largely applied in different contexts [15–22], they still lack a clear spectral description, which would be useful for the interpretation of the frequency-specific content of the investigated processes. The most popular information-theoretic measure of self-predictability is the IS defined in (4.3), which quantifies, for a scalar random process Y, the amount of information contained in the present state Y_n that can be predicted by the knowledge of its past states, Y_n^- [19]. The linear formulation of this quantity allows to expand it in the frequency domain, as we did for the first time in [11].

To this aim, we exploit the spectral representation of the AR model (2.2) and the method of spectral decomposition described in Sect. 4.1.2.2. We start noting that the TF $H^{Y(ar)}(z)$ contains spectral information about the predictable dynamics of Y, as it is directly derived from the Z-domain representation of the AR model coefficients $a_{Y,k}^{(ar)}$, $k = 1, \ldots, p$, which in turn describe these dynamics in the time domain. Then, we exploit the fact that such frequency-specific information can be particularized to each oscillatory component considering the TF factor $|H^{(\kappa)}(\bar{f})|^2$, so as to retrieve information on the pole-specific self-dynamics of the AR process. This factor is the squared TF associated to a real pole or the squared product of the two TFs associated to a pair of complex conjugated poles. We expect that stronger self-dynamics of Y at the frequency f_κ are reflected by higher values of $|H^{(\kappa)}(\bar{f})|^2$, which indeed shows a positive peak at that frequency. Therefore, we define the frequency-specific spectral LSP measure as

$$s_Y^{(\kappa)}(\bar{f}) = \frac{1}{2}\log\left(\frac{\sigma_Y^2|H^{(\kappa)}(\bar{f})|^2}{\sigma_{U_Y}^{2(ar)}}\right). \tag{4.9}$$

The spectral LSP (4.9) can be written also as $s_Y^{(\kappa)}(\bar{f}) = \frac{1}{2}\log\left(\frac{\sigma_Y^2}{\sigma_{U_Y}^{(ar)}}\right) + \frac{1}{2}\log\left(|H^{(\kappa)}(\bar{f})|^2\right) = S_Y + \bar{s}_Y^{(\kappa)}(\bar{f})$, and satisfies the spectral integration property [9], i.e., it is such that its integral extended over all frequencies returns the time domain LSP measure:

$$S_Y = \int_0^{\frac{1}{2}} s_Y^{(\kappa)}(\bar{f})\mathrm{d}\bar{f}. \tag{4.10}$$

Note that (4.10) is satisfied because the full-frequency integral of the term $\bar{s}_Y^{(\kappa)}(\bar{f})$ is null, i.e., $\int_0^{\frac{1}{2}} \log\left(|H^{(\kappa)}(\bar{f})|^2\right)\mathrm{d}\bar{f} = 0$ [2, 23]. Therefore, the spectral LSP consists of a frequency-independent part equal to S_Y and a frequency-specific part quantified by $\bar{s}_Y^{(\kappa)}(\bar{f})$, which takes negative values at some frequencies, depending on where the informative content is located.

The spectral decomposition of the pole-specific LSP into terms related to the K oscillations of the AR process, depicted in (4.9), allows to locate the self-dynamics of Y in specific spectral bands with given frequency, as well as to compute their strength as the integral of these profiles within the investigated bands. Remarkably, the spectral LSP profiles exhibit peaks similarly to the PSD, since both are derived from adaptations of the TF of the AR model describing the data. The difference between the two resides in the logarithmic formulation of the LSP in the framework of information theory. Indeed, being a measure of information shared between the present and past states of the investigated processes according to its mathematical definition, it can be quantified in natural units (*nats*) in the time domain, and in nats/Hz in the spectral domain, thus acquiring a clear meaning in the context of information theory. A simulation example showing the time and frequency domain behavior of the LSP is provided in Sect. 4.1.3.

4.1.3 Simulation Example: Spectral Decomposition of the Linear Self-predictability of a Simulated AR Process

In this section, we report the theoretical simulation proposed in [11] to study the behavior of the LSP measure (4.4) using a simulated AR process, where the exact profiles of the spectral LSP (4.9) are computed (with sampling frequency $f_s = 1$ Hz) from the true values imposed for the AR parameters. The process Y, exhibiting autonomous oscillations at different frequencies, is defined as in (2.2) with $p = 4$. The autonomous oscillations of Y are obtained placing pairs of complex-conjugate poles, with modulus ρ and phase $2\pi f$, in the complex plane representation of the process; the AR coefficients resulting from this setting are $a_{Y,1} = 2\rho\cos(2\pi f)$ and $a_{Y,2} = -\rho^2$ [8]. Here, we imposed autonomous oscillations for the process Y in the low-frequency (LF, [0.04–0.15] Hz) and high-frequency (HF, [0.15–0.4] Hz) bands of the spectrum, setting $\rho_{HF} = 0.9$, $f_{HF} = 0.3$ Hz, so that the dynamics of Y in the HF band are determined by the fixed coefficients $a_{Y,1} = -0.556$, $a_{Y,2} = -0.81$, and $\rho_{LF} = b \cdot 0.8$, $f_{LF} = 0.1$ Hz, so that the strength of the dynamics of Y in the LF band, which are determined by the coefficients $a_{Y,3}$, $a_{Y,4}$, depends on the parameter b varying in the range [0, 1]. The theoretical values of the time domain and the spectral LSP measures are computed for each value of the parameter b. Furthermore, $N_b = 100$ bootstrap realizations of the process Y, each of length $L = 1000$ points, are then generated $\forall b$ by feeding (2.2) with N_b block bootstrap versions of a single realization of the white noise process U_Y, using the theoretical coefficients $a_{Y,k}$, $k = 1, \ldots, 4$ (see BOOTSTRAP DATA ANALYSIS for technical details). The time domain and spectral LSP measures are then estimated after identifying the AR model fitting the bootstrap time series of Y; the model order was set using the AIC.

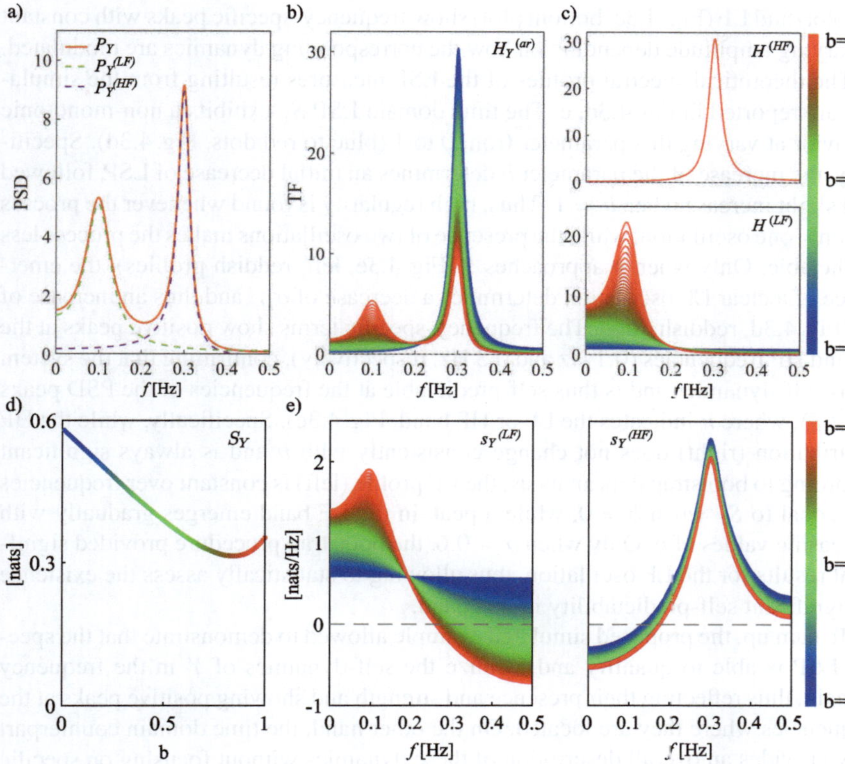

Fig. 4.3 **a** Exemplary linear spectral decomposition of the process Y into LF and HF components. **b** Transfer function of the process and its **c** HF- and LF-specific components. The time domain LSP **d** is decomposed into **e** LF- and HF-specific components. Profiles are computed at varying the parameter $b = [0, 1]$ (blue to red). The figure is adapted from [11]

Statistical significance of the spectral LSP measures in a given frequency band is then assessed exploiting the method described in SPECTRAL MEASURES OF SELF-PREDICTABILITY OF SINGLE NETWORK NODES.

The spectral decompositions of the PSD and the TF of the simulated AR process are reported in Fig. 4.3a–c. Figure 4.3a shows the theoretical PSD profile (orange continuous line) of the process Y when $b = 1$, decomposed into its two spectral components, LF (green dashed line) and HF (purple dashed line). Panel *b* shows the theoretical profiles of the TF of the AR process, $H^{Y(ar)}(f)$, at varying the parameter b from 0 (blue) to 1 (red). The TF shows only a positive peak in the HF band in absence of LF dynamics, i.e., when $b = 0$, and a positive peak in LF with amplitude increasing with the parameter b. This demonstrates that the TF is sensitive to the oscillatory content of the AR process, and it peaks wherever its self-dynamics are located in the frequency domain. Indeed, the two TF contributions in HF (Fig. 4.3c, top plot) and LF (Fig. 4.3c, bottom plot) show frequency-specific peaks with constant or varying amplitude depending on how the corresponding dynamics are modulated.

The theoretical spectral profiles of the LSP measures resulting from the simulation are reported in Fig. 4.3d, e. The time domain LSP S_Y exhibits a non-monotonic behavior at varying this parameter from 0 to 1 (blue to red dots, Fig. 4.3d). Specifically, the increase of the parameter b determines an initial decrease of LSP, followed by a slight increase when $b = 1$. Thus, high regularity is found whenever the process has only one oscillation, while the presence of two oscillations makes the process less predictable. Only when b approaches 1 (Fig. 4.3e, left, reddish profiles), the emergence of a clear LF oscillation determines a decrease of $\sigma^2_{U_Y}$ and thus an increase of S_Y (Fig. 4.3d, reddish dots). The frequency-specific terms show positive peaks at the LF and HF frequencies (0.1 Hz and 0.3 Hz, respectively), confirming that the system owns self-dynamics and is thus self-predictable at the frequencies of the PSD peaks ($s_Y^{(\kappa)}(\bar{f})$, where κ indicates the LF or HF band, Fig. 4.3e). Specifically, while the HF contribution (right) does not change consistently with b and is always significant according to bootstrap data analysis, the LF profile (left) is constant over frequencies and equal to S_Y when $b = 0$, while a peak in the LF band emerges gradually with increasing values of b. Only when $b > 0.6$, the bootstrap procedure provided significant results for the LF oscillation, thus allowing to statistically assess the existence of significant self-predictability in this band.

To sum up, the proposed simulated example allowed to demonstrate that the spectral LSP is able to quantify and localize the self-dynamics of Y in the frequency domain, thus reflecting their presence and strength and showing positive peaks at the frequencies where they are located. On the other hand, the time domain counterpart only provides an overall description of these dynamics without focusing on specific oscillatory rhythms. Remarkably, the utilization of surrogate and bootstrap data analyses allows to assess the statistical significance of the proposed measures in the time and frequency domain, thus enabling to get more confidence in the significance of the findings ad draw meaningful conclusions.

4.2 Beyond the Single Network Unit: The Interactions Between Two Nodes

4.2.1 *Time Domain*

In this section, we will characterize the time domain behavior of the two generic ith and jth network nodes taken from the set $\mathcal{X}_1, \ldots, \mathcal{X}_M$, $i = 1, \ldots, M$, in the framework of a BA, i.e., assuming that $M = 2$; the same information-theoretic measures can be defined in the scalar case, i.e., when each block is composed of 2 scalar processes. To this end, we refer the reader to the mathematical definitions provided in Sect. 2.3.2 related to the time domain representation of ARX linear models, from which several time- and frequency-domain measures can be defined with meaning in the framework of information theory to explore the coupled (causal) interactions between the two processes, as well as their self-dependencies. In the following subsections, we will introduce the mathematical definitions of coupling, causality and autonomy measures, as well as their formulation in the context of linear Gaussian processes.

4.2.1.1 Coupling and Causality Measures

The interactions of order two (i.e., pairwise interactions) between two network units can be assessed by measures of coupling and causality. *Non-directional coupling* relations between time series refer to associations which do not specify the direction of influence. This type of coupling does not assume causality between the time series, but rather looks for symmetrical statistical dependencies between them [24, 25]. Nevertheless, the principle of *causality* is fundamental in time series analysis to identify driver-response (i.e., time-lagged) relations between the processes. In the linear signal processing framework, this principle can be explored with reference to the concept of *Granger causality* (GC), which has been one of the most relevant approaches exploited by modern time series analysis. The concept of GC was originally developed by Wiener [26] and then made operative by Granger in the context of linear regression models [27]. In particular, GC relates the presence of a cause-effect relation to two aspects: the cause must precede the effect in time and must carry unique information about the present value of the effect. This relationship is not symmetrical and can be bidirectional, thus enabling the detection of both directional and reciprocal influences. Differently from non-directional measures, causality approaches exploiting this concept allow focusing on specific directional pathways of interactions within the investigated network [28].

Let us consider the bivariate vector process $\mathbf{X} = \{\mathbf{X}_1, \mathbf{X}_2\}$, where the two blocks have dimensions M_1 and M_2, respectively. In the frame of information theory, a measure of non-directional coupling is the *mutual information rate* (MIR) computed between the processes $\mathbf{X}_1$ and $\mathbf{X}_2$, quantifying the information shared by the two processes per unit of time [2, 3, 9, 29]:

$$I_{\mathbf{X}_1;\mathbf{X}_2} = \lim_{m\to\infty} \frac{1}{m} I(\mathbf{X}_{1,n:n+m}; \mathbf{X}_{2,n:n+m}) = I(\mathbf{X}_{1,n}, \mathbf{X}^-_{1,n}; \mathbf{X}_{2,n}, \mathbf{X}^-_{2,n}) - I(\mathbf{X}^-_{1,n}, \mathbf{X}^-_{2,n}). \tag{4.11}$$

The MIR is a symmetric measure (i.e., $I_{\mathbf{X}_1;\mathbf{X}_2} = I_{\mathbf{X}_2;\mathbf{X}_1}$) which can be decomposed evidencing information quantities with meaningful interpretations (see, e.g., [30, 31]):

$$I_{\mathbf{X}_1;\mathbf{X}_2} = H_{\mathbf{X}_1} + H_{\mathbf{X}_2} - H_{\mathbf{X}_1,\mathbf{X}_2}, \tag{4.12a}$$

$$I_{\mathbf{X}_1;\mathbf{X}_2} = T_{\mathbf{X}_1\to\mathbf{X}_2} + T_{\mathbf{X}_2\to\mathbf{X}_1} + I_{\mathbf{X}_1\cdot\mathbf{X}_2}. \tag{4.12b}$$

The first decomposition of the MIR given in (4.12a) evidences how it can be formulated comparing the sum of the individual ERs of the processes $\mathbf{X}_1$ and $\mathbf{X}_2$,

$H_{\mathbf{X}_1}$ and $H_{\mathbf{X}_2}$, with their joint entropy rate, $H_{\mathbf{X}_1,\mathbf{X}_2} = H(\mathbf{X}_{1,n}, \mathbf{X}_{2,n}|\mathbf{X}_{1,n}^-, \mathbf{X}_{2,n}^-)$. Conversely, the decomposition in (4.12b) expresses the MIR as the sum of the two transfer entropies from $\mathbf{X}_1$ to $\mathbf{X}_2$ and from $\mathbf{X}_2$ to $\mathbf{X}_1$ [32]

$$T_{\mathbf{X}_1\to\mathbf{X}_2} = I(\mathbf{X}_{2,n}; \mathbf{X}_{1,n}^-|\mathbf{X}_{2,n}^-) = H(\mathbf{X}_{2,n}|\mathbf{X}_{2,n}^-) - H(\mathbf{X}_{2,n}|\mathbf{X}_{1,n}^-, \mathbf{X}_{2,n}^-), \tag{4.13a}$$

$$T_{\mathbf{X}_2\to\mathbf{X}_1} = I(\mathbf{X}_{1,n}; \mathbf{X}_{2,n}^-|\mathbf{X}_{1,n}^-) = H(\mathbf{X}_{1,n}|\mathbf{X}_{1,n}^-) - H(\mathbf{X}_{1,n}|\mathbf{X}_{2,n}^-, \mathbf{X}_{1,n}^-), \tag{4.13b}$$

and a term quantifying the instantaneous information shared by $\mathbf{X}_1$ and $\mathbf{X}_2$ at zero lag

$$I_{\mathbf{X}_1\cdot\mathbf{X}_2} = I(\mathbf{X}_{1,n}; \mathbf{X}_{2,n}|\mathbf{X}_{1,n}^-, \mathbf{X}_{2,n}^-). \tag{4.14}$$

The *transfer entropy* (TE) is a well-known measure of directional information transfer related to the concept of GC, while the *instantaneous transfer* (IT) is a symmetric measure related to the concept of instantaneous causality (IC) [2].

Remarkably, the information-theoretic measure of MIR (4.11) and its decomposition terms (4.12) can be straightforwardly defined in the case of two interacting scalar processes $\{Y_1, Y_2\}$, with $Y_{1,n}$, $Y_{2,n}$, $Y_{1,n}^-$ and $Y_{2,n}^-$ taking the place of $\mathbf{X}_{1,n}$, $\mathbf{X}_{2,n}$, $\mathbf{X}_{1,n}^-$ and $\mathbf{X}_{2,n}^-$ in (4.11)–(4.14), respectively.

Linear parametric formulation. In the linear signal processing framework, a straightforward formulation of the MIR can be retrieved combining (4.2), (4.12a) and $H_{\mathbf{X}_1,\mathbf{X}_2} = \frac{1}{2}\log\left((2\pi e)^{M_1+M_2}|\Sigma_{\mathbf{U_X}}^{(arx)}|\right)$, where $\Sigma_{\mathbf{U_X}}^{(arx)}$ is the $(M_1 + M_2) \times (M_1 + M_2)$ covariance matrix of the block ARX model (2.5), thus yielding:

$$I_{\mathbf{X}_1;\mathbf{X}_2} = \frac{1}{2}\log\left(\frac{|\Sigma_{\mathbf{U}_{\mathbf{X}_1}}^{(ar)}||\Sigma_{\mathbf{U}_{\mathbf{X}_2}}^{(ar)}|}{|\Sigma_{\mathbf{U_X}}^{(arx)}|}\right), \tag{4.15}$$

where $\Sigma_{\mathbf{U}_{\mathbf{X}_1}}^{(ar)}$ and $\Sigma_{\mathbf{U}_{\mathbf{X}_2}}^{(ar)}$ are the covariance matrices of the residuals in the block AR models of $\mathbf{X}_1$ and $\mathbf{X}_2$ in the form of (2.3), respectively. In a similar manner, we can retrieve the MIR of $\{Y_1, Y_2\}$ as $I_{Y_1;Y_2} = \frac{1}{2}\log\left(\frac{\sigma_{U_{Y_1}}^{2(ar)}\sigma_{U_{Y_2}}^{2(ar)}}{|\Sigma_{\mathbf{U_Y}}^{(arx)}|}\right)$, where $\sigma_{U_{Y_1}}^{2(ar)}$ and $\sigma_{U_{Y_2}}^{2(ar)}$ are the variances of the residuals of the AR models of Y_1, Y_2 in the form of (2.2), while $\Sigma_{\mathbf{U_Y}}^{(arx)}$ is the 2×2 covariance matrix of the residuals in the ARX model (2.4).

The linear formulation of the two TEs requires the identification of the full block ARX model of $\{\mathbf{X}_1, \mathbf{X}_2\}$ (2.5) and of the restricted block AR models of $\mathbf{X}_1$ and $\mathbf{X}_2$ in the form of (2.3), to yield:

$$T_{\mathbf{X}_1\to\mathbf{X}_2} = \frac{1}{2}\log\left(\frac{|\Sigma_{\mathbf{U}_{\mathbf{X}_2}}^{(ar)}|}{|\Sigma_{\mathbf{U}_{\mathbf{X}_2}}^{(arx)}|}\right), \; T_{\mathbf{X}_2\to\mathbf{X}_1} = \frac{1}{2}\log\left(\frac{|\Sigma_{\mathbf{U}_{\mathbf{X}_1}}^{(ar)}|}{|\Sigma_{\mathbf{U}_{\mathbf{X}_2}}^{(arx)}|}\right), \tag{4.16}$$

where $\Sigma_{\mathbf{U}_{\mathbf{X}_1}}^{(arx)}$ and $\Sigma_{\mathbf{U}_{\mathbf{X}_2}}^{(arx)}$ are the block matrices of dimensions $M_1 \times M_1$ and $M_2 \times M_2$, respectively, on the main diagonal of the block covariance matrix $\Sigma_{\mathbf{U_X}}^{(arx)}$ in (2.5). In a similar manner, we can retrieve the TEs of the network $\{Y_1, Y_2\}$ as $T_{Y_1\to Y_2} =$

$\frac{1}{2}\log\left(\frac{\sigma^{2(ar)}_{U_{Y_2}}}{\sigma^{2(arx)}_{U_{Y_2}}}\right)$ and $T_{Y_2\to Y_1} = \frac{1}{2}\log\left(\frac{\sigma^{2(ar)}_{U_{Y_1}}}{\sigma^{2(arx)}_{U_{Y_1}}}\right)$, where $\sigma^{2(arx)}_{U_{Y_1}}$ and $\sigma^{2(arx)}_{U_{Y_2}}$ are the diagonal elements of the 2×2 covariance matrix $\Sigma^{(arx)}_{\mathbf{U_Y}}$ in (2.4).

The linear formulation of the quantity $I_{\mathbf{X}_1\cdot\mathbf{X}_2}$ in (4.12b) can be directly retrieved from (4.15) and (4.16) as:

$$I_{\mathbf{X}_1\cdot\mathbf{X}_2} = \frac{1}{2}\log\left(\frac{|\Sigma^{(arx)}_{\mathbf{U}_{\mathbf{X}_1}}||\Sigma^{(arx)}_{\mathbf{U}_{\mathbf{X}_2}}|}{|\Sigma^{(arx)}_{\mathbf{U_X}}|}\right). \tag{4.17}$$

A similar definition can be found in the case of the scalar processes $\{Y_1, Y_2\}$, i.e., $I_{Y_1\cdot Y_2} = \frac{1}{2}\log\left(\frac{\sigma^{2(arx)}_{U_{Y_1}}\sigma^{2(arx)}_{U_{Y_2}}}{|\Sigma^{(arx)}_{\mathbf{U_Y}}|}\right)$. The instantaneous transfer quantifies instantaneous interactions between the processes, and reflects the presence of correlations occurring between the variables that sample the two processes at the same temporal index; since they cannot be captured by the model coefficients describing time-lagged effects, these correlations are found in the model innovations. Indeed, this measure is zero when the innovations of the full ARX model are uncorrelated (i.e., when $\sigma^{2(arx)}_{U_{Y_1Y_2}} = \sigma^{2(arx)}_{U_{Y_2Y_1}} = 0$ in the scalar case) and positive otherwise. Then, in the case of a strictly causal ARX model whose covariance matrix $\Sigma^{(arx)}_{\mathbf{U_Y}}$ is diagonal, the IT vanishes by construction. However, strictly causal models are often used to fit a vector process whose scalar constituents interact also at lag 0, thus introducing zero-lag correlations among the residuals and non-zero off-diagonal elements in the input covariance matrix. In this case, the measure does not vanish in the time domain, reflecting the presence of instantaneous effects which however have not been modelled.

In this context, the time domain bivariate dependencies assessed via the MIR measure in dynamic networks of multiple interconnected nodes have been investigated, e.g., in [33], with a specific application to cardiovascular, cardiorespiratory and cerebrovascular interactions that will be discussed in Sect. 5.2.

Geweke decomposition of total dependence. In the case of Gaussian processes, it can be demonstrated that the the TEs (4.13) and the IT (4.14) are equivalent to the following Geweke logarithmic measures of GC and IC, respectively, up to a factor 2 [2, 9]:

$$T_{\mathbf{X}_1\to\mathbf{X}_2} = \frac{F_{\mathbf{X}_1\to\mathbf{X}_2}}{2}, \tag{4.18a}$$

$$T_{\mathbf{X}_2\to\mathbf{X}_1} = \frac{F_{\mathbf{X}_2\to\mathbf{X}_1}}{2}, \tag{4.18b}$$

$$I_{\mathbf{X}_1\cdot\mathbf{X}_2} = \frac{F_{\mathbf{X}_1\cdot\mathbf{X}_2}}{2}, \tag{4.18c}$$

with

$$F_{\mathbf{X}_1;\mathbf{X}_2} = F_{\mathbf{X}_1\to\mathbf{X}_2} + F_{\mathbf{X}_2\to\mathbf{X}_1} + F_{\mathbf{X}_1\cdot\mathbf{X}_2} \tag{4.19}$$

being referred to as the *Geweke decomposition* of the *total dependence* (TD) $F_{\mathbf{X}_1;\mathbf{X}_2}$, which in turn is straightforwardly related to the MIR (4.11) according to

$$I_{\mathbf{X}_1;\mathbf{X}_2} = \frac{F_{\mathbf{X}_1;\mathbf{X}_2}}{2}. \tag{4.20}$$

The same equations can be retrieved for the process $\mathbf{Y} = \{Y_1, Y_2\}$ in the case of unitary blocks, leading to $F_{Y_1;Y_2} = F_{Y_1 \to Y_2} + F_{Y_2 \to Y_1} + F_{Y_1 \cdot Y_2}$. The time domain dependencies between two network nodes assessed via the Geweke measures of TD, GC and IC have been investigated, e.g., in [34–36], with specific applications to cardiovascular, cardiorespiratory and cerebrovascular interactions reviewed and discussed in Sects. 5.3, 5.4 and 5.5.

4.2.1.2 Autonomy Measures

In parallel to exploring the concept of GC, the analysis of coupled processes has evolved also studying the role of *autonomous dynamics*, i.e. interactions that occur internally in a process independently of its link with other processes. In this context, a so-called *Granger autonomy* (GA) measure has been defined to quantify how much its own internal dynamics contribute to determine its predictability more than the dynamics of other processes potentially connected to it [37], based on a previously defined notion of degree of self-determination of a system [38, 39]. The idea behind the concept is that a system is autonomous if it is not controlled by external influences but rather it self-determines its states. This measure has been developed also in the context of information theory and applied to gain insight about the physiological mechanisms governing the autonomous dynamics of a process [6, 8, 40].

In analogy with GC, the concept of GA was formalized for the bivariate process $\mathbf{Y} = \{Y_1, Y_2\}$ assessing the predictability improvement brought to the present state of the target Y_2 by its own past states above and beyond the predictability brought by the past states of the driver Y_1 [37]. Operationally, GA is quantified comparing the full ARX model (2.4) with a restricted X model where $Y_{2,n}$ is described only from the past of $Y_{1,n}$ (2.9). Then, the predictability improvement is quantified by the logarithmic measure of GA given by [37, 41]

$$A_{Y_2} = \log\left(\frac{\lambda^2_{W_{Y_2|Y_1}}}{\sigma^{2(arx)}_{U_{Y_2}}}\right), \tag{4.21}$$

which quantifies the strength of the autonomous dynamics in the target process Y_2 comparing the error variances of the models (2.4), $\sigma^{2(arx)}_{U_{Y_2}}$, and (2.9), $\lambda^2_{W_{Y_2|Y_1}}$. In the case of Gaussian processes, the GA measure (4.21) is equivalent, up to a factor 2, to the information-theoretic measure of conditional self-entropy [6, 8, 40], i.e. $S_{Y_2|Y_1} = \frac{A_{Y_2}}{2}$.

Two notes here are of remarkable importance. First, we point out that the time domain formulation of the GA measure (4.21) was early proposed in [37] and then applied in [41] for the two scalar processes Y_1, Y_2, whose dynamics are described by the ARX model (2.4); the extension to the vector case (2.5) should be straightforward though not formulated yet. Further, we emphasize that, differently from GC, the concept of GA as expressed by [37] lacked of a valid spectral representation, as spectral measures able to identify the autonomous oscillations in a process had not been defined at that time. Our goal in [41] is to fill this gap, as we will show in Sect. 4.2.2.2.

4.2.2 *Frequency Domain*

In this section, we will derive definitions of spectral coupling, causality (Sect. 4.2.2.1) and autonomy (Sect. 4.2.2.2) measures in the case of two scalar processes $\{Y_1, Y_2\}$ starting from the decomposition of the power spectra of the processes into causal and non-causal contributes. The spectral formulations of the discussed coupling and causality measures can be extended intuitively to the vector case, i.e., when the two block processes $\{\mathbf{X}_1, \mathbf{X}_2\}$ are observed, as shown at the end of Sect. 4.2.2.1; we refer the reader to Refs. [25, 42, 43] for further mathematical details.

4.2.2.1 Coupling and Causality Measures

Let us consider the spectral representation of the ARX model (2.4). The model is first represented in the Z-domain through its Z-transforms yielding $\mathbf{Y}(z) = \mathbf{H}^{\mathbf{Y}(arx)}(z)\mathbf{U}_{\mathbf{Y}}^{(arx)}(z)$, where $\mathbf{H}^{\mathbf{Y}(arx)}(z) = [\mathbf{I} - \sum_{k=1}^{p} \mathbf{A}_{\mathbf{Y},k}^{(arx)} z^{-k}]^{-1} = \bar{\mathbf{A}}_{\mathbf{Y}}^{(arx)}(z)^{-1}$ is the 2×2 TF matrix modelling the relationships between the input $\mathbf{U}_{\mathbf{Y}}^{(arx)}(z)$ and the output $\mathbf{Y}(z)$. Computing $\mathbf{H}^{\mathbf{Y}(arx)}(z)$ on the unit circle in the complex plane ($z = e^{\mathrm{j}2\pi\bar{f}}$) and exploiting spectral factorization, it is possible to derive the PSD of the analysed stationary random process $\mathbf{Y}$, thus yielding $\mathbf{P}_{\mathbf{Y}}^{(arx)}(\bar{f}) = \mathbf{H}^{\mathbf{Y}(arx)}(\bar{f})\Sigma_{\mathbf{U}_{\mathbf{Y}}}^{(arx)}\mathbf{H}^{\mathbf{Y}(arx)*}(\bar{f})$. The PSD matrix $\mathbf{P}_{\mathbf{Y}}^{(arx)}(\bar{f})$ contains information related to the spectral properties of the two processes, i.e., to their own dynamics, through the 2 diagonal elements $P_{Y_i}^{(arx)}(\bar{f})$, $i = 1, 2$, and to the causal interactions between Y_1 and Y_2, through the off-diagonal elements $P_{Y_iY_j}^{(arx)}(\bar{f})$, $i, j \in \{1, 2\}, i \neq j$.

The pairwise interactions between the units Y_1, Y_2 can be assessed by measures of spectral coupling and causality in the frequency domain, which can be directly derived from different combinations of the elements of the 2×2 PSD matrix $\mathbf{P}_{\mathbf{Y}}^{(arx)}(\bar{f})$, of the 2×2 TF matrix $\mathbf{H}^{\mathbf{Y}(arx)}(\bar{f})$ and of the innovations of the full ARX and the restricted AR models. We remark that these measures have a straightforward interpretation in the information-theoretic framework thanks to the well-known property of spectral integration [9]. For details, please refer to [34, 35, 41], where the theoretical notions of spectral coupling and causality have been widely unravelled and applied to closed-loop physiological interactions, and to [36], the latter exploiting the concept of spectral GC to assess cardiorespiratory interactions through photoplethysmography (see Sect. 5.3 for the related applications to real data).

Measures of coherence and directed coherence. Let us consider the full ARX model. Under the assumption that the input white noises are uncorrelated at lag 0 leading to diagonality of $\Sigma_{\mathbf{U_Y}}^{(arx)}$ (i.e., the model is strictly causal) [24, 44], the ijth elements of $\mathbf{P}_{\mathbf{Y}}^{(arx)}(\bar{f})$ are factorized into (henceforth, we omit the superscript $^{(arx)}$ for clarity):

$$P_{Y_iY_j}(\bar{f}) = \sum_{q=1}^{Q} \sigma^2_{U_{Y_q}} H_{Y_iY_q}(\bar{f}) H^*_{Y_jY_q}(\bar{f}), \tag{4.22}$$

where $Q = 2$ for a bivariate process, and $\sigma^2_{U_{Y_q}}$, $q = 1, 2$, are the diagonal elements of $\Sigma_{\mathbf{U_Y}}$. This factorization allows to decompose the frequency domain measures of coupling and causality into terms eliciting the directional information from one process to another. Indeed, the *coherence* (Coh) between the two processes Y_1 and Y_2, defined as the ratio between the cross-spectrum $P_{Y_1Y_2}(\bar{f})$ and the squared root of the product between the autospectra of Y_1 and Y_2

$$\Gamma_{Y_1;Y_2}(\bar{f}) = \frac{P_{Y_1Y_2}(\bar{f})}{\sqrt{P_{Y_1}(\bar{f}) P_{Y_2}(\bar{f})}}, \tag{4.23}$$

is decomposed as follows, exploiting (4.22):

$$\Gamma_{Y_1;Y_2}(\bar{f}) = \sum_{q=1}^{2} \frac{\sigma^2_{U_{Y_q}} H_{Y_1Y_q}(\bar{f}) \sigma^2_{U_{Y_q}} H^*_{Y_2Y_q}(\bar{f})}{\sqrt{P_{Y_1}(\bar{f})}\sqrt{P_{Y_2}(\bar{f})}}. \tag{4.24}$$

Since this function is complex-valued, its squared modulus is commonly used to measure the strength of coupling in the frequency domain. Indeed, the magnitude squared coherence, $|\Gamma_{Y_1;Y_2}(\bar{f})|^2$, has a meaningful physical interpretation, since it measures the strength of the linear, non-directional coupled interactions between the processes Y_1 and Y_2 as a function of frequency, being 0 in case of uncoupling and 1 in case of full coupling. Exploiting (4.23), it can be shown that the squared Coh is related to the logarithmic spectral measure of total dependence between Y_1 and Y_2, defined by Geweke as [9]

$$f_{Y_1;Y_2}(\bar{f}) = \log\left(\frac{P_{Y_1}(\bar{f}) P_{Y_2}(\bar{f})}{|\mathbf{P_Y}(\bar{f})|}\right), \tag{4.25}$$

through the relationship $f_{Y_1;Y_2}(\bar{f}) = -\log(1 - |\Gamma_{Y_1;Y_2}(\bar{f})|^2)$. The spectral TD defined in (4.25) is non-negative and linked to the time domain measure (4.20) by the spectral integration property

$$F_{Y_1,Y_2} = 2\int_0^{\frac{1}{2}} f_{Y_1;Y_2}(\bar{f})\,\mathrm{d}\bar{f}. \tag{4.26}$$

Due to the symmetrical nature of this measure, it cannot provide information about causality; such an information may be described by means of the *directed coherence* (DC) $\gamma_{Y_1\to Y_2}(\bar{f}) = \gamma_{Y_2Y_1}(\bar{f})$ from the driver process Y_1 to the target process Y_2, interpreted as a measure of the influence of Y_1 onto Y_2, as opposed to $\gamma_{Y_1Y_2}(\bar{f})$ which measures the interaction occurring over the opposite direction, from Y_2 to Y_1. The squared DC

$$|\gamma_{Y_2Y_1}(\bar{f})|^2 = \frac{\sigma^2_{U_{Y_1}}|H_{Y_2Y_1}(\bar{f})|^2}{\sigma^2_{U_{Y_1}}|H_{Y_2Y_1}(\bar{f})|^2 + \sigma^2_{U_{Y_2}}|H_{Y_2Y_2}(\bar{f})|^2} \tag{4.27}$$

measures a normalized coupling strength, being 0 in the absence of directional coupling from Y_1 to Y_2 at the frequency f, and 1 in the presence of full coupling. Moreover, $|\gamma_{Y_2Y_1}(\bar{f})|^2$ measures the coupling strength from Y_1 to Y_2 as the normalized proportion of $P_{Y_2}(\bar{f})$ which is causally due to Y_1.

Similarly to (4.27), it is possible to define the normalized portion of $P_{Y_2}(\bar{f})$ which arises from the target process Y_2 itself as

$$|\gamma_{Y_2Y_2}(\bar{f})|^2 = \frac{\sigma^2_{U_{Y_2}}|H_{Y_2Y_2}(\bar{f})|^2}{\sigma^2_{U_{Y_1}}|H_{Y_2Y_1}(\bar{f})|^2 + \sigma^2_{U_{Y_2}}|H_{Y_2Y_2}(\bar{f})|^2}. \tag{4.28}$$

Hence, it is easy to show that the spectrum of the target process Y_2 can be decomposed as:

$$P_{Y_2}(\bar{f}) = \sum_{q=1}^{2} P_{Y_2|Y_q}(\bar{f}) = P_{Y_2|Y_1}(\bar{f}) + P_{Y_2|Y_2}(\bar{f}), \tag{4.29}$$

where $P_{Y_2|Y_1}(\bar{f}) = |\gamma_{Y_2Y_1}(\bar{f})|^2 P_{Y_2}(\bar{f})$ is the part of $P_{Y_2}(\bar{f})$ due to Y_1, which is usually referred to as the *causal* part of $P_{Y_2}(\bar{f})$; $P_{Y_2|Y_2}(\bar{f}) = |\gamma_{Y_2Y_2}(\bar{f})|^2 P_{Y_2}(\bar{f})$ measures the part of $P_{Y_2}(\bar{f})$ due to none of the other processes, but to the process Y_2 itself, which may be thus referred to as the *isolated* part of the target spectrum. The concept of isolation, related to the spectrum decomposition into causal and non-causal contributions, has been proposed for the first time in [41] with reference to bivariate processes.

Importantly, the DC defined in (4.27) has a meaningful physical interpretation, since it measures causality as the amount of signal power transferred from one process to another and is a measure of causal coupling in the frequency domain. Moreover, it can be regarded as a measure of GC from Y_1 to Y_2 thanks to its relation with logarithmic spectral measure of GC defined by Geweke and referred to as *linear feedback* [9]

$$f_{Y_1\to Y_2}(\bar{f}) = \log\left(\frac{P_{Y_2}(\bar{f})}{\sigma^2_{U_{Y_2}}|H_{Y_2Y_2}(\bar{f})|^2}\right), \tag{4.30}$$

which is non-negative and linked to the time domain GC measure (4.18a) by the spectral integration property

$$F_{Y_1 \to Y_2} = 2 \int_0^{\frac{1}{2}} f_{Y_1 \to Y_2}(\bar{f})\, \mathrm{d}\bar{f}. \tag{4.31}$$

In fact, combining (4.27), (4.29) and (4.30) one can easily show that the DC and the spectral GC are linked by the relation $f_{Y_1 \to Y_2}(\bar{f}) = -\log(1 - |\gamma_{Y_2 Y_1}(\bar{f})|^2)$ [2, 9, 24, 34, 41]. We note that, while the total coupling measure (4.25) is always non-negative, the two causal measures $f_{Y_i \to Y_j}(\bar{f})$, $i, j = 1, 2$, can take negative values at some frequencies if the process $\mathbf{Y}$ is not strictly causal (i.e. if the innovation covariance $\Sigma_{\mathbf{U}_\mathbf{Y}}$ is not diagonal). To overcome this issue, we recently proposed extended linear models to capture zero-lag interactions known a priori in physiological time series analysis and thus included them in the Geweke logarithmic spectral measures of GC (see [34, 35]).

Geweke decomposition of the total dependence in the frequency domain. To satisfy a spectral decomposition similar to (4.19), with Y_1 and Y_2 in place of $\mathbf{X}_1, \mathbf{X}_2$, i.e.,

$$f_{Y_1;Y_2}(\bar{f}) = f_{Y_1 \to Y_2}(\bar{f}) + f_{Y_2 \to Y_1}(\bar{f}) + f_{Y_1 \cdot Y_2}(\bar{f}), \tag{4.32}$$

the spectral measure of instantaneous interactions was chosen ad hoc [9]

$$f_{Y_1 \cdot Y_2}(\bar{f}) = \log \left(\frac{\sigma^2_{U_{Y_1}} |H_{Y_1 Y_1}(\bar{f})|^2 \sigma^2_{U_{Y_2}} |H_{Y_2 Y_2}(\bar{f})|^2}{|\mathbf{P}_\mathbf{Y}(\bar{f})|} \right) \tag{4.33}$$

in such a way to be linked to the time domain measure in (4.18a), right side, by the spectral integration property, i.e.,

$$F_{Y_1 \cdot Y_2} = 2 \int_0^{\frac{1}{2}} f_{Y_1 \cdot Y_2}(\bar{f})\, \mathrm{d}\bar{f}. \tag{4.34}$$

This spectral measure does not fulfill all the requirements of Geweke, different from what occurs in the time domain. Indeed, it may be negative for some frequencies and has no clear physical meaning [9]. The lack of non-negativity can be seen by considering that the integration over frequencies of the decomposition in (4.32) has to be consistent with the decomposition in (4.19). In absence of instantaneous causality, i.e., in the case of a strictly causal ARX model implying diagonality of $\Sigma_{\mathbf{U}_\mathbf{Y}}$, $F_{Y_1 \cdot Y_2} = 0$; however, $f_{Y_1 \cdot Y_2}(\bar{f})$ is generally found to be non-zero for some frequencies. Since the integral of $f_{Y_1 \cdot Y_2}(\bar{f})$ has to be null when $F_{Y_1 \cdot Y_2} = 0$, not being zero for all frequencies, this implies the violation of non-negativity.

The Geweke decomposition of the total dependence in the frequency domain, as well as the issue of zero-lag effects in physiological time series analysis, have been largely discussed in [34], and then applied to cardiorespiratory interactions during spontaneous and controlled breathing [35].

Granger Isolation. In analogy with the derivations in (4.30), the new spectral logarithmic measure of *Granger isolation* (GI) of the target Y_2 linked to the isolated part of the target spectrum was proposed for the first time in [41] and defined as

$$f_{Y_2}(\bar{f}) = \log\left(\frac{P_{Y_2}(\bar{f})}{\sigma^2_{U_{Y_1}}|H_{Y_2Y_1}(\bar{f})|^2}\right) = -\log(1-|\gamma_{Y_2Y_2}(\bar{f})|^2). \tag{4.35}$$

Moreover, following (4.31), the new time domain measure of GI integrating the spectral measure in (4.35) over all frequencies is provided, yielding

$$F_{Y_2} = 2\int_0^{\frac{1}{2}} f_{Y_2}(\bar{f})\,\mathrm{d}\bar{f}. \tag{4.36}$$

Intuitively, one might think that the GI measure (4.36) reflects the concept of GA [37], given that it is derived from the isolated (i.e., non-causal) part of the target spectrum. However, the GI behaves differently than the known GA measure defined from the error variances of full and restricted linear regression models (4.21) [6, 8, 37, 40], as discussed in [41] and displayed through different simulation examples in Sect. 4.2.3.

Spectral measures of coupling and causality for block processes. We report here the formulations of the information-theoretic spectral measures of coupling and causality for the two block processes $\{\mathbf{X}_1, \mathbf{X}_2\}$. To this end, let us consider the block ARX model (2.5) and its spectral representation. The model is first represented in the Z-domain through its Z-transforms yielding $\mathbf{X}(z) = \mathbf{H}^{\mathbf{X}(arx)}(z)\mathbf{U}_{\mathbf{X}}^{(arx)}(z)$, where $\mathbf{H}^{\mathbf{X}(arx)}(z) = [\mathbf{I} - \sum_{k=1}^{p}\mathbf{A}_{\mathbf{X},k}^{(arx)}z^{-k}]^{-1} = \bar{\mathbf{A}}_{\mathbf{X}}^{(arx)}(z)^{-1}$ is the TF matrix modelling the relationships between the input $\mathbf{U}_{\mathbf{X}}^{(arx)}(z)$ and the output $\mathbf{X}(z)$. Computing $\mathbf{H}^{\mathbf{X}(arx)}(z)$ on the unit circle in the complex plane ($z = e^{\mathrm{j}2\pi\bar{f}}$) and exploiting spectral factorization, it is possible to derive the PSD of the random process $\mathbf{X}$, thus yielding $\mathbf{P}_{\mathbf{X}}^{(arx)}(\bar{f}) = \mathbf{H}^{\mathbf{X}(arx)}(\bar{f})\Sigma_{\mathbf{U}_{\mathbf{X}}}^{(arx)}\mathbf{H}^{\mathbf{X}(arx)*}(\bar{f})$. Now, let us omit the superscript $^{(arx)}$ for clarity. The matrix $\mathbf{P}_{\mathbf{X}}(\bar{f})$ can be factorized in blocks to make explicit the power spectral densities of $\mathbf{X}_1$ and $\mathbf{X}_2$, $\mathbf{P}_{\mathbf{X}_1}(\bar{f})$ and $\mathbf{P}_{\mathbf{X}_2}(\bar{f})$, as diagonal blocks, and the cross-spectral densities between $\mathbf{X}_1$ and $\mathbf{X}_2$, $\mathbf{P}_{\mathbf{X}_1\mathbf{X}_2}(\bar{f})$ and $\mathbf{P}_{\mathbf{X}_2\mathbf{X}_1}(\bar{f})$, as off-diagonal blocks. From this factorization, a logarithmic spectral measure of the interdependence between $\mathbf{X}_1$ and $\mathbf{X}_2$ is defined by [9]

$$f_{\mathbf{X}_1;\mathbf{X}_2}(\bar{f}) = \log\left(\frac{|\mathbf{P}_{\mathbf{X}_1}(\bar{f})||\mathbf{P}_{\mathbf{X}_2}(\bar{f})|}{|\mathbf{P}_{\mathbf{X}}(\bar{f})|}\right); \tag{4.37}$$

this measure quantifies the total (symmetric) coupling between $\mathbf{X}_1$ and $\mathbf{X}_2$ and is related to the so-called *block coherence* [25, 42], extending to block processes the Coh (4.23) defined for scalar processes. Moreover, after factorizing in $M_i \times M_i$ diagonal blocks and $M_i \times M_j$ off-diagonal blocks also the transfer and innovation covariance matrices $\mathbf{H}(\bar{f})$ and $\Sigma_{\mathbf{U_X}}$, logarithmic spectral measures of the causal effect of $\mathbf{X}_j$ on $\mathbf{X}_i$ $(i, j = 1, 2)$ can be computed as [9]

$$f_{\mathbf{X}_j \to \mathbf{X}_i}(\bar{f}) = \log\left(\frac{|\mathbf{P}_{\mathbf{X}_i}(\bar{f})|}{|\mathbf{H}_{\mathbf{X}_i\mathbf{X}_i}(\bar{f})\Sigma_{\mathbf{U}_{\mathbf{X}_i}}\mathbf{H}^*_{\mathbf{X}_i\mathbf{X}_i}(\bar{f})|}\right), \tag{4.38}$$

where $\mathbf{H}_{\mathbf{X}_i\mathbf{X}_i}$ describes the transfer from $\mathbf{U}_{\mathbf{X}_i}$ to $\mathbf{X}_i$ in the frequency domain and $\Sigma_{\mathbf{U}_{\mathbf{X}_i}} = \mathbb{E}[\mathbf{U}_{\mathbf{X}_i,n}\mathbf{U}^{\intercal}_{\mathbf{X}_i,n}]$; these measures quantify the causal (asymmetric) coupling from $\mathbf{X}_1$ to $\mathbf{X}_2$ and vice-versa, and are related to the so-called *block directed coherence* [25], extending to block processes the DC (4.27) defined for scalar processes. To complete the representation of the pairwise interactions between $\mathbf{X}_1$ and $\mathbf{X}_2$, a spectral measure $f_{\mathbf{X}_1 \cdot \mathbf{X}_2}(\bar{f})$ can be defined subtracting the sum of the two causal measures (4.38) from the coupling measure (4.37) to get

$$f_{\mathbf{X}_1 \cdot \mathbf{X}_2}(\bar{f}) = \log\left(\frac{|\mathbf{H}_{\mathbf{X}_1\mathbf{X}_1}(\bar{f})\Sigma_{\mathbf{U}_{\mathbf{X}_1}}\mathbf{H}^*_{\mathbf{X}_1\mathbf{X}_1}(\bar{f})||\mathbf{H}_{\mathbf{X}_2\mathbf{X}_2}(\bar{f})\Sigma_{\mathbf{U}_{\mathbf{X}_2}}\mathbf{H}^*_{\mathbf{X}_2\mathbf{X}_2}(\bar{f})|}{|\mathbf{P}_{\mathbf{X}}(\bar{f})|}\right), \tag{4.39}$$

so as to satisfy in the frequency domain a decomposition similar to the time domain decomposition (4.19):

$$f_{\mathbf{X}_1;\mathbf{X}_2}(\bar{f}) = f_{\mathbf{X}_1 \to \mathbf{X}_2}(\bar{f}) + f_{\mathbf{X}_2 \to \mathbf{X}_1}(\bar{f}) + f_{\mathbf{X}_1 \cdot \mathbf{X}_2}(\bar{f}). \tag{4.40}$$

Importantly, the spectral measures in (4.40) are tightly linked to the similar measures given in the time domain decomposition of total dependence (4.19). In fact, it can be shown that integration over the whole frequency axis of the spectral coupling measure (4.37) returns the TD between the two processes [2], i.e.,

$$F_{\mathbf{X}_1;\mathbf{X}_2} = 2\int_0^{\frac{1}{2}} f_{\mathbf{X}_1;\mathbf{X}_2}(\bar{f})\mathrm{d}\bar{f}, \tag{4.41}$$

and that the same relation holds integrating $f_{\mathbf{X}_1 \to \mathbf{X}_2}(\bar{f})$, $f_{\mathbf{X}_2 \to \mathbf{X}_1}(\bar{f})$ and $f_{\mathbf{X}_1 \cdot \mathbf{X}_2}(\bar{f})$ to get respectively $F_{\mathbf{X}_1 \to \mathbf{X}_2}$, $F_{\mathbf{X}_2 \to \mathbf{X}_1}$, and $F_{\mathbf{X}_1 \cdot \mathbf{X}_2}$.

The Geweke measures of TD, GC and IC related to block processes have been abundantly deepened in [43], as well as applied to investigate the pairwise interactions among large-scale brain networks in subacute stroke patients after motor rehabilitation [45]. Theoretical simulations showing the behavior of the proposed measures of coupling and causality for block processes in networks of multiple nodes will be presented in Sect. 4.3.3.

Meaning of the spectral measures of coupling and causality in the framework of information theory. The Geweke spectral measures of coupling and causality defined above for scalar and block processes have a straightforward meaning in the framework of information-theory. Indeed, it is easy to show that the MIR in (4.12b) can be expanded in the frequency domain as follows:

$$I_{\mathbf{X}_1;\mathbf{X}_2} = 2\int_0^{\frac{1}{2}} i_{\mathbf{X}_1;\mathbf{X}_2}(\bar{f})\,\mathrm{d}\bar{f}, \tag{4.42}$$

and that the same relation holds integrating $i_{\mathbf{X}_1\to\mathbf{X}_2}(\bar{f})$, $i_{\mathbf{X}_2\to\mathbf{X}_1}(\bar{f})$ and $i_{\mathbf{X}_1\cdot\mathbf{X}_2}(\bar{f})$ to get respectively $I_{\mathbf{X}_1\to\mathbf{X}_2}$, $I_{\mathbf{X}_2\to\mathbf{X}_1}$, and $I_{\mathbf{X}_1\cdot\mathbf{X}_2}$, where $i_{(\cdot)} = \frac{f_{(\cdot)}}{2}$. Remarkably, the spectral integration property gives to the logarithmic measures $f_{\mathbf{X}_1;\mathbf{X}_2}(\bar{f})$ (4.37) and $f_{\mathbf{X}_j\to\mathbf{X}_i}(\bar{f})$ (4.38), $i, j = 1, 2$, the information-theoretic meaning of density of information shared between the two scalar processes, or transferred from one process to the other, at the normalized frequency $\bar{f}$. We note that the same relationships are valid for scalar processes.

The spectral integration property has been largely applied in our works (see, e.g., [34–36, 41, 43, 45]), to establish a link between the information-theoretic measures of coupling and causality and the spectral representation of linear parametric models, essential to capture the richness of oscillatory content typically observed in physiological networks.

4.2.2.2 Autonomy Measures

While causality measures have been proposed and extensively applied in the frequency domain, measures quantifying self-dependencies were still limited to the time domain formulation and lacked of a clear spectral representation [37] before the introduction of the new spectral measure of GA [41]. To retrieve its formulation, let us consider the full ARX (2.4) and the restricted X (2.9) models. The autonomous dynamics of the target process Y_2, conditioned to the knowledge of the history of the driver Y_1, can be assessed by a measure of spectral autonomy in the frequency domain, which can be directly derived from a meaningful combination of the elements of the 2×2 TF matrices $\mathbf{H}^{\mathbf{Y}(arx)}(\bar{f})$ and $\mathbf{G}^{\mathbf{Y}}(\bar{f})$ of the full ARX (2.4) and restricted X (2.9) models, respectively (see Sect. 2.4.1). First, we note that in (2.9) the removal of the predictable autonomous dynamics of the target process Y_2 makes them likely to be contained in the residual $W_{Y_2|Y_1}$, and thus not modelled by the element $G_{Y_2Y_2}(\bar{f})$ of the TF matrix $\mathbf{G}^{\mathbf{Y}}(\bar{f})$. Since in (2.4b) these autonomous dynamics are instead modelled by $H^{\mathbf{Y}(arx)}_{Y_2Y_2}(\bar{f})$, they can be emphasized comparing the two transfer functions of the full and restricted models. Accordingly, in [41], we proposed to assess the strength and frequency-specific location of the target internal dynamics through the spectral function (we remove the superscript $^{\mathbf{Y}(arx)}$ for brevity):

$$\bar{a}_{Y_2}(\bar{f}) = \log\left(\frac{|H_{Y_2Y_2}(\bar{f})|^2}{|G_{Y_2Y_2}(\bar{f})|^2}\right), \tag{4.43}$$

which captures the balance between the transfer of information within the target quantified when the self-dependencies are modelled and when they are not. We expect that stronger internal dynamics at the frequency $\bar{f}$ are reflected by higher values of $|H_{Y_2Y_2}(\bar{f})|^2$ compared with $|G_{Y_2Y_2}(\bar{f})|^2$, and thus to higher values of

$\bar{a}_{Y_2}(\bar{f})$. However, since the full-frequency integral of both $\log\left(|H_{Y_2Y_2}(\bar{f})|^2\right)$ and $\log\left(|G_{Y_2Y_2}(\bar{f})|^2\right)$ is null [23], we have that $2\int_0^{\frac{1}{2}} \bar{a}_{Y_2}(\bar{f})\mathrm{d}\bar{f} = 0$, and thus $\bar{a}_{Y_2}(\bar{f})$ will take negative values at some frequencies, and its full-frequency integral will not return the time domain GA. To counteract these issues, the spectral GA measure was introduced and defined as

$$a_{Y_2}(\bar{f}) = \log\left(\frac{\lambda^2_{W_{Y_2|Y_1}}}{\sigma^2_{U_{Y_2}}} \frac{|H_{Y_2Y_2}(\bar{f})|^2}{|G_{Y_2Y_2}(\bar{f})|^2}\right), \tag{4.44}$$

which can be written also as $a_{Y_2}(\bar{f}) = A_{Y_2} + \bar{a}_{Y_2}(\bar{f})$, showing that it consists of a frequency-independent part equal to the time domain GA (4.21) and of a frequency-specific part quantified by (4.43). Remarkably, the spectral GA measure (4.44) is zero over all frequencies in the absence of internal dynamics in the target process, i.e. $a_{Y_2}(\bar{f}) = 0\ \forall \bar{f}$ if $a^{(arx)}_{Y_2Y_2,k} = 0\ \forall k$ (see (2.4)), and satisfies the spectral integration property, i.e.

$$A_{Y_2} = 2\int_0^{\frac{1}{2}} a_{Y_2}(\bar{f})\,\mathrm{d}\bar{f}, \tag{4.45}$$

which gives this measure a straightforward interpretation in the information-theoretic framework [2, 9]. In [41], the time (4.21) and spectral (4.44) measures of GA have been applied, together with measures of causality and isolation, to study the closed-loop beat-to-beat variability of cerebrovascular interactions (see Sect. 5.6). Simulation examples showing their behaviors in different settings are shown in Sect. 4.2.3.

4.2.3 Simulation Examples: Pairwise Measures of Causality, Isolation and Autonomy

In this section, we study the behavior of the measures of GC, GI and GA presented in Sect. 4.2 using simulated AR processes (see [41]). First, we simulate open-loop (Sect. 4.2.3.1) and closed-loop (Sect. 4.2.3.2) bivariate processes where the exact profiles of the spectral measures are computed (with sampling frequency $f_s = 1$ Hz) from the true values imposed for the AR parameters. Then, we consider a multivariate system where the dynamics of two interacting processes are perturbed by a third process which is not modelled in the calculation of GC, GI and GA (Sect. 4.2.3.3); in this case, estimations are performed from finite-length realizations of the three processes. Finally, in Sect. 4.2.3.4 we discuss the results of the simulations, using them to support the comparison and interpretation of the time domain and spectral measures of GC, GI and GA.

4.2.3.1 Open-Loop System

The first simulation reproduces a bivariate process where the driver Y_1 and the target Y_2 exhibit autonomous oscillations at different frequencies, and where a causal interaction from Y_1 to Y_2 is simulated:

$$\begin{aligned} Y_{1,n} &= a_{Y_1Y_1,1}Y_{1,n-1} + a_{Y_1Y_1,2}Y_{1,n-2} + U_{Y_1,n} \\ Y_{2,n} &= a_{Y_2Y_2,1}Y_{2,n-1} + a_{Y_2Y_2,2}Y_{2,n-2} - cY_{1,n-1} + U_{Y_2,n} \end{aligned} \tag{4.46}$$

where U_{Y_1} and U_{Y_2} are independent Gaussian white noises with zero mean and unit variance. The autonomous oscillations in the two processes are obtained placing a pair of complex-conjugate poles, with modulus ρ and phase $2\pi f$, in the complex plane representation of each process [8]. Here, we set $\rho_{Y_1} = 0.9$, $f_{Y_1} = 0.3$ Hz, and $\rho_{Y_2} = b \cdot 0.8$, $f_{Y_2} = 0.1$ Hz, so that the strength of the autonomous dynamics of Y_2 depends on the parameter b. Moreover, causal interactions are set from Y_1 to Y_2 at lag $k = 1$, with strength modulated by the parameter c.

We consider the two following settings: *(i)* progressive strengthening of the internal dynamics in the process Y_2 with stable causal interaction from Y_1 to Y_2, obtained varying b from 0 to 1 with fixed $c = 0.5$; *(ii)* progressive strengthening of the causal interaction from Y_1 to Y_2 with stable internal dynamics of Y_2, obtained varying c from 0 to 1 with fixed $b = 1$. The time domain values and spectral profiles of the measures of GC from driver to target ($F_{Y_1 \to Y_2}$, $f_{Y_1 \to Y_2}(\bar{f})$), GI of the target (F_{Y_2}, $f_{Y_2}(\bar{f})$) and GA of the target (A_{Y_2}, $a_{Y_2}(\bar{f})$, $\bar{a}_{Y_2}(\bar{f})$) resulting from the two simulations are reported in panels **(A)** and **(B)** of Fig. 4.4, respectively. The GA measure A_{Y_2} reflects exclusively the presence and strength of the autonomous dynamics in the target process Y_2, as it is null when $b = 0$ and rises proportionally to b in the first setting and is constant at varying the coupling from Y_1 to Y_2 in the second setting (panels *a* of Fig. 4.4A, B, circles). Analogously, the GC measure $F_{Y_1 \to Y_2}$ reflects exclusively the presence and strength of the causal coupling from Y_1 to Y_2, being constant in case of fixed coupling $c = 0.5$ (Fig. 4.4A, a, triangles) and increasing with c when b is kept constant (Fig. 4.4B, a, triangles). The GI measure F_{Y_2} is also affected only by the causal coupling (it is constant when b varies with $c = 0.5$, Fig. 4.4A, a, squares), and is complementary to the GC measure, as it exhibits an opposite trend (it decreases at increasing c, Fig. 4.4B, a, squares).

The spectral measures of GC, GI and GA localize within specific frequency bands, related to the oscillations of the two processes, the effects described in the time domain. Indeed, the GC and GI measures $f_{Y_1 \to Y_2}(\bar{f})$ and $f_{Y_2}(\bar{f})$ exhibit respectively a peak and a valley at the frequency of the oscillation of Y_1 that is transmitted to Y_2 (i.e., $f = 0.3$ Hz, Fig. 4.4A, b, c). The shape of the spectral profile is modulated in both functions by the coupling parameter c: when $c = 0$ the GC is null at all frequencies and the GI takes the highest values; when c rises towards 1 the GC shows a more and more prominent peak at 0.3 Hz while the GI flattens progressively. As regards the GA measure $a_{Y_2}(\bar{f})$, reported for the two simulations in panels *f*, we show its decomposition into a constant part equal to the time domain GA measure A_{Y_2}

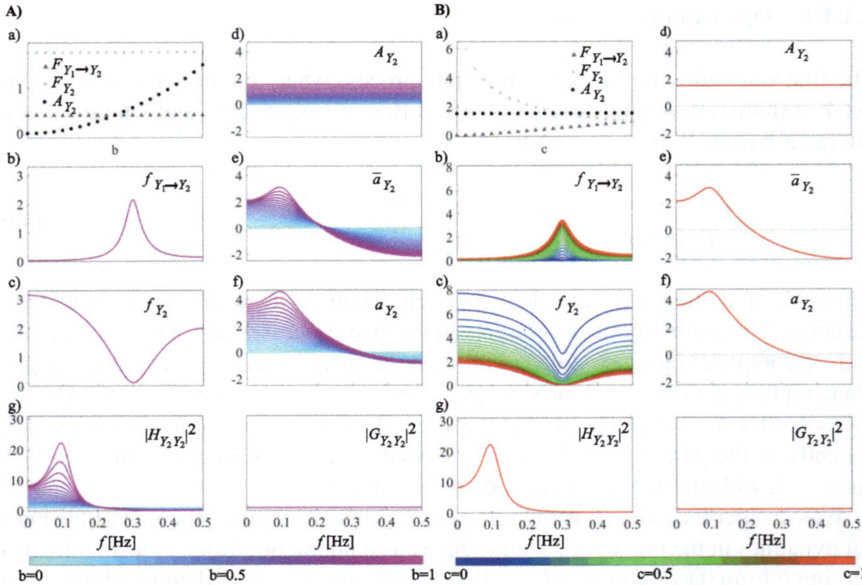

Fig. 4.4 Open-loop system: dependence of the measures of Granger Causality, Isolation and Autonomy on the strength of **A** the internal dynamics in the target process, modulated by the parameter b, and **B** the causal interaction from driver to target, modulated by the parameter c. Plots depict: **a** the time domain GC, GI, and GA; the spectral profiles of the **b** GC, **c** GI and **d–f** GA; **g** the spectral profiles of the TFs of the full and restricted models. The figure is adapted from [41]

(panels d) and a variable part $\bar{a}_{Y_2}(\bar{f})$ whose frequency average is zero (panels e). The spectral GA shows its highest values at the frequency of the autonomous oscillations imposed in the target process (i.e., $f = 0.1$ Hz); when the parameter determining the strength of this oscillation increases from $b = 0$ to $b = 1$, the spectral GA measures varies from a flat null profile up to a shape with a well-defined peak (Fig. 4.4A, f). Panels g of Fig. 4.4A, B, report the spectral profiles of the transfer functions of the full (2.4) and restricted (2.9) models, $|H_{Y_2Y_2}(\bar{f})|^2$ and $|G_{Y_2Y_2}(\bar{f})|^2$ respectively, obtained varying the parameters b and c. These profiles illustrate how the predictable autonomous dynamics of the target process, in this simulation located at 0.1 Hz, are captured in the full model by the transfer function $H_{Y_2Y_2}(\bar{f})$, but not in the reduced model by $G_{Y_2Y_2}(\bar{f})$, which indeed is flat. This corroborates the choice of the ratio between $|H_{Y_2Y_2}(\bar{f})|^2$ and $|G_{Y_2Y_2}(\bar{f})|^2$ as a meaningful index $\bar{a}_{Y_2}(\bar{f})$ displaying a peak at the frequency of the target autonomous dynamics (panels e of Fig. 4.4A, B).

4.2.3.2 Closed-Loop System

The second simulation reproduces a bivariate AR process where the driver Y_1 and the target Y_2 exhibit autonomous oscillations at different frequencies and interact in a closed loop. This type of pairwise interaction involving bidirectional causal influences from one process to the other is commonly found in physiological contexts; for instance, heart rate and arterial pressure generally interact in a closed-loop manner where the arterial pressure counteract short-term modifications of heart rate through the baroreflex (feedback mechanism), while the heart is responsible of blood pressure changes due to feedforward mechanisms such as Windkessel or Frank-Starling effects [46–48].

In our simulation, the process is defined as:

$$\begin{aligned} Y_{1,n} &= a_{Y_1Y_1,1}Y_{1,n-1} + a_{Y_1Y_1,2}Y_{1,n-2} - dY_{2,n-1} + U_{Y_1,n} \\ Y_{2,n} &= a_{Y_2Y_2,1}Y_{2,n-1} + a_{Y_2Y_2,2}Y_{2,n-2} - cY_{1,n-1} + U_{Y_2,n} \end{aligned} \tag{4.47}$$

where U_{Y_1} and U_{Y_2} are independent Gaussian white noises with zero mean and unit variance. The parameters $a_{\cdot,\cdot}$ were set as in the first simulation (Sect. 4.2.3.1) to obtain autonomous oscillations at 0.3 Hz for Y_1 and at 0.1 Hz for Y_2. In addition to the causal interaction from Y_1 to Y_2 modulated by c, a causal interaction is set from Y_2 to Y_1 at lag $k = 1$, with strength modulated by the parameter d.

We consider the three following settings: *(i)* progressive strengthening of the internal dynamics in the process Y_2 with stable causal interactions, obtained varying b from 0 to 1 with fixed $c = 0.5$ and $d = 0.2$; *(ii)* progressive strengthening of the causal interaction from Y_1 to Y_2 with stable internal dynamics of Y_2 and causal interaction from Y_2 to Y_1, obtained varying c from 0 to 1 with fixed $b = 1$ and $d = 0.2$; *(iii)* progressive strengthening of the causal interaction from Y_2 to Y_1 with stable internal dynamics of Y_2 and causal interaction from Y_1 to Y_2, obtained varying d from 0 to 1 with fixed $b = 1$ and $c = 0.5$. Figure 4.5 reports the time domain values and the frequency-domain profiles of the GA (A_{Y_2}, $a_{Y_2}(\bar{f})$) obtained in the three settings. For the sake of brevity, the trends of the GC and GI measures are not reported because they are identical to those of the first simulation (Fig. 4.4) despite the addition of the causal interaction from Y_2 to Y_1.

Figure 4.5a shows that the time domain GA increases with the strength of the internal dynamics modulated by b (circles, light blue to magenta), and remains constant increasing the strength of the interaction $Y_1 \rightarrow Y_2$ modulated by c (circles, blue to red). On the other hand, stronger interactions $Y_2 \rightarrow Y_1$ obtained increasing d determine a slight decrease of the time domain GA A_{Y_2} (circles, yellow to blue). The spectral expansion of the GA measure allows to identify the frequency bands where the internal dynamics are localized. Indeed, Fig. 4.5b–d reveal that the spectral profile of the GA measure exhibits a peak at the frequency of the autonomous oscillation imposed in the target process (i.e., 0.1 Hz). This peak is clearly modulated in amplitude by the strength of the internal dynamics of Y_2 (parameter b, Fig. 4.5b), while it changes only slightly at varying the causal interactions between Y_1 and Y_2 (parameters c and d), showing small amplitude and frequency modulations (Fig. 4.5c, d). Furthermore, the imposition of a feedback effect from Y_2 to Y_1 determines a modification of the spectral profile of $a_{Y_2}(\bar{f})$, with the emergence of a second peak around the frequency of the autonomous oscillations of Y_1 (~0.3 Hz) and of a a reverse

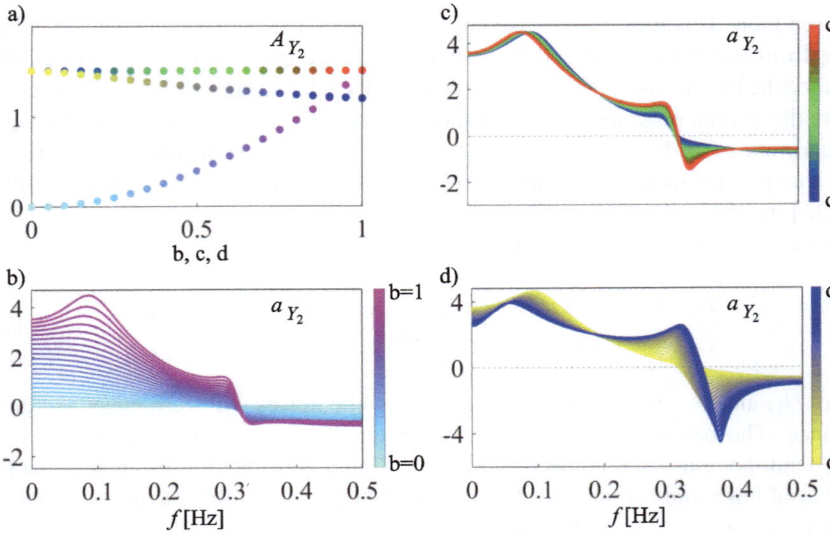

Fig. 4.5 Closed-loop system: dependence of the measure of Granger Autonomy on the parameters b, c and d, modulating respectively the strength of the internal dynamics in the target process, the causal coupling from Y_1 to Y_2 and the causal coupling from Y_2 to Y_1. Plots depict **a** the time domain GA at varying b (light blue to magenta), c (blue to red) or d (yellow to blue) in the range [0, 1], and the spectral profiles of the GA obtained varying **b** b, **c** c, and **d** d. The figure is adapted from [41]

peak at higher frequencies (Fig. 4.5b–d). The spectral integration property allows to ascribe the decrease of the time domain GA for high values of d to this behavior, as the negative peak prevails over the positive one at high frequencies, while the low-frequency peak shows preserved or slightly larger amplitude. Remarkably, we hypothesize that this effect may predominate when the frequencies at which target and driver oscillate are different; indeed, as shown in the application to cerebrovascular variables in Sect. 5.6 (see Fig. 5.8), the reverse peak is not so evident if the two processes exhibit a similar spectral behavior (panels b, d).

4.2.3.3 System with Unobserved Confounders

In this section, we study the behavior of the measures of GC, GI and GA, computed as described in Sect. 4.2 for two processes Y_1 and Y_2, when their dynamics are affected by an *unobserved* process Y_3. To do this, we simulate the three-variate process defined as

$$\begin{aligned} Y_{1,n} &= a_{Y_1Y_1,1}Y_{1,n-1} + a_{Y_1Y_1,2}Y_{1,n-2} + U_{Y_1,n} \\ Y_{2,n} &= a_{Y_2Y_2,1}Y_{2,n-1} + a_{Y_2Y_2,2}Y_{2,n-2} - 0.8Y_{1,n-1} - aY_{3,n-1} + U_{Y_2,n} \\ Y_{3,n} &= a_{Y_3Y_3,1}Y_{3,n-1} + a_{Y_3Y_3,2}Y_{3,n-2} + U_{Y_3,n}, \end{aligned} \tag{4.48}$$

where U_{Y_1}, U_{Y_2} and U_{Y_3} are independent Gaussian white noises with zero mean and unit variance. The coefficients $a_{\cdot,\cdot}$ are set to obtain autonomous oscillations in the processes depending on the modulus ρ and phase $2\pi f$ of three pairs of complex-conjugate poles. Here, we set $\rho_{Y_1} = 0.9$, $f_{Y_1} = 0.3$ Hz, $\rho_{Y_2} = b \cdot 0.8$, $f_{Y_2} = 0.1$ Hz, and $\rho_{Y_3} = 0.8$, $f_{Y_3} = 0.2$ Hz; the strength of the autonomous dynamics of Y_2 depends on the parameter b. Moreover, causal interactions are set at lag 1 both from Y_1 to Y_2, with fixed strength 0.8, and from Y_3 to Y_2, with strength weighed by the parameter a.

The analysis is performed on realizations of the three processes generated by feeding (4.48) with white noise observations, and then computing the spectral GC ($f_{Y_1 \to Y_2}(\bar{f})$), GI ($f_{Y_2}(\bar{f})$) and GA ($a_{Y_2}(\bar{f})$) measures on the time series relevant to the processes Y_1 and Y_2. We consider three parameter settings: (*I*) $a = 0, b = 0$, to simulate the absence of autonomous dynamics in the target process Y_2 and of effects from the unobserved process Y_3; (*II*) $a = 0, b = 0.8$, to simulate the presence of autonomous dynamics in Y_2 without effects from Y_3; (*III*) $a = 0.8, b = 0$, to simulate the effect of the unobserved process Y_3 on the target Y_2 in the absence of autonomous dynamics of Y_2. For each setting, 100 realizations of (4.48) were generated, each of length 500 points, and the spectral measures of GC, GI and GA were estimated after identifying the bivariate AR model fitting the time series of Y_1 and Y_2; the model order was set using the AIC [24].

The results of the analysis are reported in Fig. 4.6, showing the average spectral profiles of the GC, GI and GA measures in the three simulation conditions. The profiles of GC and GI are very similar in the three cases, revealing a clear peak of the GC, and a corresponding minimum of the GI, at the frequency of the causal interaction imposed from Y_1 to Y_2 ($f_{Y_1} = 0.3$ Hz, Fig. 4.6a, b). This documents that the presence of the unobserved confounder Y_3 acting only on the analyzed target Y_2 does not alter significantly the causal interactions from Y_1 to Y_2. On the other hand, the profiles of GA are substantially different in the three cases: Fig. 4.6c shows that the GA stays uniformly at the zero level when both autonomous target dynamics and confounding effects are absent (continuous line), peaks at ~0.1 Hz when only the autonomous dynamics are present (dashed line), and peaks at ~0.2 Hz when only the confounding effects are present (dotted line). This demonstrates that the proposed spectral measure of GA is able to capture not only the autonomous dynamics of Y_2, but also the regular dynamics simulated in Y_3 and transmitted to Y_2 via the parameter a. This may have implications in practical applications, e.g., when modifications of GA may reflect the altered strength of exogenous effects, i.e., effects acting on the target independently of the driver. For instance, in the case of the cerebrovascular closed-loop interactions between mean arterial pressure and cerebral blood flow (see Sect. 5.6), such effects might include modifications of carbon dioxide pressure, p_{CO_2}, due to internal or external stressors, which may have an impact on arteriolar vessel caliber and thus on blood flow velocity [49]. This impact, which is not observed nor quantified in the bivariate model, might alter the autonomous dynamics of the target and thus enter the computation of GA.

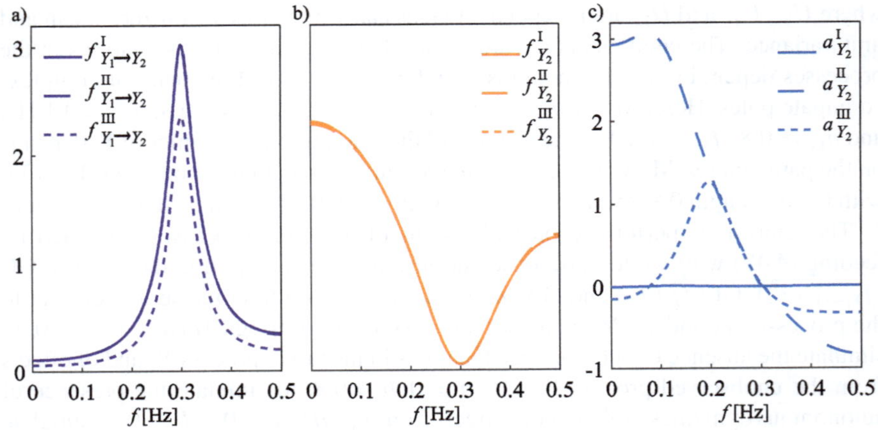

Fig. 4.6 Analysis of Granger causality, isolation and autonomy for the simulated system with unidirectional interactions and confounding effects. Plots depict the average spectral profiles of **a** GC, **b** GI and **c** GA computed in the absence of autonomous dynamics of Y_2 and confounding effects from Y_3 to Y_2 ($a = 0$, $b = 0$, continuous lines), in the presence of autonomous dynamics only ($a = 0$, $b = 0.8$, dashed lines), and in the presence of confounding effects only ($a = 0.8$, $b = 0$, dotted lines). The figure is adapted from [41]

4.2.3.4 Interpretation and Comparison of Granger Causality, Isolation and Autonomy

The reported simulations depict the theoretical properties of the measures of GC, GI and GA developed in this work. We have shown that, in a bivariate process $\{Y_1, Y_2\}$, the GC and GA measures capture selectively the causal interaction from Y_1 to Y_2 and the autonomous dynamics of Y_2, respectively, either globally or at specific frequencies when measured in the time or frequency domains. The GI measure behaves in a complementary way to the GC, decreasing with the strength of the causal interactions and thus reflecting the degree of isolation of Y_2. Importantly, the time- and frequency-domain formulations of GC, GI and GA are strictly connected by the spectral integration property, and the spectral representation allows to identify the oscillations for which causal, non-causal and autonomous effects take place. This property can be useful to detect variations in the strength of effects which are confined within specific frequency bands and can be missed if investigated in the time domain only. However, while these interpretations emerge strikingly in a bivariate system with unidirectional coupling, they can be challenged when more complex dynamics arise in the presence of closed-loop or multivariate interactions. For instance, in the second simulation (Sect. 4.2.3.2) we showed that the GA measure is influenced by the imposition of a feedback effect from the target to the driver; a similar behavior was documented in a previous work by the conditional self entropy measure [8], which is formally equivalent to the time domain GA (4.21). Herein, this dependence is localized in frequency via the proposed spectral GA measure, which exhibits an irregular profile with the appearance of a positive peak and a reverse one around the frequency of the autonomous oscillation in the driver system (see, e.g., Fig. 4.5b–d). Furthermore, the third simulation (Sect. 4.2.3.3) showed that, when an unobserved process has effects on the target, these effects may alter the spectral profile of the GA measure in a way similar to that of autonomous dynamics (see Fig. 4.6c). This may have implications in practical applications, e.g., when multiple physiological systems interact but only two of them are monitored and observed in a bivariate analysis.

Further insights on the concepts of *autonomy* and *isolation* are provided in the following, where their meanings are discussed in terms of target predictable dynamics and non-causal spectral power and the relevant measures of GA and GI are compared in an additional simulation example. Specifically, we consider the bivariate process (4.46), where the observed driver Y_1 and target Y_2 are simulated as AR processes with autonomous oscillations as specified in Sect. 4.2.3.1. A causal interaction is set from Y_1 to Y_2 at lag $k = 1$, with strength modulated by the parameter c. We consider four parameter configurations, in which the driver dynamics are fixed (AR process with an autonomous oscillation at 0.3 Hz) and the target process Y_2 is simulated as: *(i)* an isolated Gaussian white noise with zero mean and unit variance ($b = 0, c = 0$); *(ii)* an isolated AR process with an autonomous oscillation at $f_{Y_2} = 0.1$ Hz ($b = 1, c = 0$); *(iii)* an AR process with no autonomous oscillations but influenced by Y_2 at lag $k = 1$ ($b = 0, c = 1$); *(iv)* an AR process with an autonomous oscillation at $f_{Y_2} = 0.1$ Hz and causally driven by Y_1 at lag $k = 1$ ($b = 1, c = 1$). For each setting, a pair of time series Y_1, Y_2 of length 500 is generated from the AR model parameters feeding the model with white noise realizations. Then, model identification is performed for the bivariate process $\{Y_1, Y_2\}$, setting the model order at the value $p = 2$. The spectral profiles of the measures of GC, GI and GA resulting from the generated time series are reported in Fig. 4.7.

Results confirm that GI and GA measures reflect different behaviors of the target system. Figure 4.7a reports the results relevant to the setting *(i)*; the absence of internal dynamics for Y_2 (i.e., AR model coefficients relating the history of the target to its present state) is reflected by the flat spectrum of $a_{Y_2}(\bar{f})$, while the spectral profile of the GI, together with the GC, ensures the *isolation* of the target with respect to the driver. Indeed, while the GC is uniformly zero at all frequencies, thus reflecting the absence of causal interactions from Y_1 to Y_2, the spectral profile of the GI is characterized by a reverse peak at the frequency of the autonomous oscillation of the driver. These findings hold again in case of an AR target process Y_2 whose dynamics are only self-determined (Fig. 4.7b, setting *(ii)*). On the other hand, in case of a well-established causal interaction from Y_1 to Y_2 (Fig. 4.7c, setting *(iii)*, and Fig. 4.7d, setting *(iv)*), the GC shows a clear peak at the frequency of the autonomous oscillation of the driver, and the GI still owns a reverse peak at the same frequency. This finding is quite interesting, since it emerges that the profile of the GI may be related to the time- and frequency-domain behavior of the driver, which indeed remains unaltered in all the settings (AR process with an autonomous oscillation at 0.3 Hz). In another simulation, which has not been shown here for brevity, we found that the spectral profile of the GI is flat and uniform at all frequencies when

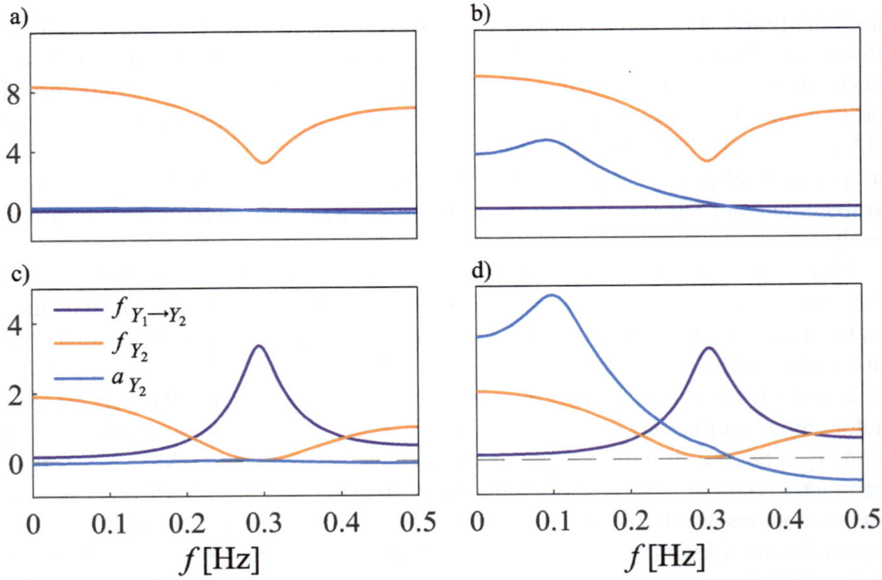

Fig. 4.7 Spectral behavior of GC, GI and GA in different simulation settings: **a** setting *(i), isolated process without autonomous dynamics*; **b** setting *(ii), isolated process with autonomous dynamics*; **c** setting *(iii), non-isolated process without autonomous dynamics*; **d** setting *(iv), non-isolated process with autonomous dynamics*. The figure is adapted from [41]

Y_1 is an isolated Gaussian noise. This suggests that the reverse peak of the GI, when present, may give indications on the frequency-specific location of a possible driving effect reducing the degree of isolation of the target process. These results seem to confirm that the measure of GI has a dual role. First, it quantifies that part of the target spectrum which cannot be explained by the driver, i.e., how much the target dynamics are *isolated* with respect to the driver process. Moreover, it gives also an information on the frequency-specific location of the *isolation* phenomenon, thus allowing to identify the spectral bands where this is less/more accentuated. On the other hand, the spectral profile of the GA is in line with what we expect to find, i.e. a positive peak in presence of an autonomous oscillation of the target process whereby it is located (Fig. 4.7b, d), or conversely a flat spectrum for $a_{Y_2}(\bar{f})$ whether the target has no internal dynamics.

In conclusion, the differences between the two measures come out clearly. The two concepts of *isolation* and *internal dynamics* reflect different mechanisms occurring in the target process, as one does not rule out the other. They can coexist in the same process (Fig. 4.7d), exist independently of each other (Fig. 4.7b, c), or not exist at all (Fig. 4.7a). Still, it is fundamental to investigate the pairwise interactions between the target and the driver through the three discussed measures, in order to have a complete overview of how two processes interact and/or oscillate independently of each other.

As regards the relation between the concepts of *isolation* and *causality*, quantified respectively by the GI and GC measures, we evidence that they are clearly complementary, as an increase in the causal part of the spectrum implies a decrease of the isolated part and vice versa (see (4.29), (4.30) and (4.35)). However, differently from the corresponding non-logarithmic DC measures (4.27) and (4.28) which sum to 1 at each frequency, the relation between the GC and GI measures is not trivial. In fact, the logarithmic transformation, which provides information-theoretic meanings to GC and GI, makes their sum to vary across frequencies, and this aspect may differentiate their behavior in practical computations; we show an example in Sect. 5.6 regarding cerebrovascular interactions in healthy individuals and subjects prone to develop postural-related syncope.

4.3 Beyond Pairwise Interactions: High-Order Interactions of Multiple Network Units

While the standard network description of complex systems is based on quantifying the link between pairs of system units, high-order interactions involving three or more units often play a major role in governing the collective network behavior. Indeed, there is mounting evidence that such measures cannot fully capture the interplay among the multiple units of a complex system [50], since they very often exhibit collective behaviors which are integrated at different hierarchical levels, thus displaying interactions that involve more than two network nodes. These high-order interactions occur for instance when brain dynamics require the joint examination of multiple units to be predicted accurately [51], or when cardiovascular interactions are influenced by the effects of the respiratory activity [52]. In this section, we will describe various information-theoretic metrics proposed to assess the emergence of HOI patterns at increasing orders, i.e., interactions involving the role of all network nodes at different levels of resolution, in both time (Sects. 4.3.1.1 and 4.3.1.2) and frequency domains (Sect. 4.3.2.1).

4.3.1 *Time Domain*

In this section, we will characterize the time domain behavior of the network $\mathcal{X}$ composed of M nodes, whose activity is described by the vector processes $\mathbf{X}_1, \ldots, \mathbf{X}_M$, each comprising M_i scalar processes, $i = 1, \ldots, M$. To this end, we refer the reader to the mathematical definitions provided in Sect. 2.3 related to the time domain representation of VAR linear models. When more than two units are considered, suitable approaches are needed to examine single-node and pairwise activity within a complex network of multiple nodes, as well as HOIs for multiplets of order $N \leq M$ at different levels of resolution, i.e., characterizing the role of the single node within the selected subset (*node-specific analysis*), the nature of the link between two nodes with respect to the remaining nodes (*link-specific analysis*), as well as the overall activity of groups of nodes or the whole network (*network-specific analysis*). To this end, information-theoretic measures of dynamic HOIs which extend to random processes the concepts of first-order OI gradient, OI, PID and local OI defined for random variables in Sect. 3.2 are required to provide a more comprehensive and detailed description of the analysed dynamic network system. In this section, we will describe these measures in the context of linear Gaussian processes. For further details, we refer the reader to [3, 43, 53–56].

4.3.1.1 Network-Specific and Node-Specific Analysis of HOIs

O-Information Rate. The MIR (4.11) is a dynamic measure of pairwise interdependence between two random processes, and can be used as a building block for the assessment of HOIs generalizing to multiple random processes the OI measure appearing in (3.12) for multiple random variables.

To this end, let us consider the Q scalar processes $\{Y_1, \dots, Y_Q\}$ grouped in the M blocks $\{\mathbf{X}_1, \dots, \mathbf{X}_M\}$, each of dimension $M_i, i = 1, \dots, M$. The dynamic interaction of order three among the processes $\mathbf{X}_i$, $\mathbf{X}_j$ and $\mathbf{X}_k$, $i, j, k \in \{1, \dots, M\}, i \neq j \neq k$, can be quantified by the *interaction information rate* (IIR) [3, 57] by using MIR terms in a formulation similar to (3.8):

$$I_{\mathbf{X}_i;\mathbf{X}_j;\mathbf{X}_k} = I_{\mathbf{X}_i;\mathbf{X}_j} + I_{\mathbf{X}_i;\mathbf{X}_k} - I_{\mathbf{X}_i;\mathbf{X}_j,\mathbf{X}_k}. \tag{4.49}$$

Moreover, exploiting the same expression valid for the OI, based on the two concepts of total correlations and dual total correlation (3.11), where the ER of random processes is used in place of the entropy of random variables, it is possible to define a so-called *O-information rate* (OIR). Specifically, the OIR of a group of random processes $\mathbf{X}^N = \{\mathbf{X}_1, \dots, \mathbf{X}_N\}$, $3 \leq N \leq M$, is computed as

$$\Omega_{\mathbf{X}^N} = (N-2)H_{\mathbf{X}^N} + \sum_{i=1}^{N} \Big[H_{\mathbf{X}_i} - H_{\mathbf{X}^N_{-i}} \Big], \tag{4.50}$$

where $\mathbf{X}^N_{-i}$ denotes the set of all the processes in $\mathbf{X}^N$ but $\mathbf{X}_i$. The OIR quantifies collective interactions among all the analyzed processes, and can be also computed in an iterative way from the OIR of a subset including $N-1$ processes, e.g., $\mathbf{X}^N_{-i}$, summing a *gradient* (referred to as Δ) which quantifies the informational increment obtained when the process $\mathbf{X}_i$ is appended to $\mathbf{X}^N_{-i}$ [3, 43]:

$$\Omega_{\mathbf{X}^N} = \Omega_{\mathbf{X}^N_{-i}} + \Delta_{\mathbf{X}_i;\mathbf{X}^N_{-i}}, \tag{4.51}$$

$$\Delta_{\mathbf{X}_i;\mathbf{X}^N_{-i}} = \sum_{\substack{i=1 \\ i\neq j}}^{N} I_{\mathbf{X}_i;\mathbf{X}^N_{-ij}} + (2-N) I_{\mathbf{X}_i;\mathbf{X}^N_{-i}}, \tag{4.52}$$

with $\mathbf{X}^N_{-ij} = \mathbf{X}^N \backslash \{\mathbf{X}_i, \mathbf{X}_j\}$. According to this definition, the OIR is zero for any two processes ($\Omega_{\mathbf{X}_i,\mathbf{X}_j} = 0$), and is equal to the gradient for three processes, i.e. $\Omega_{\mathbf{X}_i,\mathbf{X}_j,\mathbf{X}_k} = \Delta_{\mathbf{X}_i;\mathbf{X}_j,\mathbf{X}_k}$, which in turns corresponds to the IIR (4.49) with $N = 3$. Both the IIR of three processes and the OIR computed for $N > 3$ processes are symmetric, i.e. do not change if the order of the processes is swapped in the computation. Importantly, these measures, as well as the gradients, can be either positive or negative, with the sign reflecting the redundant or synergistic nature of the interactions in groups of random processes; specifically, positive values of $\Omega_{\mathbf{X}^N}$ or $\Delta_{\mathbf{X}_i;\mathbf{X}^N_{-i}}$ denote redundancy, while negative values denote synergy. While the OIR $\Omega_{\mathbf{X}^N}$ can be considered as an information-theoretic measure of the overall dynamic activity of the observed network of N processes, the OIR gradient $\Delta_{\mathbf{X}_i;\mathbf{X}^N_{-i}}$ characterizes the redundant/synergistic role of the single node $\mathbf{X}_i$ when it is appended to the subset $\mathbf{X}^N_{-i}$.

Remarkably, combinatorial explosion in the context of HOIs in physiological and neural networks, which refers to the rapid increase in the number of possible node combinations as the size of the network grows, is an open challenge in the field of Network Science. For instance, in a network with $M = 10$ nodes, the activity of the single entities is described by means of $M = 10$ values of ER, and $\frac{M(M-1)}{2} = 45$ values of MIR. However, when we start looking at interactions beyond just pairs of nodes (i.e., HOIs), the complexity skyrockets. In general, for the M-node network, the number of HOIs assessed via the OIR measure scales factorially according to the binomial coefficient $C_{M,N} = \binom{M}{N} = \frac{M!}{N!(M-N)!}$, where N is the number of nodes taken into account ($3 \leq N \leq M$). Hence, in the case of triplets ($N = 3$ nodes), the number of HOIs amounts to 120 different possible combinations among the 10 nodes. This combinatorial increase poses computational challenges and often necessitates efficient algorithms or resourceful approximations to handle the explosive growth in complexity. Moreover, issues arise also regarding the interpretation of the vast number of HOIs in multiplets of different order: understanding large networks is not just about scaling up, but also about dealing with the intricate web of interactions that emerge.

Simulation examples of the proposed measures of network- and node-specific analysis of complex networks can be found in Sects. 4.3.3.1 and 4.3.3.2. Applications to cardiovascular, cardiorespiratory and cerebrovascular interactions, as well as to brain networks, have been discussed in [3, 43, 45, 58, 59], and herein presented in Sects. 5.7, 5.9, and6.2.4–6.2.5.

Causal decomposition of the O-Information rate. The OIR increment (4.52) can be decomposed into causal and instantaneous contributions arising from the MIR decomposition (4.12b), as explained in [43]. To this aim, we note that $\Delta_{\mathbf{X}_i;\mathbf{X}^N_{-i}}$ is obtained inserting N different MIR values in (4.52), i.e. the MIRs between the processes $\mathbf{Z}_1 = \mathbf{X}_i$ and $\mathbf{Z}_2 = \mathbf{X}^N_{-i}$ where $\mathbf{Z}_1$ is fixed and $\mathbf{Z}_2$ varies with $i = 1, \ldots, N$, excluding $\mathbf{X}_i$. Then, using $\mathbf{Z}_1$ and $\mathbf{Z}_2$ in the MIR expansion [2]

$$I_{\mathbf{Z}_1;\mathbf{Z}_2} = T_{\mathbf{Z}_1 \to \mathbf{Z}_2} + T_{\mathbf{Z}_2 \to \mathbf{Z}_1} + I_{\mathbf{Z}_1 \cdot \mathbf{Z}_2}, \tag{4.53}$$

and substituting into (4.52) allows to decompose the OIR gradient as

$$\Delta_{\mathbf{X}_i;\mathbf{X}^N_{-i}} = \Delta_{\mathbf{X}_i \to \mathbf{X}^N_{-i}} + \Delta_{\mathbf{X}^N_{-i} \to \mathbf{X}_i} + \Delta_{\mathbf{X}_i \cdot \mathbf{X}^N_{-i}}, \tag{4.54}$$

where the three terms

$$\begin{aligned}
\Delta_{\mathbf{X}_i \to \mathbf{X}^N_{-i}} &= (2-N) T_{\mathbf{X}_i \to \mathbf{X}^N_{-i}} + \sum_{i=1}^{N-1} T_{\mathbf{X}_i \to \mathbf{X}^N_{-i}} \\
\Delta_{\mathbf{X}^N_{-i} \to \mathbf{X}_i} &= (2-N) T_{\mathbf{X}^N_{-i} \to \mathbf{X}_i} + \sum_{i=1}^{N-1} T_{\mathbf{X}^N_{-i} \to \mathbf{X}_i} \\
\Delta_{\mathbf{X}_i \cdot \mathbf{X}^N_{-i}} &= (2-N) I_{\mathbf{X}_i \cdot \mathbf{X}^N_{-i}} + \sum_{i=1}^{N-1} I_{\mathbf{X}_i \cdot \mathbf{X}^N_{-i}}
\end{aligned} \tag{4.55}$$

quantify the informational character of the directional information transfer from $\mathbf{X}_i$ to $\mathbf{X}^N_{-i}$, of the directional information transfer from $\mathbf{X}^N_{-i}$ to $\mathbf{X}_i$, and of the instantaneous information shared between $\mathbf{X}^N_{-i}$ and $\mathbf{X}_i$, respectively; the informational character of each term is redundant when the term is positive, and synergistic when the term is negative.

Simulation examples of the proposed measures of network- and node-specific analysis of complex networks can be found in Sect. 4.3.3.3.

Linear parametric formulation. The calculation of the high-order information-theoretic measures defined above requires an approach to compute the MIR between vector random variables. While the VAR model (2.6) provides a global representation of the overall multivariate process, to describe the linear interactions relevant to the subset of processes $\mathbf{Z} = \{\mathbf{Z}_1, \mathbf{Z}_2\} = \{\mathbf{X}_i, \mathbf{X}^N_{-i}\}$, for which the MIR terms in (4.52) and the MIR decomposition (4.54) are sought, a restricted VAR model involving only those processes should be defined as in (2.7). The restricted model parameters can be derived from the parameters of the full model (2.6) through a procedure that exploits SS models (Sect. 2.3.5.1). The computation of the OIR, the OIR gradient and its decomposition terms goes through the representation of the SS model in the frequency domain (see Sect. 2.4.1) and will be shown in Sect. 4.3.2. The procedure has been thoroughly described in [43].

Partial Information Rate Decomposition. Besides the novel concept of the OIR and its decomposition into causal and non-causal terms, which however do not put in evidence multiplets of processes which are both redundant and synergistic with equal strength, different approaches to the study of the mutual influence among many signals in dynamic networks have been proposed throughout the past decades. A common technique is the conditional/partial transfer entropy [40, 51, 60], which measures the influence of one time series on another time series in the presence of a third, a condition translated by Geweke into the frequency domain [61]. However, although the method is simple and intuitive, it has been demonstrated that fully conditioned causality analysis may encounter challenges in the presence of redundancy in time series data, failing to reveal redundant effects [62]. Approaches which separately evaluate redundant and synergistic patterns of dynamic interactions are then needed to deal with these issues.

Generally, the PID framework (Sect. 3.2) has been largely applied to random processes with temporal statistical structure. In fact, although specifically defined for random variables, PID is often needed in practice to analyze multivariate time series whose most proper statistical representation is the vector random process. Then, since both the target random process and the set of source processes constitute collections of random variables, utilization of the PID in such dynamic case is not straightforward as it implies an arbitrary selection of the variables to be extracted from each process. A common approach is to apply PID to the variables sampling the processes at the same time, performing a so-called *static PID*, under the implicit assumption that the processes are stationary and memoryless (i.e., composed by i.i.d. variables). However, the i.i.d. assumption is typically not tested in practice and is often violated in applications of information decomposition where the analyzed data exhibit nontrivial temporal correlations [63–65]. Alternatively, the PID has been applied to random processes by selecting the variables to use in order to decompose the joint information transferred from all sources to the target [66–68], yielding the PID of the popular TE measure of information transfer [32]. Nevertheless, although considering the temporal statistical structure of the multivariate process, the PID of the TE cannot account for instantaneous interactions among the processes, nor for causal interactions occurring in the causal direction from the target to the sources. Hence, the current applications of PID to random processes provide only a partial, often misleading view of the interactions among the units of dynamic network systems.

To deal with these limitations, we present the novel framework described in [55, 56] for the decomposition of the information shared dynamically between the target and the source units composing the analyzed network of random processes. The idea is to shift the paradigm of PID from the use of random variables to the use of random processes as building blocks of information decomposition: leveraging the use of information rate quantities in place of standard information quantities, we replace the MI between random variables with the MIR between random processes, and use the same lattice backbone of PID to implement the so-called *partial information rate decomposition* (PIRD). PIRD dissects the information shared per unit of time between the designed target random process and the set of source processes, quantified by the MIR, into PI rate atoms adopting a full PID perspective, or into unique, redundant and synergistic information rates adopting a coarse-graining perspective. The PIRD is solved introducing the new information-theoretic measure of *redundancy rate* which generalizes the MIR over the lattice. Among several possible definitions, the redundancy rate is formulated following a pointwise approach implemented in the frequency domain, i.e., defining a so-called spectral redundancy rate which quantifies the concept of redundancy among iso-frequency oscillatory components of the analyzed processes. This allows to achieve a non-negative decomposition of the spectral MIR between target and source processes at each specific frequency. Moreover, in the case of linear Gaussian processes [2, 9], the use of the MMI principle [69] applied to the MIR between the target and each source process leads us to retrieve a non-negative time-domain PIRD through integration of the spectral PI rate atoms over the whole frequency spectrum. While in this section we discuss the time domain formulation of the PIRD framework, the spectral PIRD and its practical implementation for linear Gaussian processes will be presented in Sect. 4.3.2.1.

Let us now consider a generic dynamic network system composed of $M + 1$ nodes, $\mathcal{Z} = \{\mathcal{Y}, \mathcal{X}_1, \ldots, \mathcal{X}_M\}$, where the activity at each node is described in terms of the vector random process $\mathbf{Z} = \{Y, X_1, \ldots, X_M\}$, with Y assumed as *target* and $\mathbf{X} = \{X_1, \ldots, X_M\}$ as the vector of *sources*. Since each random process is a collection of random variables, the application of the PID in the dynamic case implies an arbitrary selection of the variables to be extracted from each process. The most intuitive choice is to apply the PID to the variables sampling the processes at the same nth time point, setting $T = Y_n$ and $\mathbf{S} = \mathbf{X}_n$ as target and source variables and thus decomposing the static MI $I(Y_n; \mathbf{X}_n)$ through a redundancy function based on zero-lag MI terms. However, the zero-lag PID of $I(Y_n; \mathbf{X}_n)$ presupposes to work with memoryless processes, a condition that is typically not satisfied in practice as the presence of a temporal statistical structure is inherently expected in time series data. To overcome this limitation, we propose a framework for the decomposition of the information shared dynamically between the target and the source processes which makes use of the concepts of ER (4.1) and MIR (4.11). The latter can be expressed in terms of entropy rates as in (4.12a), evidencing the analogy between the concepts of entropy and MI for random variables and the concepts of ER and MIR for random processes. Hence, we use the MIR as a building block for assessing and decomposing the dynamic information shared between the scalar target process Y and the vector source process $\mathbf{X} = \{X_1, \ldots, X_M\}$ of the analyzed network system. Specifically, we formalize the PIRD which makes use of the same lattice structure of the PID [70] to expand the MIR between target and sources as:

$$I_{Y;\mathbf{X}} = \sum_{\alpha \in \mathcal{A}} I^{\delta}_{Y;\mathbf{X}_\alpha}, \tag{4.56}$$

where $I^{\delta}_{\cdot;\cdot}$ is a PI rate function defined for each atom $\alpha = \{\alpha_1, \ldots, \alpha_J\}$ of the lattice, and $\mathbf{X}_\alpha$ denotes the αth set of subsets of source processes, with $\mathbf{X}_{\alpha_j} \subseteq X$. As happens for the PID, to solve the PIRD it is necessary to define a so-called *redundancy rate* function, here denoted as $I^{\cap}_{\cdot;\cdot}$, which generalizes the MIR over the lattice and replaces the concept of redundancy function generalizing the MI. The redundancy rate of the

αth atom is obtained summing the PI rate of the same atom to the PI rates of the atoms positioned downwards in the lattice:

$$I^{\cap}_{Y;\mathbf{X}_\alpha} = \sum_{\beta \preceq \alpha} I^{\delta}_{Y;\mathbf{X}_\beta}; \tag{4.57}$$

then, once the redundancy rate is known, the information rate associated to all atoms can be retrieved via Möbius inversion of (4.57). Importantly, since the PIRD is built over the same lattice backbone as the PID, several concepts and relations defined for the PID still hold in the dynamic case (see Fig. 3.3, with the shrewdness of considering information rates in place of the static information distributed over the lattice). For instance, as the PIRD satisfies the same consistency equations valid for the PID, the redundancy function computed for an atom composed by one single source reduces to the MIR between the target and that source, i.e. $I^{\cap}_{Y;\mathbf{X}_\alpha} = I_{Y;X_i}$ when $\alpha = \{i\}$. Moreover, the rate of information shared between the M source processes $X_1, \ldots, X_M$ taken together and the target process Y can be expanded in analogy to (3.24) as the sum of $M + 2$ contributions:

$$I_{Y;\mathbf{X}} = \sum_{m=1}^{M} \mathcal{U}_{Y;X_m} + \mathcal{R}_{Y;\mathbf{X}} + \mathcal{S}_{Y;\mathbf{X}}; \tag{4.58}$$

in (4.58), we achieve a so-called *coarse-grained* PIRD, whereby each of the M terms $\mathcal{U}_{Y;X_m} = I_{Y;X_m} - \mathcal{R}_{Y;\mathbf{X}}$ identifies the *unique rate* of information produced by Y that is shared exclusively with X_m ($m = 1, \ldots, M$), the term $\mathcal{R}_{Y;\mathbf{X}}$ identifies the *redundant rate* of information produced by Y that is shared simultaneously with all the source processes in $\mathbf{X}$, and the term $\mathcal{S}_{Y;\mathbf{X}}$ identifies the *synergistic rate* of information produced by Y that only arises from knowing all the sources $X_1, \ldots, X_M$. These coarse-grained information rates correspond for the case of two sources to the PI rates of the four atoms of the redundancy rate lattice (Fig 3.3a, right), while they can be obtained for the case $M = 3$ by summing the PI rates of the atoms with redundant, unique or synergistic character (respectively, magenta, grey and blue shades in Fig. 3.3b, right).

The linear parametric formulation of the PIRD measures will be presented and detailed in Sect. 4.3.2.1 in the context of a spectral representation of the framework.

4.3.1.2 Link-Specific Analysis of HOIs

B-index Rate. The concept of B-index introduced in Sect. 3.2 for random variables can be expanded to random processes, where the MIR and the conditional MIR are used in place of the MI and CMI, respectively. To this end, let us consider the dynamic system $\mathcal{X}$ described by the vector random process $Y_n = [Y_{1,n}, \ldots, Y_{Q,n}]$. The equivalent notation evidencing the dynamics of the units $\{\mathbf{V}_i, \mathbf{V}_j, \mathbf{V}_z\}$ (see Sect. 3.2) is $\mathbf{Y} = [Y_i, Y_j, \mathbf{Y}_z]$, where $\mathbf{Y}_z = \mathbf{Y} \setminus [Y_i, Y_j]$, $i, j = 1, \ldots, Q$, $i \neq j$. In this case, the information-theoretic measure that is typically used to analyze the temporal evolution of the unit Y_i is the entropy rate (4.1), quantifying the rate of generation of new information in the process Y_i; when combined with the entropy rate of Y_j, H_{Y_j}, and with the joint entropy rate of Y_i and Y_j, H_{Y_i,Y_j}, it can be used to assess the iS of the two processes through the MIR (4.12a) [29]. Then, based on this definition in the analyzed network of random processes, the ciS becomes the conditional MIR (cMIR) $I_{Y_i;Y_j|\mathbf{Y}_z}$ obtained as

$$I_{Y_i;Y_j|\mathbf{Y}_z} = H_{Y_i|\mathbf{Y}_z} - H_{Y_i|Y_j,\mathbf{Y}_z}, \tag{4.59}$$

the niS becomes the niRS $I_{Y_i;Y_j;\mathbf{Y}_z}$, quantifying the balance between redundancy and synergy as the difference between the MIR (i.e., the iS) and the CMIR (i.e., the ciS), and the B-index becomes a *B-index rate* [53]:

$$B_{Y_i;Y_j} = \frac{I_{Y_i;Y_j;\mathbf{Y}_z}}{\max\{I_{Y_i;Y_j}, I_{Y_i;Y_j|\mathbf{Y}_z}\}}. \tag{4.60}$$

In (4.60), the B-index rate is computed dividing the difference between the MIR and the CMIR to their maximum, so as to obtain a measure ranging between -1 and 1. The limit values highlight a full imbalance between redundancy and synergy that relates to specific network topologies, specified and discussed in Sect. 3.2.

A simulated example showing the theoretical behavior of the proposed B-index rate measure quantifying link-specific interactions in complex networks can be found in Sect. 4.3.3.4. Applications to cardiovascular networks are described in Sect. 5.11.

Linear parametric formulation. The individual dynamics of the processes Y_i and Y_j and their joint dynamics described by the bivariate process $\{Y_i, Y_j\}$ are captured by three restricted VAR models in the form of (2.7), where $Y_{i,n}$, $Y_{j,n}$ and $\left[Y_{i,n}Y_{j,n}\right]^\intercal$ are used in place of $\mathbf{Z}_n$, while $Y_{i,n-k}$, $Y_{j,n-k}$ and $\left[Y_{i,n-k}Y_{j,n-k}\right]$ in place of $\mathbf{Z}_{n-k}$. The parameters of these restricted models, i.e. the coefficients $\mathbf{B}_{\mathbf{Z},k}$, and the covariance of the residuals $\Sigma_{\mathbf{W}_\mathbf{Z}}$, can be derived from the parameters of the full model (2.6) $\mathbf{A}^{(var)}_{\mathbf{Y},k}$ and $\Sigma^{(var)}_{\mathbf{U}_\mathbf{Y}}$, through a procedure that solves the YW equations to derive the covariance structure of $\mathbf{Y}$ and then reorganizes such structure to relate it to the covariances of Y_i, Y_j or $\mathbf{Y}_z$ (Sect. 2.3.5.1). Then, under the assumption of joint Gaussianity for the overall process $\mathbf{Y}$, the information measures capturing the network interactions can be derived straightly from the covariances of the residuals of the restricted models. Specifically, the entropy rates of Y_i, Y_j or $\{Y_i, Y_j\}$ are obtained as in (4.2), from which the MIR is computed as in (4.15) (see the formulations for scalar processes).

The procedure described above can be repeated to define restricted models capturing the dynamics of the vector process $\mathbf{Y}_z$ as well as of the joint processes $\mathbf{Y}^i = \{Y_i, \mathbf{Y}_z\}$ and $\mathbf{Y}^j = \{Y_j, \mathbf{Y}_z\}$ using VAR formulations as in (2.7), and then to compute the entropy rates of $\mathbf{Y}_z$, $\mathbf{Y}^i$ and $\mathbf{Y}^j$ as in (4.2). This allows to obtain formulations of the MIR terms $I_{Y_i;\mathbf{Y}_z} = H_{Y_i} + H_{\mathbf{Y}_z} - H_{\mathbf{Y}^i}$ and $I_{Y_i;\mathbf{Y}^j} = H_{Y_i} + H_{\mathbf{Y}^j} - H_{\mathbf{Y}}$ following (4.15), from which the ciS $I_{Y_i;Y_j|\mathbf{Y}_z}$ is computed as in (4.59) subtracting

$I_{Y_i;\mathbf{Y}_z}$ from $I_{Y_i;\mathbf{Y}^j}$, and the niRS $I_{Y_i;Y_j;\mathbf{Y}_z}$ and B-index rate $B_{Y_i;Y_j}$ are computed as in (4.60).

4.3.2 Frequency Domain

In this section, starting from the mathematical definitions provided in Sect. 2.4 related to the frequency domain representation of VAR linear models, we will characterize the frequency domain behavior of the network $\mathcal{X}$ composed of M nodes, whose activity is described by the vector processes $\mathbf{X}_1, \ldots, \mathbf{X}_M$, each comprising M_i scalar processes, $i = 1, \ldots, M$. While information-theoretic node- and network-specific measures of dynamic HOIs which extend to the frequency domain the concepts defined in the time domain in Sect. 4.3.1 have been validated through the utilization of the OIR [3, 43] and PIRD framework [55, 56], measures characterizing the link-specific spectral interactions in networks of multiple nodes are still lacking, though they can be derived straightforwardly exploiting the concepts of spectral MIR and CMIR to get a spectral B-index rate as an expansion of (4.60).

Furthermore, we remark that single-node and pairwise activities involving only one or two (groups of) node(s) in the observed network can be investigated through a very simple approach, based on representing the VAR model (2.6) in the frequency domain thus yielding (2.20), which in turn can be partitioned in $M \times M$ blocks if the subdivision of $\mathbf{Y}$ in M blocks $\mathbf{X} = \{\mathbf{X}_1, \ldots, \mathbf{X}_M\}$ is considered (2.21). Then, the spectral ERs h_{Y_i} ($h_{\mathbf{X}_i}$) and spectral MIRs $i_{Y_i;Y_j}$ ($i_{\mathbf{X}_i;\mathbf{X}_j}$) can be computed by taking the corresponding elements of $\mathbf{P}^{(var)}_{\mathbf{Y}}(\bar{f})$ ($\mathbf{P}^{(var)}_{\mathbf{X}}(\bar{f})$), i.e., $P^{(var)}_{Y_i}(\bar{f})$ ($\mathbf{P}^{(var)}_{\mathbf{X}_i}(\bar{f})$) to compute the ER as in (4.5), and $\{P^{(var)}_{Y_i,Y_j}(\bar{f}), \mathbf{P}^{(var)}_{[Y_iY_j]}(\bar{f})\}$ ($\{\mathbf{P}^{(var)}_{\mathbf{X}_i,\mathbf{X}_j}(\bar{f}), \mathbf{P}^{(var)}_{[\mathbf{X}_i\mathbf{X}_j]}(\bar{f})\}$) to compute the MIR as in (4.37), where $\mathbf{P}^{(var)}_{[Y_iY_j]}(\bar{f}) = \begin{bmatrix} P^{(var)}_{Y_i}(\bar{f}) & P^{(var)}_{Y_iY_j}(\bar{f}) \\ P^{(var)}_{Y_jY_i}(\bar{f}) & P^{(var)}_{Y_j}(\bar{f}) \end{bmatrix}$ (the same holds for $\mathbf{P}^{(var)}_{[\mathbf{X}_i\mathbf{X}_j]}(\bar{f})\}$). Theoretical simulations exploiting this approach will be presented in Sects. 4.3.3.1 and 4.3.3.2 [3].

4.3.2.1 Network-Specific and Node-Specific Analysis of HOIs

O-Information Rate. The parametric implementation of the OIR decomposition, which relies on the mathematical definitions provided in Sect. 2.4 related to the frequency domain representation of VAR linear models, allows to expand the OIR and its decomposition terms in the frequency domain, thus obtaining spectral patterns of HOIs between the analysed processes. These concepts have been investigated and described with fine detail in [3, 43].

As a first step, we consider the full VAR model (2.6) and the restricted VAR model (2.7), whose identification procedures have been described in Sects. 2.3.5 and 2.3.5.1, respectively; details on their spectral representations can be found in Sects. 2.4 and

2.4.1, respectively. The PSD matrix of the restricted process $\mathbf{Z} = \{\mathbf{Z}_1, \mathbf{Z}_2\}$ (2.24) can be factorized in blocks to make explicit the power spectral densities of $\mathbf{Z}_1$ and $\mathbf{Z}_2$, $\mathbf{S}_{\mathbf{Z}_1}(\bar{f})$ and $\mathbf{S}_{\mathbf{Z}_2}(\bar{f})$, as diagonal blocks, and the cross-spectral densities between $\mathbf{Z}_1$ and $\mathbf{Z}_2$, $\mathbf{S}_{\mathbf{Z}_1\mathbf{Z}_2}(\bar{f})$ and $\mathbf{S}_{\mathbf{Z}_2\mathbf{Z}_1}(\bar{f})$, as off-diagonal blocks. From this factorization, the logarithmic spectral measure of the total (symmetric) interdependence between $\mathbf{Z}_1$ and $\mathbf{Z}_2$ is defined as in (4.37), yielding $f_{\mathbf{Z}_1;\mathbf{Z}_2}(\bar{f}) = \log\left(\frac{|\mathbf{S}_{\mathbf{Z}_1}(\bar{f})||\mathbf{S}_{\mathbf{Z}_2}(\bar{f})|}{|\mathbf{S}_{\mathbf{Z}}(\bar{f})|}\right)$ [9]. Moreover, after factorizing in $R_i \times R_i$ diagonal blocks and $R_i \times R_j$ off-diagonal blocks also the transfer and innovation covariance matrices $\tilde{\mathbf{H}}^{\mathbf{Z}}(\bar{f})$ and $\Sigma_{\mathbf{W}_\mathbf{Z}}$, logarithmic spectral measures of the causal (asymmetric) effect of $\mathbf{Z}_j$ on $\mathbf{Z}_i$ $(i, j = 1, 2)$ can be computed as in (4.38). This yields $f_{\mathbf{Z}_j \to \mathbf{Z}_i}(\bar{f}) = \log\left(\frac{|\mathbf{S}_{\mathbf{Z}_i}(\bar{f})|}{|\tilde{\mathbf{H}}^{\mathbf{Z}}_{\mathbf{Z}_i\mathbf{Z}_i}(\bar{f})\Sigma_{\mathbf{W}_{\mathbf{Z}_i}}\tilde{\mathbf{H}}^{\mathbf{Z}*}_{\mathbf{Z}_i\mathbf{Z}_i}(\bar{f})|}\right)$ [9], where $\tilde{\mathbf{H}}^{\mathbf{Z}}_{\mathbf{Z}_i\mathbf{Z}_i}$ describes the transfer from $\mathbf{W}_{\mathbf{Z}_i}$ to $\mathbf{Z}_i$ in the frequency domain and $\Sigma_{\mathbf{W}_{\mathbf{Z}_i}} = \mathbb{E}[\mathbf{W}_{\mathbf{Z}_i,n}\mathbf{W}^\intercal_{\mathbf{Z}_i,n}]$. To complete the representation of the pairwise interactions between $\mathbf{Z}_1$ and $\mathbf{Z}_2$, a spectral measure instantaneous causality can be defined as $f_{\mathbf{Z}_1\cdot\mathbf{Z}_2}(\bar{f}) = \log\left(\frac{|\tilde{\mathbf{H}}^{\mathbf{Z}}_{\mathbf{Z}_1\mathbf{Z}_1}(\bar{f})\Sigma_{\mathbf{W}_{\mathbf{Z}_1}}\tilde{\mathbf{H}}^{\mathbf{Z}*}_{\mathbf{Z}_1\mathbf{Z}_1}(\bar{f})||\tilde{\mathbf{H}}^{\mathbf{Z}}_{\mathbf{Z}_2\mathbf{Z}_2}(\bar{f})\Sigma_{\mathbf{W}_{\mathbf{Z}_2}}\tilde{\mathbf{H}}^{\mathbf{Z}*}_{\mathbf{Z}_2\mathbf{Z}_2}(\bar{f})|}{|\mathbf{S}_{\mathbf{Z}}(\bar{f})|}\right)$ so as to satisfy in the frequency domain a decomposition similar to the time domain decomposition (4.53):

$$f_{\mathbf{Z}_1;\mathbf{Z}_2}(\bar{f}) = f_{\mathbf{Z}_1\to\mathbf{Z}_2}(\bar{f}) + f_{\mathbf{Z}_2\to\mathbf{Z}_1}(\bar{f}) + f_{\mathbf{Z}_1\cdot\mathbf{Z}_2}(\bar{f}). \tag{4.61}$$

Importantly, the spectral measures in (4.61) are tightly linked to the corresponding measures computed in time domain in (4.53), as specified in Sect. 4.2.2.1.

The spectral integration property can be exploited not only to compute the time domain measures in (4.53) as the integral of the corresponding spectral measures in (4.61), but also to achieve a causal decomposition of the OIR formulated for spectral functions. Indeed, it is easy to show that the frequency-specific OIR increment defined in analogy to (4.52) as

$$\delta_{\mathbf{X}_i;\mathbf{X}^N_{-i}}(\bar{f}) = \sum_{\substack{i=1\\ i\neq j}}^{N} i_{\mathbf{X}_i;\mathbf{X}^N_{-ij}}(\bar{f}) + (2-N)i_{\mathbf{X}_i;\mathbf{X}^N_{-i}}(\bar{f}), \tag{4.62}$$

where $i_{(\cdot)} = \frac{f_{(\cdot)}}{2}$, satisfies the spectral integration property, i.e. $\Delta_{\mathbf{X}_i;\mathbf{X}^N_{-i}} = \int_0^{\frac{1}{2}} \delta_{\mathbf{X}_i;\mathbf{X}^N_{-i}}(\bar{f})\mathrm{d}\bar{f}$, and can also be expanded through a causal decomposition similar to (4.54) as

$$\delta_{\mathbf{X}_i;\mathbf{X}^N_{-i}}(\bar{f}) = \delta_{\mathbf{X}_i\to\mathbf{X}^N_{-i}}(\bar{f}) + \delta_{\mathbf{X}^N_{-i}\to\mathbf{X}_i}(\bar{f}) + \delta_{\mathbf{X}_i\cdot\mathbf{X}^N_{-i}}(\bar{f}), \tag{4.63}$$

where the three terms on the r.h.s. of (4.63) are obtained expanding $i_{\mathbf{X}_i;\mathbf{X}^N_{-i}}(\bar{f})$ and $i_{\mathbf{X}_i;\mathbf{X}^N_{-ij}}(\bar{f})$ in (4.62) according to (4.61). Moreover, the spectral OIR increment (4.62) can be used to compute recursively a frequency-domain version of the OIR, in analogy to (4.51), as

$$\nu_{\mathbf{X}^N}(\bar{f}) = \nu_{\mathbf{X}^N_{-i}}(\bar{f}) + \delta_{X_i;\mathbf{X}^N_{-i}}(\bar{f}), \tag{4.64}$$

which again satisfies the spectral integration property, i.e. $\Omega_{\mathbf{X}^N} = \int_0^{\frac{1}{2}} \nu_{\mathbf{X}^N}(\bar{f})\mathrm{d}\bar{f}$.

Therefore, the spectral versions of the HOI measures defined in this section can be meaningfully interpreted as densities of the synergistic/redundant character of the information shared between multiple stochastic processes. To conclude this section it is worth noting that, in the case of $N = 3$ processes, the spectral OIR (4.64) is a frequency-domain analogous of the IIR defined in (4.49), which can be recovered through whole-band integration. This measure has been recently defined for triplets of random processes [71], and also extended to the spectral computation of separate measures of redundancy and synergy within the PID framework [57]. As shown in the simulation examples of Sects. 4.3.3.1, 4.3.3.2 and 4.3.3.3 and practical applications of Sects. 5.9, and 6.2.4–6.2.5 the evaluation of the spectral IIR of three processes, and more generally of the spectral OIR of multiple processes, allows to assess the informational character of specific oscillations within circuits of nodes of the analyzed network.

Frequency-Specific Partial Information Rate Decomposition. To solve the PIRD identified by (4.56), it is necessary to define a redundancy rate function taking values over the lattice underlying the decomposition. In principle, the redundancy rate can be defined following any of the several approaches formulated for the PID, adapting it to the calculation of the MIR between random processes in place of the MI between random variables. A particularly useful approach is to derive redundancy measures working on the specific realizations of the random variables at hand, rather than on the variable themselves, and then compute redundancy via statistical expectation. This *pointwise* approach has been followed to put forth PID methods that define *local* redundancy measures computed for single realizations of the variables, from which a *global* redundancy is obtained taking the ensemble average over all possible realizations [72–74]. The same approach can be followed for the PIRD, e.g., working on the local version of the MIR to formalize the notion of pointwise redundancy rate; the redundancy rate among processes could be then retrieved by ensemble averaging, which for stationary processes corresponds to time-domain averaging. Here, we propose an approach that is conceptually similar, but is implemented through a pointwise representation in frequency rather than in time. Specifically, we characterize the analyzed network of random processes $\mathbf{Z}$ in the frequency domain, considering the information provided about a particular oscillatory component of the target process Y by the iso-frequency oscillatory components of the source processes collected in $\mathbf{X}$. The idea is to perform the entire PIRD on the pointwise level for a particular frequency, i.e., to decompose the spectral (frequency-specific) MIR denoted as $i_{Y;\mathbf{X}}(\bar{f})$ and identified from the expansion of the MIR in the frequency domain as in (4.42). While different integral transforms (e.g., the wavelet transform) could in principle be used to expand the MIR, we use a definition of spectral MIR which satisfies (4.42) for Gaussian processes to formalize the PIRD in the frequency domain. Crucially, (4.42) connects the time- and frequency-domain representations of information-theoretic quantities for random processes, and is exploited here to relate the PIRD (4.56) to its frequency-domain extension. Such an extension is denoted as *spectral PIRD* and is formulated, for the oscillatory components of the network process assessed at the frequency $\bar{f}$, expressing the spectral MIR as:

$$i_{Y;\mathbf{X}}(\bar{f}) = \sum_{\alpha \in \mathcal{A}} i^{\delta}_{Y;\mathbf{X}_\alpha}(\bar{f}), \tag{4.65}$$

where $i^{\delta}_{\cdot;\cdot}(\bar{f})$ is the spectral PI rate function defined over a lattice specifically identified on the oscillations with frequency $\bar{f}$, and $\mathbf{X}_\alpha$ denotes the set of subsets of source processes indexed by the atom α. As happens with PID, to solve (4.65) we need to identify a *spectral redundancy rate* through the imposition of a certain number of reasonable axioms or constraints on the spectral MIR, and then follow the PID formalism to derive the pointwise (spectral) atoms of information at the frequency $\bar{f}$. If $i^{\cap}_{Y;\mathbf{X}_\alpha}(\bar{f})$ denotes the spectral redundancy rate computed for the αth node of the spectral lattice defined at the frequency $\bar{f}$, the corresponding spectral PI rate is computed recursively, in analogy to (3.22), as:

$$i^{\delta}_{Y;\mathbf{X}_\alpha}(\bar{f}) = i^{\cap}_{Y;\mathbf{X}_\alpha}(\bar{f}) - \sum_{\beta \prec \alpha} i^{\delta}_{Y;\mathbf{X}_\beta}(\bar{f}). \tag{4.66}$$

Furthermore, once the atoms are identified via (4.66), they can be properly grouped to obtain a coarse-grained representation of the spectral PIRD which takes the form:

$$i_{Y;\mathbf{X}}(\bar{f}) = \sum_{m=1}^{M} u_{Y;X_m}(\bar{f}) + r_{Y;\mathbf{X}}(\bar{f}) + s_{Y;\mathbf{X}}(\bar{f}), \tag{4.67}$$

where the $M+2$ atoms reflect the M unique contributions of each source process, as well as the redundant and synergistic contributions of all sources, to the rate of information produced by the target process at the specific frequency $\bar{f}$; the coarse-grained spectral PIRD is obtained adopting the same criteria already described for the PID [75], leading to the coarse-grained atoms illustrated in Fig. 3.3 (right panels). In (4.66), we provide a solution for the spectral PIRD up to the definition of a proper spectral redundancy rate function. Here, we propose to assess spectral redundancy following the MMI principle [69] applied to the spectral MIR computed between the target and each source process at the frequency of interest. Specifically, we define the *spectral MMI redundancy rate* function of the atom $\alpha = \{\alpha_1, \ldots, \alpha_J\}$ of the spectral redundancy lattice as:

$$i^{\cap}_{Y;\mathbf{X}_\alpha}(\bar{f}) = i^{\mathrm{SMMI}}_{Y;\mathbf{X}_\alpha}(\bar{f}) := \min_{j=1,\ldots,J} i_{Y;\mathbf{X}_{\alpha_j}}(\bar{f}), \tag{4.68}$$

where $J = |\alpha|$ indicates the cardinality of the atom; e.g., $J = 1$ if $\alpha = \{1\}$ or $\alpha = \{12\}$, while $J = 2$ if $\alpha = \{\{3\}, \{12\}\}$. Note that, in contrast with the classical MMI formulation performed for random variables [69], here the minimum MIR is searched at the pointwise (frequency-specific) level. While a pointwise implementation of the MMI criterion would be cumbersome if performed in the time domain, and it is indeed avoided by the existing pointwise approaches [72, 73] because the local MI or the local MIR can take negative values, in our case is favored by the non-negativity of the spectral MIR. Importantly, from the perspective of information decomposition based on lattice structures [70], the spectral PIRD (4.65, 4.66) is conceptually equivalent to the PIRD (4.56, 4.57) and to the PID (3.20, 3.21); what changes in the three formulations is only the quantity to be decomposed, from the MI $I(T; \mathbf{S})$ for the case of random variables to the MIR $I_{Y;\mathbf{X}}$ for the case of random processes, and to the spectral MIR $i_{Y;\mathbf{X}}(\bar{f})$ for the case of oscillatory components of random processes. Since these quantities maintain the same meaning and properties (i.e., they are non-negative measures of shared information), the three formulations of PID, PIRD and spectral PIRD will lead to a unique solution for the atoms once a redundancy function (respectively, the redundancy among collections of random variables, the redundancy rate among collections of random processes, and the spectral redundancy rate among collections of oscillatory components of random processes) is fixed over a lattice structure like that in Fig. 3.3. Therefore, moving from the PIRD defined on the pointwise level (frequency-specific) to that defined on the process-level (time domain) is straightforward exploiting spectral integration applied to the redundancy rate and PI rate functions:

$$I^{\cap}_{Y;\mathbf{X}_\alpha} = 2\int_0^{\frac{1}{2}} i^{\cap}_{Y;\mathbf{X}_\alpha}(\bar{f})\mathrm{d}\bar{f}, \tag{4.69}$$

$$I^{\delta}_{Y;\mathbf{X}_\alpha} = 2\int_0^{\frac{1}{2}} i^{\delta}_{Y;\mathbf{X}_\alpha}(\bar{f})\mathrm{d}\bar{f}; \tag{4.70}$$

the same spectral integration property holds to relate the unique, redundant and synergistic atoms of the coarse-grained spectral PIRD (4.67) to their corresponding time-domain atoms of the PIRD (4.58). Remarkably, solving the frequency-specific PIRD to obtain the PI rate $i^{\delta}_{Y;\mathbf{X}_\alpha}(\bar{f})$ for each atom α via (4.66), and then integrating these contributions along the whole frequency axis according to (4.70) to get time domain values, is equivalent to integrating the spectral redundancy rate functions $i^{\cap}_{Y;\mathbf{X}_\alpha}(\bar{f})$ in the range $[0, \frac{1}{2}]$ via (4.69) and then applying the PIRD in the time domain to get the contributions $I^{\delta}_{Y;\mathbf{X}_\alpha}$. This is guaranteed by the property of linearity of the definite integrals, and allows also to develop band-specific PIRD schemes whereby the decomposition of multivariate information rates is achieved in the time domain but is limited to oscillatory components whose frequencies are confined within a specific band of the spectrum.

Benchmark simulations are displayed in Sects. 4.3.3.5 and 4.3.3.6, while an exemplary application to network physiology is presented in Sect. 5.10.

Linear parametric formulation. In the linear signal processing framework, the analyzed set of stochastic processes $\mathbf{Z} = \{Y, X_1, \ldots, X_M\}$ can be described in terms of its PSD matrix expressed as in (2.21), which is the central element for the implementation of the spectral PIRD. In fact, it is well-known that, for jointly Gaussian processes $\mathbf{X}$ and Y, the spectral MIR decomposed by PIRD corresponds to half of the spectral TD in (4.37). Moreover, for any given atom $\alpha = \{\alpha_1, \ldots, \alpha_J\}$ of the spectral redundancy lattice, the application of (4.37) can particularized to the element α_j as follows:

$$i_{Y;\mathbf{X}_{\alpha_j}}(\bar{f}) = \frac{1}{2}\log\left(\frac{|\mathbf{P}_{\mathbf{X}_{\alpha_j}}(\bar{f})|P_Y(\bar{f})}{|\mathbf{P}_{[Y\mathbf{X}_{\alpha_j}]}(\bar{f})|}\right), \tag{4.71}$$

where

$$\mathbf{P}_{[Y\mathbf{X}_{\alpha_j}]}(\bar{f}) = \begin{bmatrix} P_Y(\bar{f}) & \mathbf{P}_{Y\mathbf{X}_{\alpha_j}}(\bar{f}) \\ \mathbf{P}_{\mathbf{X}_{\alpha_j}Y}(\bar{f}) & \mathbf{P}_{\mathbf{X}_{\alpha_j}}(\bar{f}) \end{bmatrix}; \tag{4.72}$$

the computation of (4.71) for each $j = 1, \ldots, J$ yields the spectral MIR terms to be used in (4.68) for the computation of the spectral redundancy function. Remarkably, the spectral redundancy rate defined by (4.68, 4.71) is a proper redundancy function as it satisfies the axioms originally proposed by Williams and Beer, i.e.,

- *symmetry*: $i^{\cap}_{Y;\mathbf{X}_{\{\alpha_1,\ldots,\alpha_J\}}}(\bar{f})$ is symmetric w.r.t. the α_j, $j = 1, \ldots, J$;
- *self-redundancy*: $i^{\cap}_{Y;\mathbf{X}_{\alpha}}(\bar{f}) = i_{Y;X_{\alpha}}(\bar{f})$ when $|\alpha| = 1$;
- *monotonicity*: $i^{\cap}_{Y;\mathbf{X}_{\{\alpha_1,\ldots,\alpha_{J-1},\alpha_J\}}}(\bar{f}) \leq i^{\cap}_{Y;\mathbf{X}_{\{\alpha_1,\ldots,\alpha_{J-1}\}}}(\bar{f})$, with equality if $\alpha_{J-1} \subseteq \alpha_J$.

These properties are derived straightforwardly from the minimum MIR definition (4.68) of the spectral redundancy rate, and from the properties of non-negativity and monotonicity of the spectral MIR (4.71) [42].

The practical computation of the PSD matrix, whose elements are exploited to estimate all the MIR terms entering the PIRD, is performed inducing a linear parametric representation of the observed dynamics. Specifically, the analyzed stochastic process $\mathbf{Z} = \{Y, X_1, \ldots, X_M\}$ is described as a VAR process analyzed in the frequency domain as described in Sect. 2.4; the parameters of the VAR model (in the form of (2.6)) can be easily estimated from realizations of the process $\mathbf{Z}$ available in the form of multivariate time series. Model identification is based on classical OLS estimation implemented optimizing the model order p through the AIC, as described in Sect. 2.3.5 with fine details.

4.3.3 Simulation Examples

4.3.3.1 Gaussian Processes Interacting in Star Structures

In this example, taken from [3], we show the time domain and spectral behavior of the information-theoretic measures characterizing single-node (ER), pairwise (MIR) and high-order (OIR) interactions in a network of multiple nodes. Specifically, we consider $Q = 5$ scalar stationary Gaussian stochastic processes, each describing

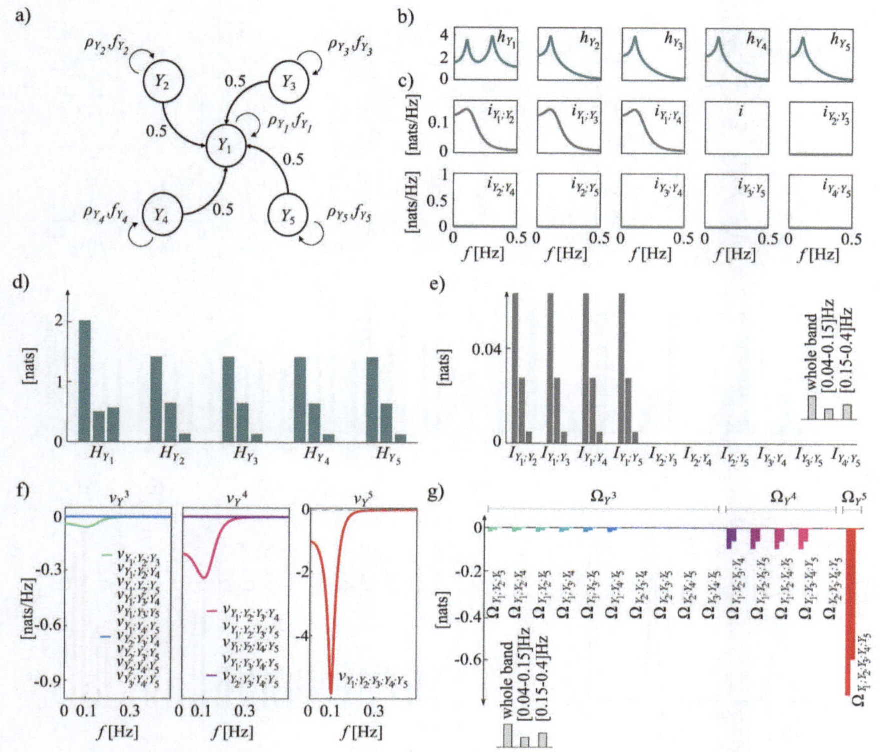

Fig. 4.8 **a** Simulation design. Spectral **b** ERs and **c** MIRs. **d** ER and **e** MIR values integrated in the whole band (left bars), the range [0.04–0.15] Hz (middle bars) and the range [0.15–0.4] Hz (right bars) of the spectrum. **f** Spectral OIRs of orders 3–5. **g** OIRs of orders 3–5 integrated as in **d**, **e**. The figure is adapted from [3]

the dynamic activity of the network nodes $\mathcal{X}_i, i = 1, \ldots, M$, with $M = Q$. Their lagged interactions are mapped by a VAR process of order $p = 5$ (2.6). Here, we imposed autonomous oscillations for the process Y_1, setting $\rho_{Y_1} = 0.95$, $f_{Y_1} = 0.3$ Hz, so that the dynamics of Y_1 are determined by the fixed coefficients $a_{Y_1Y_1,1} = -0.587$, $a_{Y_1Y_1,2} = -0.9$. Conversely, the remaining processes exhibit autonomous oscillations imposed setting $\rho_{Y_i} = 0.95$, $f_{Y_i} = 0.1$ Hz, so that the dynamics of $Y_i, i = 2, \ldots, Q$, are determined by the fixed coefficients $a_{Y_iY_i,1} = 1.537$, $a_{Y_iY_i,2} = -0.9$. The sampling frequency was fixed to $f_s = 1$ Hz. To analyze network interactions, we considered a *star* structure in two different configurations, where the process Y_1 is *(i)* a *receiver* (Fig. 4.8a) or *(ii)* a *sender* (Fig. 4.9a) of information for the remaining processes $Y_i, i = 2, \ldots, Q$. The model parameters imposed for these settings are all zero except for *(i)* $a_{Y_1Y_j,1} = 0.5, j = 2, \ldots, Q$, and *(ii)* $a_{Y_iY_1,1} = 0.5, i = 2, \ldots, Q$, respectively.

The VAR model (2.6) is studied in the frequency domain by deriving the 5×5 PSD $\mathbf{P_Y}(\bar{f})$ (2.20). This leads to compute the exact values of the time and frequency

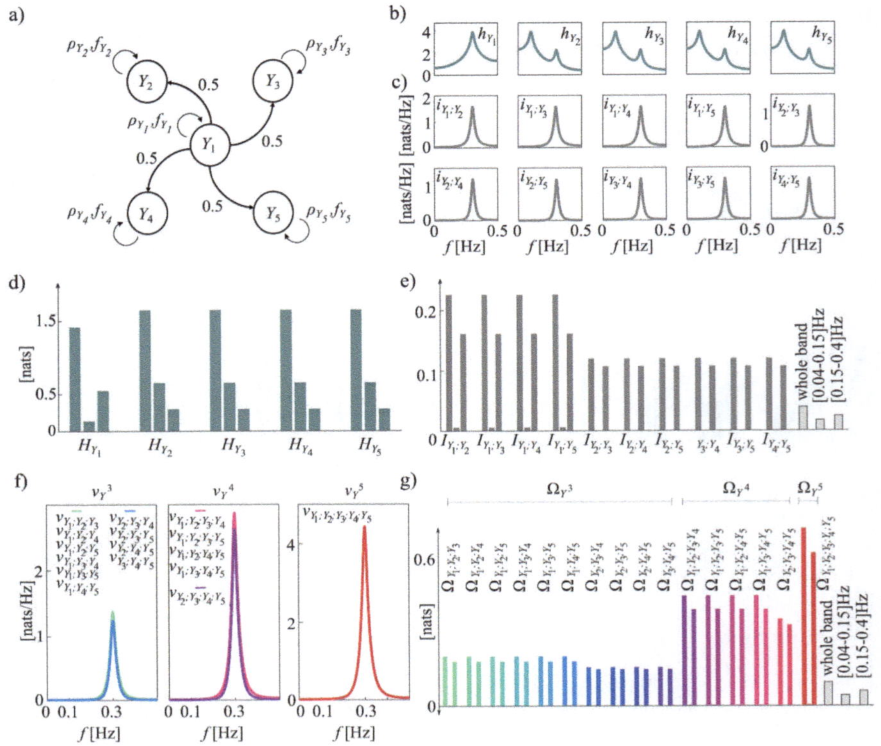

Fig. 4.9 **a** Simulation design. Spectral **b** ERs and **c** MIRs. **d** ER and **e** MIR values integrated in the whole band (left bars), the range [0.04–0.15] Hz (middle bars) and the range [0.15–0.4] Hz (right bars) of the spectrum. **f** Spectral OIRs of orders 3–5. **g** OIRs of orders 3–5 integrated as in **d**, **e**. The figure is adapted from [3]

domain information measures of ER, MIR and OIR for the simulated process. Results of the two simulation settings are shown in Figs. 4.8b–g and 4.9b–g, respectively. The spectral profiles of entropy rates, mutual information rates and O-information rates are shown in panels *b*, *c*, and *f*, respectively, while in panels *d*, *e*, and *g* the ER, MIR and OIR values integrated along the whole band (left bars), the range [0.04–0.15] Hz (middle bars) and range [0.15–0.4] Hz (right bars) are depicted.

The first configuration, shown in Fig. 4.8a, is predominantly synergistic since each of the source processes $Y_i, i = 2, \ldots, Q$ sends unique information to the target process Y_1 at the same frequency (0.1 Hz). In fact, as shown in panel *b*, the spectral entropy rate of the latter, i.e., h_{Y_1}, shows peaks not only at 0.3 Hz, which represents its own oscillating frequency, but also at 0.1 Hz, due to dynamic information transferred from the rest of the system. Conversely, the spectral profiles of the entropy rate of the sources, i.e., h_{Y_i}, are characterized by peaks at 0.1 Hz, showing that the information content of these processes is located around that frequency. Remarkably, these findings suggest that the spectral entropy rates are characterized by frequency-specific peaks wherever the series is more predictable and owns autonomous dynamics. These results are confirmed by the spectral behavior of the MIRs shared between Y_1 and Y_i, $i = 2, \ldots, Q$ ($i_{Y_1;Y_i}$), which show peaks at 0.1 Hz, as well as between pairs of sources ($i_{Y_{s1};Y_{s2}}$; $s_1 = 2, \ldots, Q-1$, and $s_2 = 3, \ldots, Q$, with $s_1 \neq s_2$), which are null at each frequency given the absence of coupled interactions between them (Fig. 4.8c). The same trend is visible by looking at the spectral profiles of the 3rd-order OIR ($\nu_{\mathbf{Y}^3}$, Fig. 4.8f), which display negative peaks at 0.1 Hz when the analyzed multiplet includes the target Y_1 and two of the sources, while are null at each frequency when it includes only three isolated source processes. Correspondingly, adding one source process to multiplets already including the target increases the amount of synergy in the system, as shown by the spectral profiles of the 4th- and 5th-order OIR ($\nu_{\mathbf{Y}^N}$, $N = 4, 5$, Fig. 4.8f). The synergistic behavior of the network, mainly confined to the band with central frequency 0.1 Hz, is confirmed by the integration of the spectral profiles along given frequency ranges, showing that all the information shared between the multiple interacting processes in the network is located around 0.1 Hz (Fig. 4.8g).

The second configuration, shown in Fig. 4.9a, is predominantly redundant since each of the processes Y_i, $i = 2, \ldots, Q$, receives the same information from the source process Y_1 at 0.3 Hz. The spectral entropy rates of the targets show peaks not only at 0.1 Hz, which represents their own oscillating frequency, but also at 0.3 Hz, due to dynamic information transferred from the source Y_1 (h_{Y_i}, panel *b*). This is in agreement with the spectral behavior of the MIRs shared between pairs of processes ($i_{Y_{s1};Y_{s2}}$; $s_1 = 1, \ldots, Q-1$, and $s_2 = 2, \ldots, Q$, with $s_1 \neq s_2$, panel *c*), which show peaks around 0.3 Hz, thus confirming the frequency-specific redundant character of the multiple interactions in the analyzed network. The high-order description of the VAR process Y confirms these findings, as highlighted by the positive peaks of the spectral OIR $\nu_{\mathbf{Y}^N}$, $N = 3, \ldots, 5$ (Fig. 4.9f), located around 0.3 Hz, whose amplitude increases with the size of the multiplet. Their whole-band and band-specific integration returns positive values indicating an overall prevalence of redundancy (Fig. 4.9g).

4.3.3.2 Multiple Interacting Gaussian Processes Analyzed in Blocks

In this simulation example, taken from [3], we show the time domain and spectral behavior of the information-theoretic measures characterizing single-node (ER), pairwise (MIR) and high-order (OIR) interactions in a network of multiple nodes. Specifically, we consider $Q = 6$ stationary Gaussian stochastic processes, $\mathbf{Y} = \{Y_1, \ldots, Y_Q\}$, grouped in $M = 3$ blocks, $\mathbf{X} = \{\mathbf{X}_1, \ldots, \mathbf{X}_M\}$. Network interactions are mapped by a six-variate VAR process of order 3 configured to reproduce coexisting redundant and synergistic interactions [71, 76, 77]. Following the structure of the VAR model in (2.6), autonomous oscillations in the processes Y_i, $i = 3, \ldots, 6$, are obtained placing complex-conjugate poles with radii $\rho_{Y_3} = \rho_{Y_5} = 0.85$, $\rho_{Y_4} = \rho_{Y_6} = 0.95$ and normalized frequencies $f_{Y_3}/f_s = 0.1$, $f_{Y_5}/f_s = 0.1$, $f_{Y_4}/f_s = 0.35$ and $f_{Y_6}/f_s = 0.2$ in the complex plane. Assuming a sampling frequency $f_s = 100$ Hz, the poles determine oscillations at 10, 35 and 20 Hz. We set $\mathbf{X}_1 = \{Y_1, Y_2\}, \mathbf{X}_2 = \{Y_3, Y_4\}$ and $\mathbf{X}_3 = \{Y_5, Y_6\}$, and the causal interactions between different blocks are specified to obtain the common driver effect $Y_5 \leftarrow Y_4 \rightarrow Y_3$, the common child effect $Y_3 \rightarrow Y_1 \leftarrow Y_5$ and the unidirectional couplings $Y_5 \rightarrow Y_6$, $Y_1 \rightarrow Y_2$. The model parameters imposed for these settings are all zero except for $a_{Y_1Y_5,1} = a_{Y_5Y_4,1} = a_{Y_1Y_3,2} = a_{Y_3Y_4,3} = 0.5$ and $a_{Y_2Y_1,2} = a_{Y_6Y_5,2} = 0.3$, as depicted in Fig. 4.10a.

The VAR model (2.6) is studied in the frequency domain by deriving the 6×6 PSD $\mathbf{P}_\mathbf{X}(\bar{f})$ (2.21). This leads to compute the exact values of the time and frequency domain information measures of ER, MIR and OIR for the simulated process. Results of the simulation are reported in Fig. 4.10. The spectral profiles of entropy rates, mutual information rates and O-information rate are shown in panels *b*, *c*, and *d*, respectively, while in panel *e* the ER, MIR and OIR values integrated along the whole band (left bars), the range [8–12] Hz (middle bars) and the range [30–40] Hz (right bars) are depicted. In panel *b*, the spectral ER profiles of the three blocks of processes are shown ($h_{\mathbf{X}_1}$,$h_{\mathbf{X}_2}$,$h_{\mathbf{X}_3}$). Two distinct oscillations at 10 and 35 Hz are consistently observed, indicating that the information content of the system is predominantly localized in specific bands of the spectrum. Specifically, $\mathbf{X}_1$ exhibits the highest spectral information content at 10 Hz, originating directly from the causal links $\mathbf{X}_2 \rightarrow \mathbf{X}_1$ and $\mathbf{X}_3 \rightarrow \mathbf{X}_1$. Conversely, the spectral ER profile of $\mathbf{X}_2$ exhibits the highest peak at 35 Hz, directly linked to the amplitude of the oscillatory activity of Y_4, which, in turn, is controlled by the value of ρ_{Y_4}. Lastly, $\mathbf{X}_3$ features three different oscillations at 10, 20 and 35 Hz, with the latter exhibiting the lowest amplitude since it is transferred from $\mathbf{X}_2$ and it is not representative of the autonomous oscillatory activity of $\mathbf{X}_3$. The analysis of the spectral MIRs, whose profiles are reported in panel *c*, reveals the presence of a dynamical coupling occurring at 10 and 35 Hz. In particular, the common drive role of Y_4, directed to Y_5 and Y_3, ensures the presence of a prominent peak at 35 Hz in all the spectral MIR profiles. On the other hand, the presence of dynamic coupling at 10 Hz between $\mathbf{X}_1$ and $\mathbf{X}_2$ ($i_{\mathbf{X}_1;\mathbf{X}_2}$), as well as between $\mathbf{X}_1$ and $\mathbf{X}_3$ ($i_{\mathbf{X}_1;\mathbf{X}_3}$), is driven by the presence of an oscillation at 10 Hz in both Y_3 and Y_5 directly transmitted towards Y_1. This is not the case when analysing the MIR between $\mathbf{X}_2$ and $\mathbf{X}_3$, which shows only the presence of an oscillation at 35 Hz as a result of the interaction between Y_4 and Y_5.

The complexity of the interactions in the analyzed network is well explained by the analysis of the spectral OIR of order 3 ($\nu^3_\mathbf{X}$, Fig. 4.10d), which reveals the coexistence of redundancy occurring at 35 Hz due to the cascade mechanism $\mathbf{X}_2 \rightarrow \mathbf{X}_3 \rightarrow \mathbf{X}_1$, and synergy occurring at 10 Hz due to the common child structure $\mathbf{X}_2 \rightarrow \mathbf{X}_1 \leftarrow \mathbf{X}_3$. This is evident only when analysing the spectral profile of the OIR, but it is not detectable employing the time domain measure. Indeed, the analysis of the OIR in the time domain evidences the impossibility of discriminating the coexistence of synergistic and redundant contributions. The integrated value over the whole frequency spectrum is negative, indicating synergy; redundancy emerges only when integrating within the range [30–40] Hz (Fig. 4.10e).

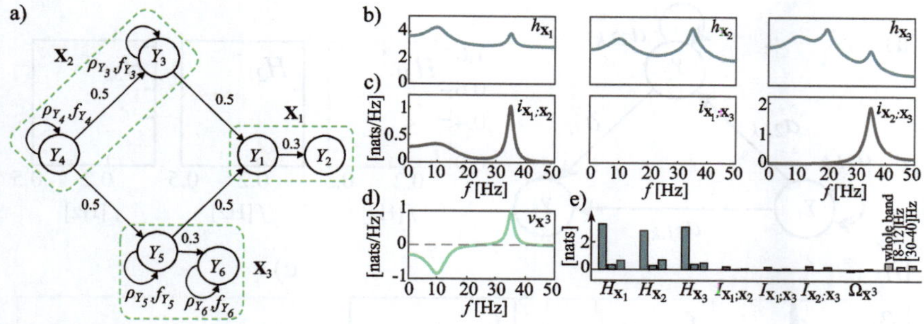

Fig. 4.10 **a** Simulation design. Spectral **b** ERs and **c** MIRs. **d** Spectral OIR of order 3. **e** ER, MIR and OIR values integrated in the whole band (left bar), the range [8–12] Hz (middle bar) and the range [30–40] Hz (right bar) of the spectrum. The figure is adapted from [3]

4.3.3.3 Causal Transfer of Information in a Simulated Cardiovascular Network

This simulation, taken from [43], reproduces the trivariate system proposed in [57], adapted to generate realistic cardiovascular and respiratory dynamics; here, we show the time domain and spectral behavior of the information measures of high-order interaction defined in Sects. 4.3.1 and 4.3.2, i.e., the OIR and its decomposition terms. The activity of this system is mapped by a trivariate VAR process defined as in (2.6) fed by independent Gaussian innovations, for which the parameters are set as illustrated in Fig. 4.11a (for clarity, we omit the subscript $Y_{(\cdot)}$ and indicate only the numbers inside round brackets; the same holds for TFs in panel *a*, right). The vector process is studied keeping the three scalar processes separate ($M = Q = 3$), and assuming sampling frequency $f_s = 1$ (spectral functions are described completely in the frequency range 0–0.5 Hz). The coefficient matrix $\mathbf{A}^{(var)}_{\mathbf{Y},k}$ is designed to mimic the dynamics of respiration (Y_1), arterial pressure (Y_2) and heart period (Y_3) variability, generating self-dependencies for the processes Y_1 and Y_2 through the coefficients $a_{Y_1Y_1,k}$ and $a_{Y_2Y_2,k}$, and imposing causal effects along the directions $Y_1 \rightarrow Y_2$, $Y_1 \rightarrow Y_3$ and $Y_2 \rightarrow Y_3$ through the coefficients $a_{Y_2Y_1,k}$, $a_{Y_3Y_1,k}$ and $a_{Y_3Y_2}$. Self-dependencies are set to induce oscillations in the respiratory band (~0.35 Hz) for Y_1 and in the low-frequency band (~0.1 Hz) for Y_1 and particularly for Y_2, while causal effects are set to realize a high-pass filter from Y_1 to Y_2, a low-pass filter from Y_1 to Y_3 and an all-pass configuration from Y_2 to Y_3 (spectral transfer functions are shown in Fig. 4.11a, right); low- and high-pass filtering are achieved through FIR filters of order 20 with cut-off frequency of 0.2 Hz.

The application of the OIR framework to the VAR parameters describing the simulated process leads to the spectral functions depicted in Fig. 4.11b, c. The PSD profiles (Fig. 4.11b, diagonal plots) highlight oscillations at ~0.1 and ~0.35 Hz for the three processes. The causal coupling between pairs of processes (Fig. 4.11b, off-diagonal plots) evidences the presence of information flows originating from the first process (nonzero profiles of $f_{Y_1\rightarrow Y_2}$, $f_{Y_1\rightarrow Y_3}$ and $f_{Y_2\rightarrow Y_3}$) and the absence of

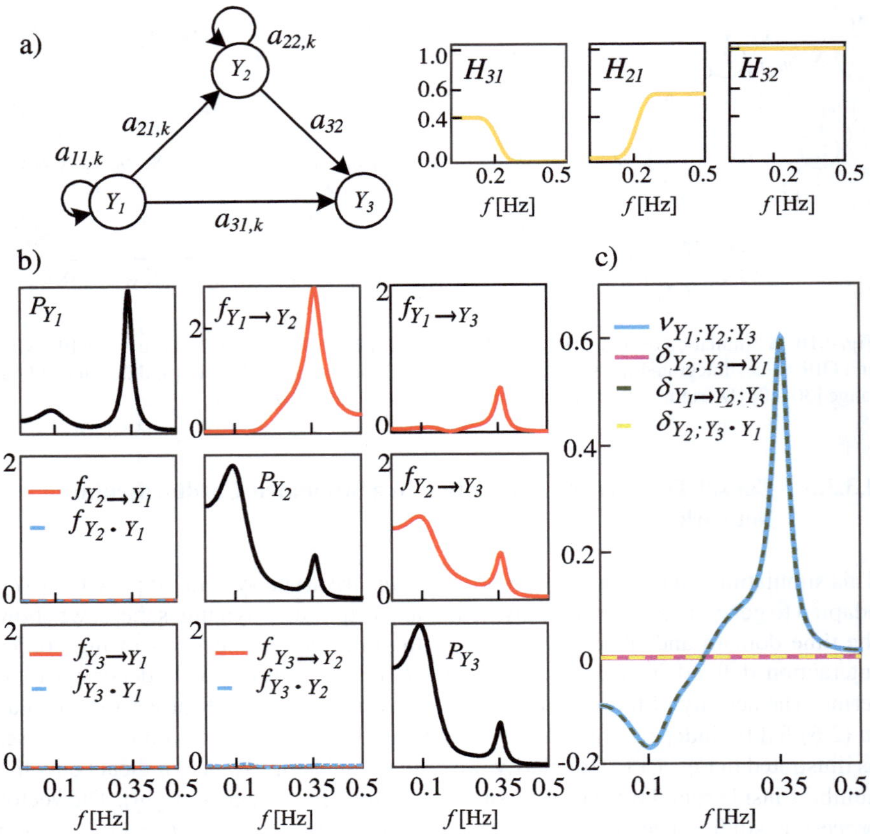

Fig. 4.11 **a** Connectivity structure of the simulated VAR process (left) and spectral TFs describing causality patterns (right); **b** PSD of the three processes (diagonal) and components of the causal decomposition of the spectral total coupling between each pair of processes (off-diagonal); **c** spectral profiles of the OIR of the three processes and of the components of its causal decomposition. The figure is adapted from [43]

information flowing back towards it ($f_{Y_3\to Y_2} = f_{Y_2\to Y_1} = f_{Y_3\to Y_1} = 0$ at each frequency). Note that, given the unidirectional coupling and the absence of instantaneous interactions, in virtue of (4.61) the three nonzero causal coupling measures are equivalent to the spectral measures of total coupling $f_{Y_1;Y_2}$, $f_{Y_1;Y_3}$ and $f_{Y_2;Y_3}$ (red curves in Fig. 4.11b); whole-band integration of such measures leads to the MIR quantifying the total information shared between pairs of processes, whose values result $I_{Y_1;Y_2} = T_{Y_1\to Y_2} = 0.28$ nats, $I_{Y_1;Y_3} = T_{Y_1\to Y_3} = 0.05$ nats and $I_{Y_2;Y_3} = T_{Y_2\to Y_3} = 0.24$ nats. Then, computation of the MIR between one process and the remaining two leads to obtain the OIR via (4.52), which for this simulation is $\Omega_{Y_1;Y_2;Y_3} = 0.019$ nats, denoting a small redundant interaction among the three processes. Importantly, the spectral expansion (Fig. 4.11c) reveals that this small OIR value is the balance between a synergistic interaction at low frequencies ($\Omega_{Y_1;Y_2;Y_3} = -0.15$ nats in the band 0.04–0.12 Hz) and a redundant interaction at higher frequencies ($\Omega_{Y_1;Y_2;Y_3} = +0.33$ nats in the band 0.31–0.39 Hz). We also highlight that the causal decomposition of the OIR $\nu_{Y_1;Y_2;Y_3} = \delta_{Y_1;Y_2;Y_3}$ reveals the unidirectional nature of the OIR increment (i.e., $\delta_{Y_1;Y_2;Y_3} = \delta_{Y_1 \to Y_2,Y_3}$ and $\delta_{Y_2,Y_3 \to Y_1} = \delta_{Y_1 \cdot Y_2,Y_3} = 0$). The opposite OIR values observed in the two frequency bands can be explained by the simulation design (see Fig. 4.11a): synergy and redundancy arise respectively because the flow of information from Y_1 to Y_3 is entirely mediated by Y_2 at the respiratory frequency (the path $Y_1 \to Y_3$ is blocked by $H_{Y_3 Y_1}$ at ~0.35 Hz), and because such flow occurs via the independent paths $Y_1 \to Y_3$ and $Y_2 \to Y_3$ at lower frequencies (the path $Y_1 \to Y_2$ is blocked by $H_{Y_2 Y_1}$ at ~0.1 Hz).

4.3.3.4 High-Order Links in a Simulated Dynamic System

In this section, we propose a simulated example taken from [53], where we characterize the B-index rate of a network comprising 6 nodes, whose activity is mapped by a multivariate random process defined by the VAR model (2.6) implemented with $\Sigma^{(var)}_{\mathbf{U_Y}} = \mathbf{I}$, $p = 2$, and time lagged coefficients reported in Fig. 4.12a, b (for brevity, the superscript $^{(var)}$ is omitted, and the subscripts of the model coefficients are numbers only). These settings produce two different connectivity structures, both characterized by two highly connected nodes (hubs, processes Y_1 and Y_6) interacting with four nodes with lower connectivity degree (leaves, processes $Y_2 - Y_5$). The hub Y_1 acts as a source sending information to the leaves; the hub Y_6 acts either as a source or as a sink, respectively sending information to the leaves which behave as receivers (Fig. 4.12a), or receiving information from the leaves which behave as mediators (Fig. 4.12b).

Figure 4.12c, d reports, for different process realizations obtained as multivariate time series with $L = 1000$ samples, the matrix of the B-index rates computed for Gaussian processes as detailed in Sect. 4.3.1, where non-significant values of the estimated MIR and cMIR leading to $B = 1$, $B = -1$ or $B = \text{NaN}$ were assessed by using surrogates, as explained with fine details in LINK- SPECIFIC ANALYSIS OF HOIS, and the reconstructed network structure obtained by pruning the links with non-significant MIR and/or cMIR. The analyzed realizations reproduce the typical structures of hub-leaves interactions which are described and analyzed in the following. A single star structure (right-positioned) with the hub Y_6 connected to the leaves $Y_2 - Y_5$ is obtained when $a_{Y_1 Y_i} = 0, i = 2, \ldots, 5$; the hub acts as a source in Fig. 4.12c and as a sink in Fig. 4.12d. These two configurations are correctly detected in terms of structure (Y_1 is isolated and Y_6 is connected to all leaves), and are differentiated by the clear redundancy (red) and synergy (blue) evidenced by the B-index rate in the two cases. The configuration with hub sending information to the leaves, which results as fully redundant due to the dominance of common drive effects, is reproduced also when $a_{Y_i Y_6} = 0$ ($i = 2, \ldots, 5$, left-positioned star), as shown in Fig. 4.12e, f where Y_1 is the hub and Y_6 is isolated.

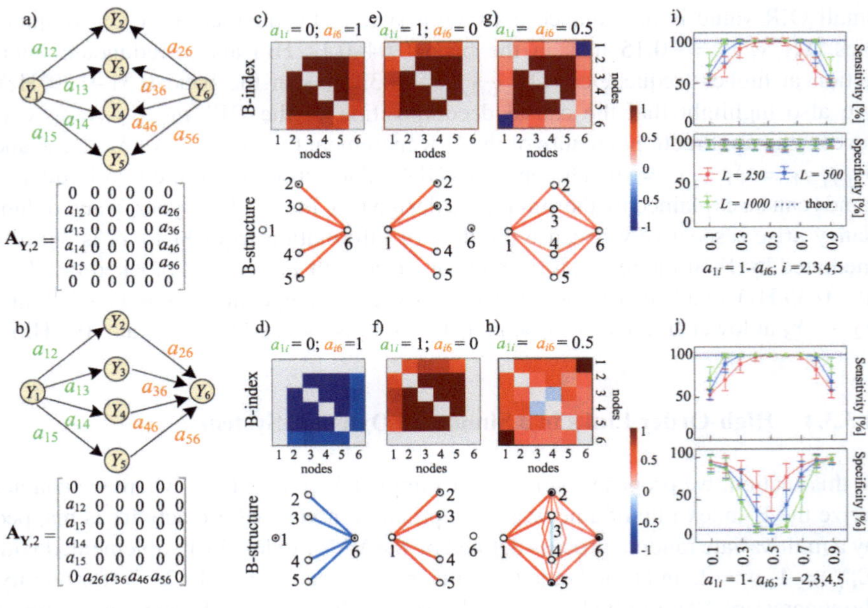

Fig. 4.12 **a, b** Simulated systems with six units mapped by Gaussian processes. The matrix of B-index rate and the corresponding reconstructed structures are investigated for the parameter settings: **c, e, f** single star structures where the hub sends information to four leaves; **d** a single star structure where the hub receives information from the leaves; **g** two competing star structures where both hubs send information to the same leaves; **h** two interacting star structures with information propagated from one hub to the other via mediation of the leaves. **i, j** Performance of network reconstruction. The figure is adapted from [53]

The most rich configurations arise when the coefficients defining the structures in Fig. 4.12a, b are all set to nonzero values. Specifically, when we set $a_{Y_1Y_i} = a_{Y_iY_6} = 0.5$ in Fig. 4.12a, a configuration with two competing star structures where two hubs send information to the same leaves is obtained; in this case, all links are redundant due to the dominance of common-drive effects (except the link between the two hubs which is fully synergistic due to a common cause effects towards all leaves), and the network is reconstructed without errors (Fig. 4.12g). On the other hand, setting $a_{Y_1Y_i} = a_{Y_iY_6} = 0.5$ in Fig. 4.12b, we obtain a configuration with propagation between two stars where one hub sends information to the other through mediation of the leaves; in this case, the links are again dominantly redundant due to the abundance of common drive and cascade effects, but the network reconstruction suffers from false-positive detections of links between the leaves caused by the simultaneous presence of common drive and common cause effects determining significant MIR and cMIR.

Finally, the accuracy of network reconstruction was investigated for different time series length L at increasing the weight, in the network structures of Fig. 4.12a, b, of the connections from the hub Y_1 to the leaves $Y_2 - Y_5$ ($a_{Y_1Y_i} \in [0.1, 0.9]$), while simultaneously decreasing the weight of the connections between the leaves and the hub Y_6 ($a_{Y_iY_6} = 1 - a_{Y_1Y_i}$), so as to gradually move from single-star to two-star structures and back. The results reported in Fig. 4.12i, j show that the sensitivity depends on the data length and on the link strength, reaching the expected value of 100% when the links are balanced in strength. The specificity was very high and substantially unaffected by the data length in the case of competing stars (Fig. 4.12i), while it was higher than expected for unbalanced link strength and approaching the expected value of 14.3% only for balanced link strength and high L (Fig. 4.12j). These results confirm the good performance of the B-index in reconstructing the statistical structure underlying directional networks, with the limitations related to its inability to resolve conditions of contemporaneous common drive and common cause effects impinging on a link.

4.3.3.5 Partial Information Rate Decomposition: Effects of Temporal Correlations

In this theoretical example, taken from [55], we characterize the PIRD in both the time and frequency domains in simulated dynamic networks involving three processes $\mathbf{Z} = \{Y, X_1, X_2\}$, comparing the time-domain measures with the standard PID decomposing instantaneous interactions among the processes. To do this we use a three-variate VAR process simulated with $f_s = 1$, in which different regimes of dynamic interaction are set by varying the parameters related to zero-lag effects, lagged interactions, and autonomous dynamics. Specifically, the 3-VAR process is defined as:

$$\begin{aligned} Y_n &= cX_{1,n-1} + cX_{2,n-2} + U_{Y,n} \\ X_{1,n} &= \sum_{k=1}^{2} a_{1,k}X_{1,n-k} + U_{X_1,n} \\ X_{2,n} &= \sum_{k=1}^{4} a_{2,k}X_{2,n-k} + U_{X_2,n} \end{aligned} \tag{4.73}$$

where U_Y, U_{X_1} and U_{X_2} are Gaussian white noises with zero mean and unit variance. The covariance matrix of the residuals,

$$\Sigma_{\mathbf{U_Z}} = \begin{pmatrix} 1 & \sigma^2_{U_{YX_1}} & \sigma^2_{U_{YX_2}} \\ \sigma^2_{U_{X_1Y}} & 1 & \sigma^2_{U_{X_1X_2}} \\ \sigma^2_{U_{X_2Y}} & \sigma^2_{U_{X_2X_1}} & 1 \end{pmatrix},$$

is built in such a way to generate zero-lag cross-correlations among the processes modulated inversely by the parameter c, imposing $\sigma^2_{U_{YX_1}} = \sigma^2_{U_{YX_2}} = \sigma^2_{U_{X_1X_2}} = 0.8 - c$. The autonomous oscillations in the two source processes X_1 and X_2 are obtained placing pairs of complex-conjugate poles, with modulus ρ and phase $2\pi f$, in the complex plane representation of each process; the AR coefficients resulting from this setting at lags $1, 2$ are $a_1 = 2\rho\cos(2\pi f)$ and $a_2 = -\rho^2$ [8]. Here, we place a pair of poles for the process X_1, setting $\rho = c$, $f = 0.1$ Hz so that the strength of the autonomous dynamics determined by the coefficients $a_{1,1}$ and $a_{1,2}$ depends on the parameter c; similarly, we place two pairs of poles for X_2, setting $\rho_1 = c$, $f_1 = 0.1$ Hz and $\rho_2 = 1.125c$, $f_2 = 0.3$ Hz so that the strength of the autonomous dynamics of X_2 determined by the coefficients $a_{2,k}$, $k = 1, \ldots, 4$, depends on the parameter c. Moreover, causal interactions are set from X_1 to Y at lag $k = 1$ and from X_2 to Y at lag $k = 2$, with strength modulated by the parameter c. The parameter c is varied in the range [0–0.8], thus allowing (i) progressive strengthening of the autonomous dynamics in the processes X_1, X_2 and of the causal interaction from X_1 and X_2 to Y, as well as (ii) progressive weakening of the zero-lag interactions among the three processes. The simulation design is shown in Fig. 4.14a (time-lagged interactions are depicted as solid black arrows; zero-lag interactions are depicted as dashed grey lines).

The PIRD was performed computing the spectral redundancy rate according to (4.68), with the spectral MIR function computed as in (4.71) after deriving the PSD matrix from the VAR parameters, and then obtaining the spectral PI rate through (4.66) which—in this case with $M = 2$ sources—yields immediately the coarse-grained terms in (4.67); all these terms were then integrated over the full frequency axis to obtain the unique $\mathcal{U}_{Y;X_1}$, $\mathcal{U}_{Y;X_2}$, redundant $\mathcal{R}_{Y;X_1;X_2}$ and synergistic $\mathcal{S}_{Y;X_1;X_2}$ information rates in the time domain. An example of how spectral redundancy is computed as the minimum MIR at each frequency f is shown in Fig. 4.13 for one specific VAR configuration ($c = 0.5$). Figure 4.14 reports the spectral MIR and redundancy rate functions, as well as the time-domain values of the PIRD terms, investigated at varying the simulation parameter c. In Fig. 4.14f, the time-domain PIRD is compared with the instantaneous PID which decomposes the MI $I(Y_n; X_{1,n}, X_{2,n})$; the latter was computed following the linear parametric formulation sketched in Sect. 3.1.1 with $\mathbf{V}_i = Y_n$ and $\mathbf{V}_j = \left[X_{1,n} X_{2,n}\right]$, after deriving the zero-lag covariance of the processes via solution of the Yule Walker equations of the VAR process (see Sect. 2.3.5.1).

The results suggest that the rate of dynamic information shared by multivariate processes is deeply affected by the balance between instantaneous and time-lagged interactions. When the three processes interact only at lag zero and do not exhibit self-dependencies ($c = 0$), the spectral profiles of the MIR and redundancy rate measures are flat (Fig. 4.14b–e, black lines) and the time-domain PIRD and zero-lg PID measures coincide (Fig. 4.14f with $c = 0$). Increasing the parameter c determines modifications of the spectral profiles of the MIR functions (Fig. 4.14b–d), due to the emergence of spectral peaks around 0.1 and 0.3 Hz induced by the self-dependencies rising in X_1 and X_2, as well as of causal interactions along the directions $X_1 \to Y$ and $X_2 \to Y$. As a result, the profile of the spectral redundancy rate is also modulated by c (Fig. 4.14e), as are the time-domain MIR and PIRD terms (Fig. 4.14f, dashed lines). The modification of the VAR parameters affects also the zero-lag PID (Fig. 4.14f, solid lines), whose information atoms were modified in a substantially different way than those of the PIRD. In fact, both redundant and synergistic contributions com-

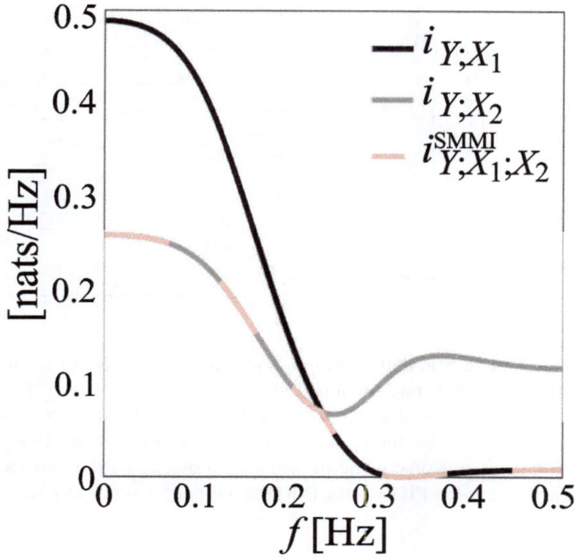

Fig. 4.13 Example of how the spectral redundancy rate function is computed as the minimum of the interaction between each individual source and the target at the specific frequency $\bar{f}$. The figure is adapted from [55]

puted via PID ($\mathcal{R}$ and $\mathcal{S}$, solid lines in Fig. 4.14f) decrease towards zero at increasing c, while we rather expect an increase of synergy due to the emerging common child structure of the simulated system (Fig. 4.14a). This effect is well evidenced by the PIRD, whose synergistic contribution increases with c becoming far higher than the redundant contribution ($\mathcal{R}$ and $\mathcal{S}$, dashed lines in Fig. 4.14f). Moreover, the PIRD detects a rise of both the unique information rates relevant to the two sources ($\mathcal{U}_1$ and $\mathcal{U}_2$, dashed lines in Fig. 4.14f), which is expected due to the emergence of causal interactions along the directions $X_1 \rightarrow Y$ and $X_2 \rightarrow Y$ with strength modulated by c. However, the same is not true if the unique information is measured by the zero-lag PID exploiting the time domain definition of redundancy (3.23) based on the MMI-PID, confirming a known limitation of such PID which always forces to zero the unique information of the source sharing the lowest information with the target (in this case $\mathcal{U}_2$, dashed line in Fig. 4.14f).

4.3.3.6 Frequency-Specific Coarse-Grained Partial Information Rate Decomposition

In this section, we provide an illustrative example of the full-frequency and band-limited PIRD applied to a network of four nodes where different high-order behaviors emerge at different frequencies. The example is adapted from [55]. We consider a VAR process simulated with $f_s = 1$, where the lagged interactions between the target Y and the three sources $\mathbf{X} = [X_1, X_2, X_3]$ are set to induce a network topology with both common drive and common child structures (Fig. 4.15a). The VAR process is defined as:

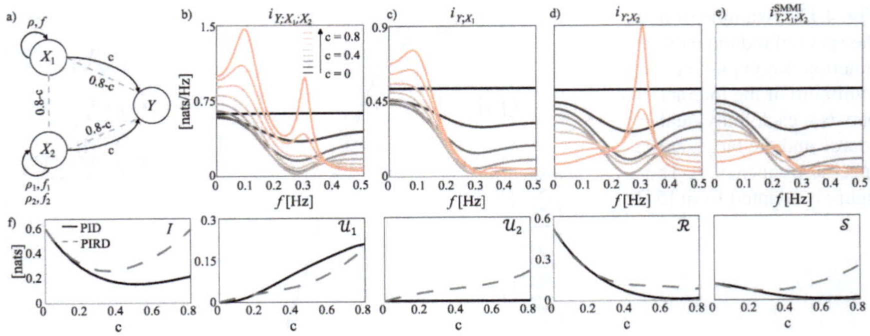

Fig. 4.14 **a** Simulation design. **b–e)** Spectral profiles of the joint MIR, the individual MIRs, and the redundancy rate, obtained at varying the parameter c from 0 (black) to 0.8 (pink). **f** From left to right: time-domain behavior of the total information shared between the target and the two sources, of the unique information shared between the target and each individual source, and of the redundant and synergistic information provided by the two sources to the target, measured using the the zero-lag PID and the PIRD as a function of the parameter c. The figure is adapted from [55]

$$
\begin{aligned}
Y_n &= X_{1,n-1} + X_{3,n-1} + U_{Y,n} \\
X_{1,n} &= \sum_{k=1}^{2} a_{1,k} X_{1,n-k} + U_{X_1,n} \\
X_{2,n} &= \sum_{k=1}^{2} a_{2,k} X_{2,n-k} + X_{1,n-1} + U_{X_2,n} \\
X_{3,n} &= \sum_{k=1}^{2} a_{3,k} X_{3,n-k} + U_{X_3,n}
\end{aligned}
\tag{4.74}
$$

where $\mathbf{U} = \left[U_Y, U_{X_1}, U_{X_2}, U_{X_3}\right]$ is a vector of four zero-mean independent Gaussian white noises with unit variance. The autonomous oscillations in the three source processes X_1, X_2, X_3 are obtained placing a pair of complex-conjugate poles, with modulus ρ and phase $2\pi f$, in the complex plane representation of each process. Here, we set $\rho = 0.8$, $f = 0.3$ Hz for the processes X_1 and X_2, so that their autonomous dynamics are determined by the coefficients $a_{1,1} = a_{2,1} = -0.494$; $a_{1,2} = a_{2,2} = -0.64$; similarly, we set $\rho = 0.9$, $f = 0.1$ Hz for X_3, so that its autonomous dynamics are determined by the coefficients $a_{3,1} = 1.456$; $a_{3,2} = -0.81$. Moreover, causal interactions are set from X_1 and X_3 to Y and from X_1 to X_2 at lag $k = 1$, with unitary strength.

The PIRD was applied computing the spectral MIRs between each individual source and the target ($i_{Y;X_i}(\bar{f}), i = 1, 2, 3$; Fig. 4.15b), as well as between groups of sources and the target (e.g., $i_{Y;X_1,X_2}(\bar{f})$); the spectral redundancy function was then computed as in (4.68), and exploited to retrieve the unique ($\mathcal{U}_{Y;X_1}, \mathcal{U}_{Y;X_2}, \mathcal{U}_{Y;X_3}$), redundant ($\mathcal{R}_{Y;\mathbf{X}}$) and synergistic ($\mathcal{S}_{Y;\mathbf{X}}$) information rates in the time domain; the

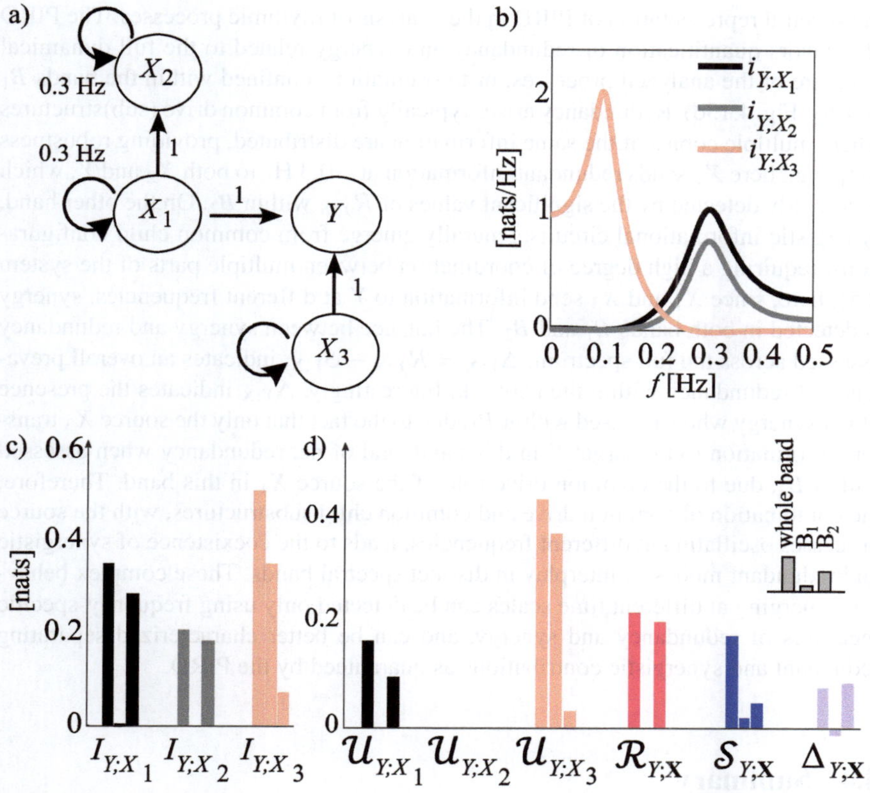

Fig. 4.15 **a** Network structure. **b** Spectral MIRs. **c, d** Time-domain values of the pairwise MIRs and of the unique, redundant and synergistic PIRD components integrated along the whole frequency axis (left), and within the bands B_1 (middle), B_2 (right); the balance $\Delta_{Y;\mathbf{X}} = \mathcal{R}_{Y;\mathbf{X}} - \mathcal{S}_{Y;\mathbf{X}}$ is also reported. The figure is adapted from [55]

latter were obtained as the whole-band integral of the correspondent spectral functions, as well as the integral taken along two spectral bands centered around the simulated stochastic oscillations (i.e., $B_1 = [0.04–0.15]$ Hz and $B_2 = [0.15–0.4]$ Hz).

The resulting time-domain values of the individual MIR terms and of the PIRD components are shown in Fig. 4.15c, d. The comparison highlights how, contrary to the pairwise MIR, the coarse-grained PIRD allows to disentangle the underlying network structure. Indeed, whilst non-zero MIR values are detected between Y and X_2 (Fig. 4.15c), the unique contribution of X_2 to Y was null, and non-zero unique contributions are correctly identified only in the presence of direct links (i.e., $\mathcal{U}_{Y;X_1}$, $\mathcal{U}_{Y;X_3}$). Remarkably, such non-zero unique contributions are mainly visible when assessed within the frequency bands for which oscillatory components are imposed (i.e., B_2 for $\mathcal{U}_{Y;X_1}$ and B_1 for $\mathcal{U}_{Y;X_3}$), thus confirming the important role played by the spectral representation of PIRD in the analysis of rhythmic processes. The PIRD also favors quantification of redundancy and synergy related to the full dynamical structure of the analyzed processes, or to oscillations confined within the bands B_1 and B_2 (Fig. 4.15d). Redundancy arises typically from common drive (sub)structures where multiple copies of the same information are distributed, providing robustness [67]; e.g., here X_1 sends redundant information at ~0.3 Hz to both X_2 and Y, which is correctly detected by the significant values of $\mathcal{R}_{Y;\mathbf{X}}$ within B_2. On the other hand, synergistic informational circuits generally emerge from common child configurations, requiring a high degree of coordination between multiple parts of the system [65]; here, since X_1 and X_3 send information to Y at different frequencies, synergy is detected in both bands B_1 and B_2. The balance between synergy and redundancy assessed across the full spectrum, $\Delta_{Y;\mathbf{X}} = \mathcal{R}_{Y;\mathbf{X}} - \mathcal{S}_{Y;\mathbf{X}}$, indicates an overall prevalence of redundancy within the network. Interestingly, $\Delta_{Y;\mathbf{X}}$ indicates the presence of net synergy when assessed within B_1, due to the fact that only the source X_3 transfers information to the target Y in this band, and of net redundancy when assessed within B_2, due to the common drive role of the source X_1 in this band. Therefore, the combination of common drive and common child substructures, with the source processes oscillating at different frequencies, leads to the coexistence of synergistic and redundant modes of interplay in distinct spectral bands. These complex behaviors emerging at different time scales can be detected only using frequency-specific measures of redundancy and synergy, and can be better characterized separating redundant and synergistic contributions as guaranteed by the PIRD.

4.4 Summary

This chapter introduced a framework to measure dynamic interactions of different orders in networks of multiple interconnected systems, expanded in the frequency domain to explore their spectral patterns. The framework has been defined for dynamic systems, whose activity is described by dynamic information measures which extend to random processes the measures defined in Chap. 3 for random variables. Its linear parametric formulation is made possible under the assumption of stationary jointly Gaussian processes.

The dynamic network system $\mathcal{X}$ composed of M nodes can be characterized in terms of the activities involving the single nodes, pairs of nodes, and groups of nodes. These activities are assessed straightforwardly using univariate, bivariate and high-order measures of dynamic interactions.

The first group of measures includes the *entropy rate* (ER) H_{X_i}, quantifying the rate of generation of new information in the random process X_i, and the information storage (IS) S_{X_i}, which is a measure of self-predictability quantifying the information stored in X_i as the MI between the present $X_{i,n}$ and the past $X_{i,n}^-$. Spectral measures of ER and IS, the latter referred to as spectral linear self-predictability (LSP), can be retrieved exploiting the power spectral densities of autoregressive models and are particularly useful to identify the oscillatory patterns of random processes in the frequency domain, as also demonstrated by the proposed simulated example in Sect. 4.1.3. The non-directional (symmetric) or directional (causal) interactions between pairs of processes, as well as the autonomous self-dependencies of one process given another, can be explored through pairwise information-theoretic measures with spectral meaning. Specifically, the *mutual information rate* (MIR) $I_{X_i;X_j}$ shared between the processes X_i and X_j is a measure of non-directional coupling quantifying the information shared by the two processes per unit of time. The MIR can be decomposed into causal terms, namely the *transfer entropy* (TE) $T_{X_i \rightarrow X_j}$, a measure of directional information transfer, and the *instantaneous transfer* (IT) $I_{X_i \cdot X_j}$, a non-directional measure of zero-lag interaction. Remarkably, under the assumption of Gaussianity, the MIR, TE and IT measures are equivalent to the Geweke measures of total dependence, Granger causality (GC) and instantaneous causality, respectively. Further, in parallel to exploring the concept of GC, we formalize the analysis of coupled processes to investigate also the role of autonomous dynamics, i.e., interactions that occur internally in a process, say X_j, independently of its link with other processes. In this context, the Granger autonomy (GA) measure A_{X_j} has been defined to quantify how much the internal dynamics of one process contribute to determine its predictability more than the dynamics of other processes potentially connected to it. Spectral measures of coupling, causality and autonomy can be retrieved exploiting the power spectral densities of full ARX and restricted AR/X models and are essential to identify the oscillatory patterns of coupled interactions and self-dependencies along given spectral bands of interest.

While the standard network description of complex systems is based on quantifying the link between pairs of system units, high-order interactions involving three or more units often play a major role in governing the collective network behavior. Inside our framework of information dynamics, we propose different measures to assess HOI patterns at increasing orders, i.e., interactions emerging from several network nodes at different levels of resolution, in both time and frequency domains. Network- and node-specific analyses of dynamic HOIs can be performed through the measures of *O-information rate* (OIR) $\Omega_{\mathbf{X}^N}$ of the multiplet $\mathbf{X}^N = \{X_1, \ldots, X_N\}$ $(N \leq M)$, and the first-order *gradient of OIR* $\Delta_{X_i;\mathbf{X}^N_{-i}}$, expanding to random processes the concepts of OI and OI gradient defined for random variables. Remarkably, the gradient of OIR can be decomposed into causal and instantaneous contributions arising from the MIR decomposition. The link-wise measure of the *B-index rate* $B_{X_i;X_j}$ highlights the balance between redundancy and synergy in the dynamic interaction among the two observed units X_i and X_j and the rest of the system, and allows to retrieve information about the network topology from such interaction. Here, it is worth mentioning the striking role of our new partial information rate decomposition (PIRD) applied to random processes and following the steps of the well-known PID for random variables. Its strength lies in providing an approach which separately evaluates redundant and synergistic patterns of interactions among random processes in both the time and the frequency domain. Indeed, the OIR and its decomposition into causal and non-causal terms do not put in evidence multiplets of variables which are both redundant and synergistic with equal strength, while pairwise and fully conditioned causality fail to reveal respectively synergistic and redundant effects [62].

All the high-order information-theoretic measures can be straightforwardly expanded in the frequency domain by exploiting the (restricted) VAR representation of the considered group of nodes, thus enabling the identification of redundant and synergistic patterns of interactions at multiple levels of resolution. Specifically, high-order spectral measures can be obtained from the power spectral density matrices of the (restricted) VAR processes; they satisfy the spectral integration property thus allowing to draw a connection with their time domain counterparts and warrant a focus on behaviors that can remain hidden in the time domain, e.g., the coexistence of redundancy and synergy in different frequency bands of the spectrum. Finally, the spectral representation allows to define a new concept of redundancy via the PIRD valid for groups of random processes.

References

1. Thomas M Cover. *Elements of information theory*. John Wiley & Sons, 1999.
2. Daniel Chicharro. "On the spectral formulation of Granger causality". In: *Biological cybernetics* 105 (2011), pp. 331–347.
3. Laura Sparacino et al. "Measuring hierarchically-organized interactions in dynamic networks through spectral entropy rates: theory, estimation, and illustrative application to physiological networks". In: *Neurocomputing* 630 (2025), p. 129675.
4. Adam B Barrett, Lionel Barnett, and Anil K Seth. "Multivariate Granger causality and generalized variance". In: *Physical Review E* 81.4 (2010), p. 041907.
5. Luca Faes et al. "Estimating the decomposition of predictive information in multivariate systems". In: *Physical Review E* 91.3 (2015), p. 032904.
6. Luca Faes et al. "Information decomposition in multivariate systems: definitions, implementation and application to cardiovascular networks". In: *Entropy* 19.1 (2016), p. 5.
7. Lionel Barnett, Adam B Barrett, and Anil K Seth. "Granger causality and transfer entropy are equivalent for Gaussian variables". In: *Physical review letters* 103.23 (2009), p. 238701.
8. Luca Faes, Alberto Porta, and Giandomenico Nollo. "Information decomposition in bivariate systems: theory and application to cardiorespiratory dynamics". In: *Entropy* 17.1 (2015), pp. 277–303.
9. John Geweke. "Measurement of linear dependence and feedback between multiple time series". In: *Journal of the American statistical association* 77.378 (1982), pp. 304–313.
10. Laura Sparacino et al. "Spectral analysis of the beat-to-beat variability of arterial compliance". In: *2022 12th Conference of the European Study Group on Cardiovascular Oscillations (ESGCO)*. IEEE. 2022, pp. 1–2.
11. Laura Sparacino et al. "A method to assess linear self-predictability of physiologic processes in the frequency domain: application to beat-to-beat variability of arterial compliance". In: *Frontiers in Network Physiology* 4 (2024), p. 1346424.
12. Giuseppe Baselli et al. "Spectral decomposition in multichannel recordings based on multivariate parametric identification". In: *IEEE transactions on biomedical engineering* 44.11 (1997), pp. 1092–1101.
13. Riccardo Pernice et al. "Comparison of frequency domain measures based on spectral decomposition for spontaneous baroreflex sensitivity assessment after Acute Myocardial Infarction". In: *Biomedical Signal Processing and Control* 68 (2021), p. 102680.
14. Jana Krohova et al. "Vascular resistance arm of the baroreflex: methodology and comparison with the cardiac chronotropic arm". In: *Journal of Applied Physiology* 128.5 (2020), pp. 1310–1320.

15. Alberto Porta et al. “Measuring regularity by means of a corrected conditional entropy in sympathetic outflow”. In: *Biological cybernetics* 78 (1998), pp. 71–78.
16. Alberto Porta et al. “Complexity and nonlinearity in short-term heart period variability: comparison of methods based on local nonlinear prediction”. In: *IEEE Transactions on Biomedical Engineering* 54.1 (2006), pp. 94–106.
17. Silvia Erla et al. “k-Nearest neighbour local linear prediction of scalp EEG activity during intermittent photic stimulation”. In: *Medical engineering & physics* 33.4 (2011), pp. 504–512.
18. Joshua S Richman and J Randall Moorman. “Physiological time-series analysis using approximate entropy and sample entropy”. In: *American journal of physiology-heart and circulatory physiology* (2000).
19. Joseph T Lizier, Mikhail Prokopenko, and Albert Y Zomaya. “Local measures of information storage in complex distributed computation”. In: *Information Sciences* 208 (2012), pp. 39–54.
20. Luca Faes et al. “Investigating the mechanisms of cardiovascular and cerebrovascular regulation in orthostatic syncope through an information decomposition strategy”. In: *Autonomic Neuroscience* 178.1-2 (2013), pp. 76–82.
21. Michael Wibral et al. “Local active information storage as a tool to understand distributed neural information processing”. In: *Frontiers in neuroinformatics* 8 (2014), p. 1.
22. Chiara Barà et al. “Local and global measures of information storage for the assessment of heartbeat-evoked cortical responses”. In: *Biomedical Signal Processing and Control* 86 (2023), p. 105315.
23. Yuo A Rozanov. *Stationary stochastic processes*. 1967.
24. Luca Faes, Silvia Erla, and Giandomenico Nollo. “Measuring connectivity in linear multivariate processes: definitions, interpretation, and practical analysis”. In: *Computational and mathematical methods in medicine* 2012 (2012).
25. Luca Faes and Giandomenico Nollo. “Measuring frequency domain Granger causality for multiple blocks of interacting time series”. In: *Biological cybernetics* 107.2 (2013), pp. 217–232.
26. Norbert Wiener. “The theory of prediction”. In: *Modern mathematics for engineers* (1956).
27. Clive WJ Granger. “Investigating causal relations by econometric models and cross-spectral methods”. In: *Econometrica: journal of the Econometric Society* (1969), pp. 424–438.
28. Alberto Porta and Luca Faes. “Wiener–Granger causality in network physiology with applications to cardiovascular control and neuroscience”. In: *Proceedings of the IEEE* 104.2 (2015), pp. 282–309.
29. Tyrone E Duncan. “On the calculation of mutual information”. In: *SIAM Journal on Applied Mathematics* 19.1 (1970), pp. 215–220.
30. Gorana Mijatovic et al. “An information-theoretic framework to measure the dynamic interaction between neural spike trains”. In: *IEEE Transactions on Biomedical Engineering* 68.12 (2021), pp. 3471–3481.
31. Chiara Barà et al. “Comparison of discretization strategies for the model-free information-theoretic assessment of short-term physiological interactions”. In: *Chaos: An Interdisciplinary Journal of Nonlinear Science* 33.3 (2023).
32. Thomas Schreiber. “Measuring information transfer”. In: *Physical review letters* 85.2 (2000), p. 461.
33. Laura Sparacino et al. “Statistical Approaches to Characterize Functional Connectivity in Brain and Physiologic Networks on a Single-Subject Basis”. In: *2023 45th Annual International Conference of the IEEE Engineering in Medicine & Biology Society (EMBC)*. IEEE. 2023, pp. 1–4.
34. Riccardo Pernice et al. “Spectral decomposition of cerebrovascular and cardiovascular interactions in patients prone to postural syncope and healthy controls”. In: *Autonomic Neuroscience* (2022), p. 103021.
35. Riccardo Pernice et al. “Assessment of Cardiorespiratory Interactions During Spontaneous and Controlled Breathing: Linear Parametric Analysis”. In: *2022 12th Conference of the European Study Group on Cardiovascular Oscillations (ESGCO)*. IEEE. 2022, pp. 1–2.

36. Gabriele Volpes et al. "A portable multisensor system to assess cardiorespiratory interactions through photoplethysmography". In: *2022 IEEE International Symposium on Medical Measurements and Applications (MeMeA)*. IEEE. 2022, pp. 1–6.
37. Anil K Seth. "Measuring autonomy and emergence via Granger causality". In: *Artificial life* 16.2 (2010), pp. 179–196.
38. Margaret A Boden. "The philosophy of artificial life". In: (1996).
39. Nils Bertschinger et al. "Autonomy: An information theoretic perspective". In: *Biosystems* 91.2 (2008), pp. 331–345.
40. Alberto Porta et al. "Conditional self-entropy and conditional joint transfer entropy in heart period variability during graded postural challenge". In: *PLoS One* 10.7 (2015), e0132851.
41. Laura Sparacino et al. "A method to assess granger causality, isolation and autonomy in the time and frequency domains: theory and application to cerebrovascular variability". In: *IEEE Transactions on Biomedical Engineering* (2023).
42. Aatira G Nedungadi, Mingzhou Ding, and Govindan Rangarajan. "Block coherence: a method for measuring the interdependence between two blocks of neurobiological time series". In: *Biological cybernetics* 104.3 (2011), pp. 197–207.
43. Luca Faes et al. "A new framework for the time-and frequency-domain assessment of high-order interactions in networks of random processes". In: *IEEE Transactions on Signal Processing* 70 (2022), pp. 5766–5777.
44. Mingzhou Ding, Yonghong Chen, and Steven L Bressler. "Granger causality: basic theory and application to neuroscience". In: *Handbook of time series analysis: recent theoretical developments and applications* (2006), pp. 437–460.
45. I Pirovano et al. "Rehabilitation modulates high-order interactions among large-scale brain networks in subacute stroke". In: *IEEE Transactions on Neural Systems and Rehabilitation Engineering* 31 (2023), pp. 4549–4560.
46. Steffen Schulz et al. "Cardiovascular and cardiorespiratory coupling analyses: a review". In: *Philosophical Transactions of the Royal Society A: Mathematical, Physical and Engineering Sciences* 371.1997 (2013), p. 20120191.
47. Michal Javorka et al. "Basic cardiovascular variability signals: mutual directed interactions explored in the information domain". In: *Physiological Measurement* 38.5 (2017), p. 877.
48. J Krohova et al. "Information domain analysis of respiratory sinus arrhythmia mechanisms". In: *Physiological Research* 67 (2018), S611–S618.
49. Simone Cencetti, Gabriele Bandinelli, and Alfonso Lagi. "Effect of PCO2 changes induced by head-upright tilt on transcranial Doppler recordings". In: *Stroke* 28.6 (1997), pp. 1195–1197.
50. Federico Battiston et al. "Networks beyond pairwise interactions: Structure and dynamics". In: *Physics reports* 874 (2020), pp. 1–92.
51. Sebastiano Stramaglia et al. "Expanding the transfer entropy to identify information circuits in complex systems". In: *Physical Review E* 86.6 (2012), p. 066211.
52. Alberto Porta et al. "Accounting for respiration is necessary to reliably infer Granger causality from cardiovascular variability series". In: *IEEE transactions on biomedical engineering* 59.3 (2011), pp. 832–841.
53. Gorana Mijatovic et al. "Assessing High-Order Links in Cardiovascular and Respiratory Networks via Static and Dynamic Information Measures". In: *IEEE Open Journal of Engineering in Medicine and Biology* (2024).
54. Gorana Mijatovic et al. "Network Representation of Higher-Order Interactions Based on Information Dynamics". In: *arXiv preprint* arXiv:2408.15617 (2024).
55. Laura Sparacino et al. "Decomposing Multivariate Information Rates in Networks of Random Processes". In: *arXiv preprint* arXiv:2502.04555 (2025). URL: https://arxiv.org/abs/2502.04555.
56. Luca Faes et al. "Partial Information Rate Decomposition". In: *arXiv preprint* arXiv:2502.04550 (2025).
57. Luca Faes et al. "Information decomposition in the frequency domain: A new framework to study cardiovascular and cardiorespiratory oscillations". In: *Philosophical Transactions of the Royal Society A* 379.2212 (2021), p. 20200250.

58. Luca Faes et al. "Quantifying High-Order Interactions in Cardiovascular and Cerebrovascular Networks". In: *2022 12th Conference of the European Study Group on Cardiovascular Oscillations (ESGCO)*. IEEE. 2022, pp. 1–2.
59. Laura Sparacino et al. "Quantifying High-Order Interactions in Complex Physiological Networks: A Frequency-Specific Approach". In: *International Conference on Complex Networks and Their Applications*. Springer. 2022, pp. 301–309.
60. Vasily A Vakorin, Olga A Krakovska, and Anthony R McIntosh. "Confounding effects of indirect connections on causality estimation". In: *Journal of neuroscience methods* 184.1 (2009), pp. 152–160.
61. John F Geweke. "Measures of conditional linear dependence and feedback between time series". In: *Journal of the American Statistical Association* 79.388 (1984), pp. 907–915.
62. Sebastiano Stramaglia et al. "Disentangling high-order effects in the transfer entropy". In: *Physical Review Research* 6.3 (2024), p. L032007.
63. Jim W Kay, Jan M Schulz, and William A Phillips. "A comparison of partial information decompositions using data from real and simulated layer 5b pyramidal cells". In: *Entropy* 24.8 (2022), p. 1021.
64. Thomas F Varley et al. "Information-processing dynamics in neural networks of macaque cerebral cortex reflect cognitive state and behavior". In: *Proceedings of the National Academy of Sciences* 120.2 (2023), e2207677120.
65. Thomas F Varley et al. "Multivariate information theory uncovers synergistic subsystems of the human cerebral cortex". In: *Communications biology* 6.1 (2023), p. 451.
66. Jana Krohova et al. "Multiscale information decomposition dissects control mechanisms of heart rate variability at rest and during physiological stress". In: *Entropy* 21.5 (2019), p. 526.
67. Andrea I Luppi et al. "A synergistic workspace for human consciousness revealed by integrated information decomposition". In: *BioRxiv* (2020).
68. Andrea I Luppi et al. "A synergistic core for human brain evolution and cognition". In: *Nature Neuroscience* 25.6 (2022), pp. 771–782.
69. Adam B Barrett. "Exploration of synergistic and redundant information sharing in static and dynamical Gaussian systems". In: *Physical Review E* 91.5 (2015), p. 052802.
70. Paul L Williams and Randall D Beer. "Nonnegative decomposition of multivariate information". In: *arXiv preprint* arXiv:1004.2515 (2010).
71. Yuri Antonacci et al. "Measuring high-order interactions in rhythmic processes through multivariate spectral information decomposition". In: *IEEE Access* 9 (2021), pp. 149486–149505.
72. Robin AA Ince. "Measuring multivariate redundant information with pointwise common change in surprisal". In: *Entropy* 19.7 (2017), p. 318.
73. Abdullah Makkeh, Aaron J Gutknecht, and Michael Wibral. "Introducing a differentiable measure of pointwise shared information". In: *Physical Review E* 103.3 (2021), p. 032149.
74. Aaron J Gutknecht, Michael Wibral, and Abdullah Makkeh. "Bits and pieces: Understanding information decomposition from part-whole relationships and formal logic". In: *Proc. Roy. Soc. A* 477.2251 (2021), p. 20210110.
75. Fernando E Rosas et al. "Reconciling emergences: An information-theoretic approach to identify causal emergence in multivariate data". In: *PLoS computational biology* 16.12 (2020), e1008289.
76. Luca Faes, Daniele Marinazzo, and Sebastiano Stramaglia. "Multiscale information decomposition: Exact computation for multivariate Gaussian processes". In: *Entropy* 19.8 (2017), p. 408.
77. Yuri Antonacci et al. "Information transfer in linear multivariate processes assessed through penalized regression techniques: validation and application to physiological networks". In: *Entropy* 22.7 (2020), p. 732.

Chapter 5
Applications to Physiological Networks

In the field of *Network Physiology*, data-driven methods for network inference can be exploited to build network models from sets of time series describing the activity of the observed network nodes [1]. Straightforward examples involve the well-known cardiovascular interactions between the cardiac and vascular subsystems [2], as well as the tangled coupled dynamics between the cardiac and the respiratory subsystems [3]. Indeed, *cardiovascular* and *cardiorespiratory networks* reflecting the modulation of heart rate, vascular and respiratory variability [4] have been largely studied, with the aim of disentangling the most relevant mechanisms occurring in diverse physiological states and conditions.

It is well known that heart rate is one of the physiological parameters characterized by the highest variability in the resting state. *Heart rate variability* (HRV) varies with age and gender [5–7], and its lack (or depression) has been described as a marker of several pathological states, e.g., nervous system disorders [8], diabetes [9], arterial hypertension [10], and myocardial infarction [11]. The physiology behind the regulation of cardiac dynamics is complex, but most studies agree that the main components of the normal sinus rhythm are related to the control exerted by the *autonomic nervous system* (ANS) [6, 7, 12]. During ventilation, the activity of the sinus atrial node is directly influenced by the modulation of vagal neurons directed to the heart, controlled by the central respiratory drives (direct communication between respiratory and cardiomotor centers), the lung inflation reflex and the changes in arterial blood pressure (ABP) transferred to heart rate via *baroreflex* [6, 13–16]. These mechanisms result in the so-called *respiratory sinus arrhytmia* (RSA), for which there is an increase of heart rate during the inspiration phase and a decrease during the expiration phase of ventilation [12, 16, 17]. Recent studies documented the underexplored complexity of cardiorespiratory interactions, highlighting the important role exerted by synchronization mechanisms [12]. Besides the HRV, the resting-state *blood pressure variability* (BPV) is also known to be strongly influenced by a relevant non-baroreflex effect (commonly known as *feedforward*) led by

L. Sparacino, *Spectral Information Dynamics in Network Neuroscience and Physiology*, Studies in Computational Intelligence 1235,
https://doi.org/10.1007/978-3-032-05416-6_5

changes in heart rate, reasonably due to the Windkessel [18] and/or Frank–Starling [19] mechanisms.

Nevertheless, these networks are extremely intricate and an appropriate characterization of their dynamics would require the involvement of many variables: cardiovascular and cardiorespiratory loops would represent only a small portion of a more complex and wider system. While heart rate, ABP and respiration still remain the most studied [4], very little is known about, e.g., the short-term-variability nature of *arterial compliance* (AC), a cardiovascular variable characterizing mechanical and structural properties of the arteries [20] expected to be affected by the sympathetic and vagal activities in different patho-physiological conditions, as well as by heart rate, blood pressure and respiratory variabilities [20–22]. In addition, *cerebrovascular regulation* is fundamental in humans for the maintenance of suitable values of cerebral blood flow in spite of internal and external disturbances. The short-term cerebrovascular regulation is related to the mechanism of *cerebral autoregulation* (CA), which has been defined as the intrinsic ability of the brain to independently regulate and thus maintain almost constant the cerebral blood flow (CBF), via changes in cerebrovascular resistance, when mean arterial pressure (MAP) fluctuates [23, 24]. An extensive study of CA in humans has been carried out by Lassen [25], who illustrated the CA curve and suggested that CBF remains almost constant within a relatively broad MAP range ($\sim$60–150 mmHg). Moreover, the advent of transcranial doppler ultrasound technology allowed to achieve a temporal resolution sufficient to assess beat-to-beat changes in the CBF velocity (CBFV), and thus to investigate the dynamic properties of CA [26]. The functioning of the closed-loop dynamic interactions between the spontaneous variability of MAP and mean CBFV (MCBFV) is generally studied non-invasively in different patho-physiological states [26–28].

It is important to highlight that all these mechanisms are challenged by a number of stressors, e.g., *postural stress*, which induces a reorganization of cardiovascular and cerebrovascular oscillations and of their coupling related to the shift in the sympatho-vagal balance towards sympathetic activation and parasympathetic withdrawal, or *controlled breathing*, which instead may have a role in modifying causality patterns between, e.g., heart rate and arterial pressure (AP). Remarkably, the orthostatic challenge is known to alter important physiological mechanisms operating in the resting state condition; orthostasis has been associated to venous pooling of the blood in the lower portion of the body, thus decreasing cardiac filling, cardiac output and stroke volume [29] in turn determining a drop of ABP sensed by baroreceptors, vagal inhibition and sympathetic activation directed to the heart and vessels [30, 31]. Hence, probing the investigated network after its modification due to a given stressor is of remarkable importance to characterize the type and modalities of network adaptation. For instance, evoking an orthostatic stress, e.g., through *head-up tilt test*, represents a useful tool to study the baroreflex and the cerebrovascular autoregulation mechanisms, both in healthy conditions and in the presence of autonomic dysfunctions.

In this chapter, we will discuss the application of the static and dynamic measures of pairwise (e.g., the MIR, the GC and the GA) and high-order (e.g., the OI/OIR and the OI gradients) connectivity defined respectively in Chaps. 3 and 4 to a variety of physiological signals, starting from univariate analysis of beat-to-beat arterial compliance time series (Sect. 5.1) and then moving towards bivariate (Sects. 5.2–5.6) and high-order (Sects. 5.7–5.11) analyses of more complex physiological systems. Our aim is to elicit non-invasively the physiological mechanisms underlying complex cardiovascular, cardiorespiratory and cerebrovascular regulation from the (joint) analysis of the spontaneous variability of the main cardiovascular, cardiorespiratory and cerebrovascular parameters. In our works, we carried out diverse systematic analyses of the (causal) interactions in physiological networks comprising a different number of variables, i.e., arterial compliance, the well-known heart rate, arterial pressure and respiration, and the cerebral blood flow beat-to-beat time series. This has been done exploiting information-theoretic approaches with meaning in the frequency domain, since parametric spectral analysis of the investigated interactions is known to provide valuable insights into the behavior of physiological time series displaying a rich oscillatory content [32, 33], typically manifested within the low-frequency (LF, [0.04–0.15] Hz) and high-frequency (HF, [0.15–0.4] Hz) bands of the spectrum in the case of, e.g., cardiovascular and respiratory variables [34, 35].

5.1 Arterial Compliance Short-Term Variability Analysis

Arterial compliance is an important cardiovascular parameter characterizing mechanical and structural properties of arteries and significantly influencing ventricular-arterial coupling [36, 37]. It is defined by the slope of pressure-volume relationship, i.e., a change in the arterial volume in relation to a given change in arterial pressure. Since this relationship is nonlinear due to complex structure of arterial wall, the value of measured AC will be different when measured around different points of the nonlinear pressure-volume relationship [38]. Furthermore, AC is influenced also by arterial size, wall thickness and smooth muscle activity [38, 39]. Decreased arterial compliance (or increased arterial stiffness) is associated with several physiological states and pathological processes, and thus its estimation has for a long time interested clinicians and cardiovascular physiologists. However, the non-invasive estimation of compliance is still a challenging task. In [37], the most often used methods of AC estimation were reviewed and classified to better clarify their applicability, necessary requirements and limitations. Remarkably, further research studies from the same group [20] allowed to develop a methodology for the non-invasive beat-to-beat estimation of compliance during standardized protocol including resting phases, orthostatic stress and cognitive load. To solve this task, the authors used a recently developed method for continuous and non-invasive determination of the time constant τ, i.e., the rate of the ABP decay during diastolic phase, using only parameters which are known to be robust against distortions due to wave reflections arising when ABP curve cannot be measured centrally [20, 40]. Together with calculated total peripheral resistance (TPR) values from the ratio of mean arterial pressure (MAP) and cardiac output (CO), the latter obtained using impedance cardiography

(ICG), compliance values were computed as the ratio of τ and TPR for each heart beat.

We exploited the novelty of this estimation method to study for the first time short-term beat-to-beat variability time series of arterial compliance acquired in healthy subjects, with the aim of investigating the time and spectral behavior of this parameter in response to the postural stress. Specifically, in this section we report the application of the time (4.10) and frequency domain (4.9) measure of linear self-predictability on beat-to-beat time series of arterial compliance obtained in young healthy subjects [20], presented in [22].

Experimental Protocol and Data Acquisition. The original study, approved by the ethical committee of the *Jessenius Faculty of Medicine, Comenius University*, included a total of 81 young and healthy Caucasians, aged 18.56 ± 2.88 years. ABP signal from finger, obtained by the photoplethysmographic volume-clamp method followed by brachial ABP reconstruction (Finometer Pro, FMS Netherlands), and electrocardiogram (ECG, CardioFax ECG-9620, NihonKohden Japan) were recorded during two phases of the experimental protocol: *(i)* the resting supine position (REST), started 8 min after the beginning of the measurement, and *(ii)* the upright position reached after passive head-up tilt (TILT), started 3 min after the position change from supine to tilt. Heart period (HP) intervals were extracted as the time distance between consecutive R peaks of the ECG. Hemodynamics parameters including CO were derived on a beat-to-beat basis exploiting the ICG (CardioScreen 2000, Medis, Germany) and exerted a main role in the subsequent determination of the AC time series. The value of AC was quantified through a recently developed method [40], based on a reliable estimation of the time constant τ for each heart beat separately, as well as on the exploitation of the relationship between τ, AC and the TPR based on the two-element Windkessel model [20, 37, 40]. Since the measurement of hemodynamic parameters using ICG is very sensitive to movement artifacts, skin condition and distribution of fat, in some cases these parameters were then not determined for each heart beat, and then only 39 subjects were selected for further analysis. All the acquired signals were digitized at a sampling rate of 1 kHz. Transient changes in cardiovascular parameters between consecutive phases of the study protocol were excluded from analysis. Then, stationary segments of 300 consecutive beats were extracted from the original recordings in the two phases of the protocol. We remark that an approximate value of 300 heart beats conforms to the *short-term* HRV standard, which allows the assessment of autonomic tone under different patho-physiological conditions. We refer the reader to [20] for further details about data acquisition and time series extraction.

Data and Statistical Analysis. The time series extracted for each subject in the two experimental conditions were regarded as realizations of the AC discrete-time process (in the following, referred to as C), assumed as uniformly sampled with a sampling frequency equal to the inverse of the mean HP. First, classical time domain markers, i.e. the mean and variance of AC ($\mu_C \left[\frac{\mathrm{mL}}{\mathrm{mmHg}}\right]$ and $\sigma_C^2 \left[\frac{\mathrm{mL}^2}{\mathrm{mmHg}^2}\right]$) were computed.

Table 5.1 Time domain indexes (mean μ_C and variance σ_C^2) of AC in the REST and TILT experimental conditions

	REST	TILT
$\mu_C \left[\frac{\text{mL}}{\text{mmHg}}\right]$	1.76 ± 0.41	1.42 ± 0.28*
$\sigma_C^2 \left[\frac{\text{mL}^2}{\text{mmHg}^2}\right]$	0.022 ± 0.015	0.015 ± 0.008*

Values are computed over 39 subjects and expressed as mean ± standard deviation. *Wilcoxon signed rank test for paired data*: *, $p < 0.05$ REST versus TILT

Then, the series were pre-processed by removing the mean value. An AR model in the form of (2.2) was fitted on each pre-processed series using OLS identification and setting the model order p according to the AIC (maximum scanned model order equal to 14). Since the use of the AIC sometimes led to duplicate peaks or negative power as a result of spectral decomposition [41] (Sect. 4.1.2.2), the model order was manually adjusted so as to detect spectral components with positive power. After AR identification and spectral decomposition of the model TF (4.7), the spectral profiles were computed according to (4.8). Moreover, the LF and HF components, i.e. $P_C^{(\mathrm{LF})}(\bar{f})$ and $P_C^{(\mathrm{HF})}(\bar{f})$ respectively, were computed from the poles with central frequency located in the ranges [0.04–0.15] Hz and [0.15–0.4] Hz, respectively, and the related variance was obtained from the pole residuals (σ_{LF}^2 and σ_{HF}^2). For some subjects, more than one peak was found in these bands; in such cases, the poles with the highest power were selected for further analysis. Finally, the spectral profiles of the LSP measure in (4.9), computed for the LF and HF oscillations, were integrated in these bands and marked as $s_C^{(LF)}$, $s_C^{(HF)}$, respectively. The time domain LSP was obtained exploiting (4.10) and marked as S_C.

To test the statistical significance of the time and frequency domain LSP measures, surrogate and bootstrap data analyses were implemented as described in THE DYNAMICS OF THE SINGLE NODE and SPECTRAL MEASURES OF SELF-PREDICTABILITY OF SINGLE NETWORK NODES, respectively. As regards statistical analysis, the distributions of the computed measures were tested for normality using the *Anderson-Darling test.* Since the hypothesis of normality was rejected for most of the distributions, and given the small sample size, non-parametric tests were employed. Specifically, the statistical significance of the difference between REST and TILT conditions, as well as between integrated PSD values in LF and HF bands in a given experimental condition, was assessed using the *Wilcoxon signed-rank test for paired data.* In this work, a significance level $\alpha = 0.05$ was used to compute confidence intervals of the surrogate and bootstrap distributions as well as to conduct statistical tests.

Results and Discussion. The results of the time domain analysis are reported in Table 5.1, revealing that both the mean μ_C and the variance σ_C^2 of the AC time series decreased significantly with head-up tilt ($p < 0.001$). This is in accordance with previous findings [20, 42, 43] and suggests that, when higher sympathetic activity is assumed, i.e., during the orthostatic challenge, the well-known changes of heart rate

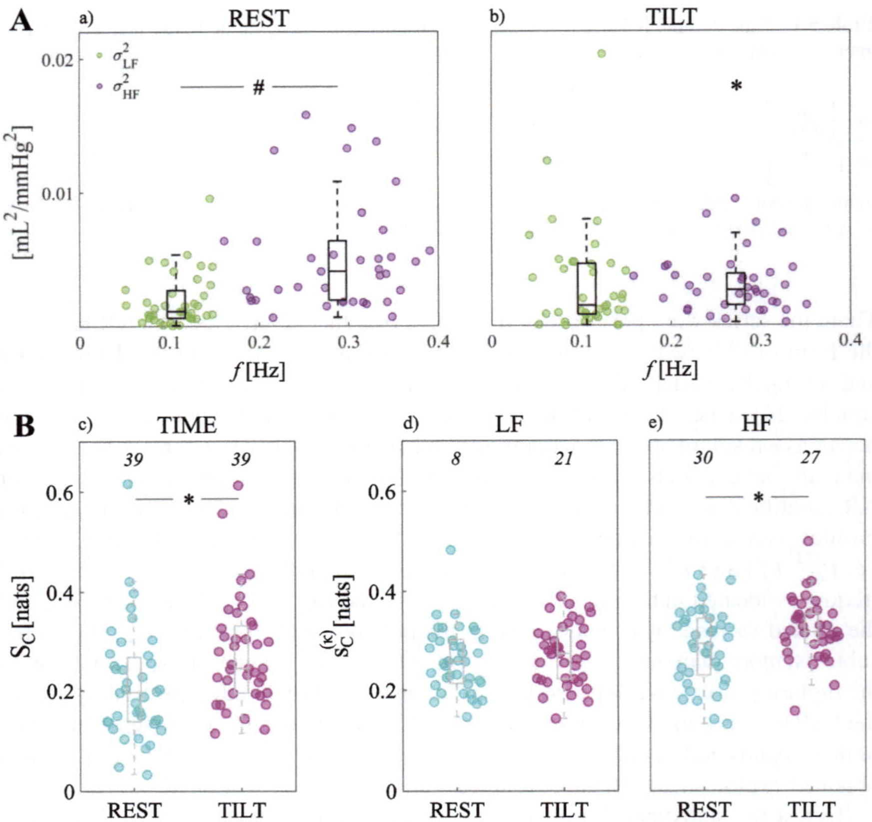

Fig. 5.1 **A** Spectral power of AC in **a** REST and **b** TILT depicted as boxplot distributions and individual values with coordinates ($f_{\mathrm{LF}}, \sigma^2_{\mathrm{LF}}$) and ($f_{\mathrm{HF}}, \sigma^2_{\mathrm{HF}}$). **B** Measures of LSP integrated in **c** time, **d** LF and **e** HF bands in REST and TILT. Numbers indicate statistically significant LSP values. *Wilcoxon signed-rank test for paired data*: $*$, $p < 0.05$ REST versus TILT; #, $p < 0.05$ LF versus HF. The figure is adapted from [22]

and total peripheral resistance occur rapidly through baroreflex mechanisms [44–46], and are accompanied by a simultaneous rise in arterial stiffness.

Figure 5.1 shows the boxplot distributions of the spectral power of AC in the REST (Fig. 5.1A, a) and TILT (Fig. 5.1A, b) conditions computed within the LF (σ^2_{LF}, green circles) and HF (σ^2_{HF}, purple circles) bands, and depicted in a way such that subject-specific information relevant to the frequency location of the LF and HF spectral peaks is also provided (each circle has coordinates ($f_\kappa, \sigma^2_\kappa$), where κ represents the LF or HF band). The total number of subjects is 39, but only the subjects for those the algorithm detected at least one pole in the LF and HF bands are shown. While the tendency of the LF power is towards an increase moving from REST to TILT ($p = 0.068$), the HF power significantly decreases ($p = 0.002$). Furthermore, the assessment of the significance of the difference between power values integrated in LF and HF bands in a given condition revealed that the latter ones are predominant ($p < 0.001$) during the supine rest (Fig. 5.1A, a, σ^2_{LF} vs. σ^2_{HF}). This finding may reflect the fact that HF oscillations of AC can be heavily affected by several respiration-related mechanisms, including *(i)* the direct mechanical effect of intrathoracic pressure oscillations on the arterial wall stretch, and *(ii)* the effect of HF oscillations in AC modulators such as heart rate and ABP, with former bringing information about the mechanisms of RSA [37, 47]. The predominance of HF withdraws with tilt due to a slight increase of LF power ($p = 0.068$) and a significant decrease of HF power ($p = 0.002$), as depicted in Fig. 5.1A, b. An increase of magnitude of LF oscillations can reflect the sympathetically-driven vasomotion as a result of its baroreflex-mediated activation associated with orthostasis [44, 45, 48]. Conversely, a decrease of magnitude of HF oscillations could be attributed to the parasympathetic inhibition during orthostasis reflected by decreased RSA magnitude [30, 49]. Quantifying the effects of potential drivers of AC oscillations, such as changes of heart rate and TPR, could improve our understanding of the observed changes in AC variability [20, 48].

In Fig. 5.1B, the spectral representation of the LSP is shown in terms of boxplot distributions of the integrated measure over all frequencies (S_C, panel *c*), as well as in the LF ($s_C^{(LF)}$, panel *d*) and HF ($s_C^{(HF)}$, panel *e*) bands, computed in the REST (left boxplots, cyan circles) and TILT (right boxplots, magenta circles) conditions. The significant increase of the time domain LSP moving from REST to TILT ($p = 0.002$) is confirmed only in the HF band of the spectrum ($p = 0.007$). This suggests that the overall increase of regularity of the process cannot be generalized to the whole frequency content of arterial compliance, but is rather confined to the HF band of the spectrum and may have different origins. One of them is related to the mathematical nature of the LSP measure $s_C^{(F)}(\bar{f})$ (F represents the LF or HF band), whose spectral profile is given by the sum of the frequency-independent term S_C, and the zero-mean term $\bar{s}_C^{(F)}(\bar{f})$ showing a peak in band F. Potential tilt-induced significant increases of S_C are thus frequency-independent and distributed uniformly along the whole frequency axis. Then, the spectral LSP $s_C^{(F)}(\bar{f})$ is influenced by this increase even in the case when there is no significant change in fluctuations of $\bar{s}_C^{(F)}(\bar{f})$; this influence has major weight in the larger spectral band due to the higher number of integrated frequency bins (i.e., the HF band), and may be thus responsible for the observed change of self-predictability in this band. From a physiological point of view, the degree of complexity of arterial compliance could be a result of the combined effects of external influences modulating its dynamic activity and operating over different temporal scales, such as direct mechanical or neural influences arising from central oscillators (respiratory and vasomotor oscillators), feedback loops (e.g. baroreceptive closed-loop circuit), and complex physiologic mechanisms adjusting TPR, ABP and heart rate.

At first sight, the unaltered regularity of AC in the LF band can be attributed to a hidden tilt-induced sympathetic activation, due to the high degree of co-ordination and synchronicity of several simultaneous control mechanisms regulating AC in both the resting state and tilt conditions (e.g., heart rate, blood pressure and TPR). However, bootstrap data analysis yielded opposite results, since we found that the significance of AC self-predictability in the LF band increased with tilt (from 8/39, i.e., 20.51%, in the supine rest to 21/39, i.e., 53.85%, in tilt), as depicted in panel *d*. This finding is of great importance and confirms the slight activation of sympathetic vasomotor control observed for the power spectral density of compliance [44, 45, 48], besides possibly reflecting an increased amplitude of LF oscillations in external modulators such as ABP or TPR [31, 50]. Noteworthy, the augmented number of significant spectral measures in this band can be explained by considering the subject-specific frequency profiles of $s_C^{(LF)}(\bar{f})$, which are likely to show more prominent peaks in LF during tilt, in accordance with wider fluctuations of $\bar{s}_C^{(LF)}(\bar{f})$ and in spite of an overall frequency-independent increase of S_C (as shown in *panel* c), i.e., the threshold for assessing significance in bootstrap data analysis (SPECTRAL MEASURES OF SELF-PREDICTABILITY OF SINGLE NETWORK NODES).

The tilt-induced physiologic responses resulting in decreased respiratory rates and increased tidal volumes [49, 51, 52], as well as in a slight diminished complexity of the respiration signal [53], may be responsible for the increase of regularity of AC in the HF band. Indeed, the increased mechanical influences on arterial vessels due to an augmented tidal volume are likely to produce an augmented coupling between arterial stiffness and respiration, reflected by an increase of AC self-predictability in the HF band. Moreover, an increased HF-related regularity in AC could be attributed also to the effect of increased magnitude of ABP variability in this band [31], probably resulting from the tilt-induced suppression of buffering effect of RSA on ABP variability at the respiratory frequency [31, 54]. Noteworthy, the latter findings highlight one important limitation of the LSP measure, which is its formulation in absence of a multivariate context taking into account potential oscillatory external drivers of AC variability, such as ABP and respiration. The spectral measure of autonomous dynamics defined as *Granger Autonomy* (Sect. 4.2.2.2) could be thought as an extension of the LSP to the bivariate case since it takes into account potential confounding mechanisms deriving from external sources [55]. It is worth noting that the significance of HF self-predictability decreases from 30/39 (76.92%) in the supine rest to 27/39 (69.23%) during tilt, as depicted in panel *e*. One more time, this result could be interpreted by looking at the spectral profiles of $s_C^{(HF)}(\bar{f})$: while the increase of self-predictability in HF may be associated to the frequency-independent increase of the term S_C, the tilt-induced diminished significance of the spectral LSP in the same band could be the result of less prominent peaks due to dampened fluctuations of $\bar{s}_C^{(HF)}(\bar{f})$. Again, if combined with the increase of significance of LF regularity, this result confirms a parasympathetic withdrawal related to heart rate control and suggests the importance of LF fluctuations when the process has to cope with the physiological perturbations due to the orthostatic challenge.

Conclusion. The application of the proposed approach to arterial compliance data has demonstrated the significance of computing frequency-specific self-predictability measures in the case of physiological variables rich of oscillatory components with different frequencies and shape, suggesting that the overall changes of self-predictability in the time domain may be confined to specific bands of the spectrum. Moreover, investigating the spectral self dynamics of poorly investigated physiological processes such as arterial compliance may have a great impact in understanding their role in multivariate contexts. Further studies from our research group are currently looking at more complex networks of physiological variables including, besides compliance, heart rate, arterial pressure and respiration, with the aim to provide a more comprehensive investigation of the regulatory mechanisms affecting cardiovascular and cardiorespiratory interactions in response to external physiological perturbations [56].

5.2 The Rate of Mutual Information Exchanged in Networks of Cardiovascular, Respiratory and Cerebrovascular Variables

This section reports the analysis of dynamic functional connectivity in physiologic networks by pairwise information-theoretic measures of total coupling between cardiovascular, respiratory and cerebrovascular time series measured in a representative subject prone to develop postural-related syncope [28, 57], published in [58]. The aim of the study consisted in drawing preliminary conclusions from descriptive indexes of physiopathological states estimated from individual recordings of biomedical signals, using statistical analyses that focus on subject-specific differences between experimental conditions. Significance and variations between conditions have been statistically validated on a single-subject basis through the use of surrogate and bootstrap data analyses (see Appendix A for mathematical details).

Experimental Protocol and Data Analysis. The subject was monitored in the resting supine position (REST) and in the upright position during the early-tilt (ET, 2 min. after tilting the bed table) and the late-tilt (LT, starting after prolonged postural stress, before the onset of the symptoms of pre-syncope), measuring the ECG, the finger AP signal, the CBFV via transcranial doppler and the respiration signal via a thoracic impedance belt. In each phase, synchronous time series of $L = 250$ consecutive values of the HP, systolic AP (SAP), MAP, MCBFV and respiration amplitude (RESP) were measured from the acquired signals on a beat-to-beat basis [28, 57]. For further details about the experimental protocol, signal acquisition and time series extraction, see [28, 57] and Sect. 5.5. From the $Q = 5$ time series $\mathbf{Y} = \{Y_1, \ldots, Y_5\}$, dynamic functional connectivity was assessed computing the MIR $I_{Y_i;Y_j}$, $i, j = 1, \ldots, 5, i \neq j$ (4.15, scalar version) between each pair of series (model order p set according to the AIC for the full model comprising the 5 processes (2.6), $q = 20$ for the reduced model comprising the processes Y_i and Y_j (2.7) whose model parameters were derived solving the YW equations as described in Sect. 2.3.5.1), and employing the surrogate and bootstrap approaches to assess the significance of the MIR in a given condition (REST, ET, or LT) (COUPLED DYNAMICS BETWEEN PAIRS OF NODES) and the significance of the MIR variation between two conditions, referred to as δ (STATISTICAL SIGNIFICANCE OF THE DIFFERENCE BETWEEN EXPERIMENTAL CONDITIONS).

Results and Discussion. The results shown in Fig. 5.2 display the maps of dynamic functional connectivity in the three analyzed conditions (Fig. 5.2a), as well as the maps of pairwise differences between the MIR values computed in two conditions (Fig. 5.2b). The statistical analyses performed at the level of a single subject allow to infer, for this subject, physiologically expected behaviors related to cardiovascular, cerebrovascular and cardiorespiratory mechanisms induced by postural stress and pre-syncope. These behaviors include: *(i)* the drop of the MIR between SAP and HP upon presyncope, suggesting an impairment of cardiovascular interaction possibly related to baroreflex impairment [57, 59]; *(ii)* the increase of the MIR between MAP and MCBFV moving from REST to ET and from ET to LT, suggesting a progressive weakening of CA causing the MCBFV to depend more and more passively on MAP [57, 59]; *(iii)* the decrease of the MIR between HP and RESP moving from REST to ET, reflecting a decrease of the strength of cardiorespiratory interactions

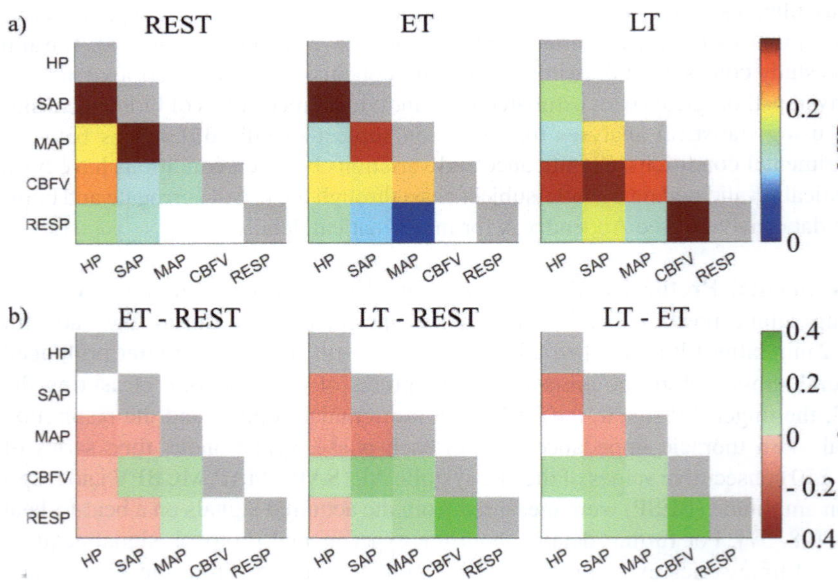

Fig. 5.2 Representative subject with previous history of postural-related syncope. **a** Pairwise connectivity assessed computing the MIR between pairs of time series; white squares: not significant **b** Differences (δ) between MIR values estimated in two conditions; green and red squares: positive (ET > REST, left; LT > REST, middle; LT > ET, right) and negative (ET < REST, left; LT < REST, middle; LT < ET, right) differences, respectively; white squares: not significant differences. The figure is adapted from [58]

due to sympathetic activation and vagal withdrawal [60]. The two former findings *(i)–(ii)* were also documented in [59] using the time domain measure of Geweke total dependence (4.20), while the latter result *(iii)* is interestingly accompanied by an increase of the MIR between RESP and the other variables (SAP, MAP, CBFV), suggesting that the respiratory activity exhibits a complex interplay with the other physiologic variables and may play a role in orthostatic intolerance [28].

Conclusion. The present work supports the use of surrogate and bootstrap data analyses for the single-subject investigation of pairwise connectivity in physiological variability studies. While the reported results need to be validated on larger datasets, the proposed approach revealed physiologically plausible patterns of dynamic functional connectivity in the application to networks of cardiovascular, cerebrovascular and respiratory variables. In perspective, the proposed single-subject analysis may have clinical relevance for subject-specific investigations and treatment planning.

5.3 A Portable Multisensor System to Assess Cardiorespiratory Interactions Through Photoplethysmography

Nowadays, the ever-growing interest to health and quality of life of individuals and the advancements in electronic devices technology are pushing the development of portable and wearable biomedical devices able to pursue a minimally invasive monitoring of physiological parameters in daily-life conditions. Such devices can now carry out a real-time assessment of the overall health status and possibly even detect ongoing diseases. In this context, in [61], we carried out a synchronous real-time acquisition of cardiorespiratory signals (electrocardiographic, photoplethysmographic, and respiration-related) using a low-invasive multisensor portable acquisition system [62, 63]. Starting from the time series extracted from the signals, the strength of causal interactions directed from the respiratory process to the heart rate has been assessed through bivariate GC measures. The aim is to prove whether and in what extent the non-invasive and cost-effective *photoplethysmographic* (PPG) *technique* can be used alone to assess cardiorespiratory interactions without acquiring breathing or ECG signals.

Experimental Protocol and Signal Processing. Details on the portable system can be found in [61–63]. Three different signals were synchronously acquired, i.e., 3-lead ECG, the PPG signal and the respiratory (RESP) signal. The four ECG electrodes were positioned according to Einthoven's triangle on the wrists and legs, the PPG probe was positioned on the left wrist, while the breath probe was placed on the nose. All the signals were sampled with a sampling frequency 500 Hz and 24-bit resolution. An appositely developed GUI was used to show in real time the acquired signals and to send the recorded data wirelessly via Bluetooth to a personal computer, in order to save them for the subsequent offline analyses (we refer to [62, 63] for further information). Measurements were carried out on 6 healthy subjects (3 males and 3 females; 24.3 $\pm$ 1.9 years) monitored in a sitting position and undergoing a two-phase protocol aimed at assessing cardiorespiratory interactions during spontaneous (SB) and controlled breathing (CB, breathing rate: 20 breaths/min, i.e., 0.33 Hz). In order to ensure a correct execution of the paced breathing, the subjects were instructed to follow an appositely developed visual metronome application showing different colors according to the current breathing phase, i.e. inhalation/expiration (see [63]). The acquired ECG, PPG and RESP signals were first filtered using a zero-phase fourth-order Butterworth bandpass filter (ECG: [0.1–20] Hz, PPG and RESP: [0.1–15] Hz). Starting from the ECG signals, the R peaks were detected using a modified version of the Pan-Tompkins algorithm [64] in order to extract the R-R interval (RRI) time series. With regard to PPG signals, a threshold-based peak detection algorithm has been employed to locate the waveform minima of the acquired signal which have been then used for pulse-pulse interval (PPI) time series. We also performed the reconstruction of the respiratory signal starting from only the PPG signals, using both a filtering-based approach and Empirical Mode Decomposition (EMD) [65]. For the first approach, a band-pass filter was applied accounting for a range of respiratory frequencies determined in accordance to the knowledge that respiration variability usually falls within the HF band, i.e., [0.15–0.4] Hz [35, 66]. Only in two of the six subjects the range was adapted to take into account the respiratory peak falling out of the HF band (we chose [0.08–0.33] Hz and [0.25–0.5] Hz to detect peaks at $\sim$0.1 Hz and $\sim$0.45 Hz, respectively). The second method has been already widely used for extracting breathing rate from PPG signals (see e.g., [65]). We have applied a simplified version of the EMD algorithm presented in [65], herein summarized: *(i)* find the local maxima (Mi) and the local minima (mi) of the PPG signal ($x(t)$); *(ii)* interpolate the maxima and minima using the same number of points of the PPG, generating thus the upper $M(t)$ and lower $m(t)$ envelopes, respectively (the MATLAB cubic spline interpolation algorithm has been employed); *(iii)* compute the average envelope as $e(t) = \frac{M(t)+m(t)}{2}$; *(iv)* subtract the average envelope to the PPG signal, $x(t) := x(t) - e(t)$. The steps *(i)–(iv)* should be repeated until the new $x(t)$ does not vary from the $x(t)$ at the previous iteration. In our approach the EMD is stopped at the first loop, and the reconstructed breathing signal has been selected as the average envelope [65]. Four different respiration time series were extracted and used for computing the cardiorespiratory interactions measures reported in the following section: *(i)* R_{RRI}: series extracted as the values of the respiration signal sampled at the times of ECG R peaks; *(ii)* R_{PPI}: series extracted as the values of the respiration signal sampled at the times of PPG peak minima; *(iii)* $R_{PPG_{filter}}$: series extracted as the values of the respiration signal reconstructed through the filtering approach sampled at the times of PPG peak minima; *(iv)* $R_{PPG_{EMD}}$: series extracted as the values of the respiration signal reconstructed through the EMD technique sampled at the times of PPG peak minima. For all the time series (RRI, PPI and *(i)–(iv)* RESP series), we have extracted 300-point stationary windows.

Data Analysis. The linear parametric analysis of cardiorespiratory interactions was performed considering both lagged and instantaneous (i.e., not delayed) effects from respiration (RESP, driver process) to the heart period (HP, target process), as the common adopted measurement convention assumes that $RESP_n$, sampled at the onset of the nth RRI, could have a role in determining HP_n variability [67, 68]. Time series were first pre-processed using a high-pass AR filter with cut-off frequency of 0.0156 times the sampling rate [69], the latter computed for each subject assuming the series as uniformly sampled with the mean HP taken as the sampling period. Each pair of stationary zero-mean HP and $RESP$ series collected in the joint bivariate process $\mathbf{Y} = [Y_1, Y_2] = [HP, RESP]$ was then fitted by an extended AR model including zero-lag effects, in the form of (7.1) where $k = 0, \ldots, p$ and $A_{\mathbf{Y},0} = \begin{bmatrix} 0 & a_{Y_1Y_2,0} \\ 0 & 0 \end{bmatrix}$. For details about the construction of extended models including zero-lag effects by previous physiological knowledge, we refer the reader to Appendix B. A reduced AR model involving only the HP process in the form of (2.2) was then formulated. Identification of the full (2.4) and restricted (2.2) models was performed via the OLS approach, setting the model order p according to the AIC for each subject (with maximum scanned model order equal to 8) (see Sect. 2.3.5 for the identification procedure). To study the causal effect of the respiratory process to the heart period, the frequency domain measure of Geweke GC was then computed along the direction $R \rightarrow H$ (in the form of (4.30), where the relevant terms of the PSD, the model TF and the variance of the AR residual are computed from the extended AR model (7.1)); then, the spectral distribution $f_{R\rightarrow H}(\bar{f})$ was integrated alongside the whole frequency axis to obtain $F_{R\rightarrow H}$ as in (4.31), and within the HF band of the spectrum to get $f_{R\rightarrow H}(HF)$. Specifically, the HF band was determined individually for each subject by first locating the respiratory peak and then selecting the band with a width of ± 0.06 Hz around such peak.

Results and Discussion. Figure 5.3 shows the subject-specific results of time domain and spectral analysis. The displayed measures are the normalized PSD of the respiratory process, computed as the ratio of the PSD in HF to the total PSD in $[0 - f_s/2]$ (panel *a*), the time domain GC (panel *b*) and the spectral GC integrated within the HF band of the spectrum (panel *c*), in both phases of the protocol (panels above: SB; panels below: CB). We have analysed the following combinations of respiratory and heart period time series: *(i)* heart period: RRI time series extracted from ECG; respiratory time series: R_{RRI}; *(ii)* heart period: PPI time series extracted from PPG; respiratory time series: R_{PPI}; *(iii)* heart period: PPI time series extracted from PPG; respiratory time series: $R_{PPG_{filt}}$; *(iv)* heart period: PPI time series extracted from PPG; respiratory time series: $R_{PPG_{EMD}}$. Two subjects presented spontaneous breathing rates higher or lower than usual (respectively, ~0.45 Hz and ~0.1 Hz), falling out of the frequency band commonly referred to as respiratory band, i.e., [0.15–0.4] Hz. Surprisingly, it has been previously found that, in many healthy subjects, breathing frequency slows down to the LF band and entrainment of the cardiovascular rhythm around 0.1 Hz often occurs [70]. In our study, slow breathing was found in 1 (17%) of the 6 subjects who performed the experimental protocol. This put a constraint in

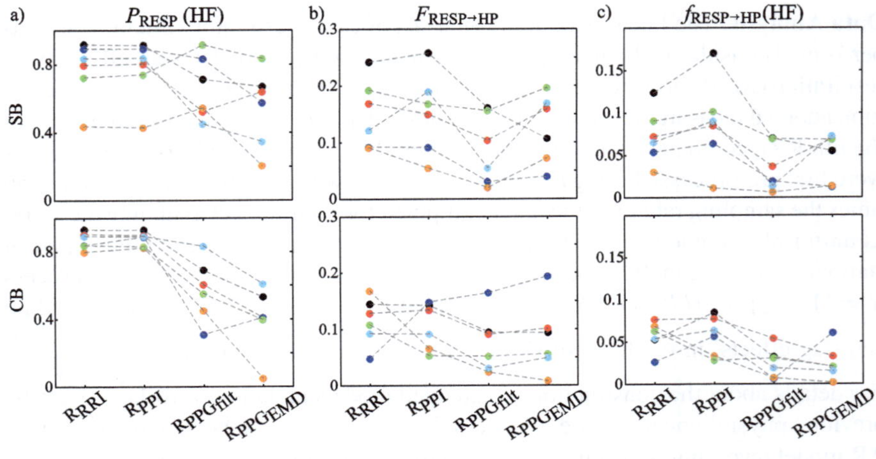

Fig. 5.3 **a** Normalized PSD of RESP in the HF band. **b** Time- and **c** frequency domain GC from RESP to HP. Measures are assessed during spontaneous breathing (SB, top row) and controlled breathing (CB, bottom row) considering the four respiration time series R_{RRI}, R_{PPI}, $R_{PPG_{filter}}$, $R_{PPG_{EMD}}$ for all 6 subjects, each represented by a different color. The figure is adapted from [61]

the selection of the HF band for the computation of spectral measures, as we have chosen subject-specific HF ranges to take into account possible outliers. However, this approach may cause to not detect all the power in HF band. Indeed, we noticed that the bandwidth around the peak is larger when reconstructing the respiration signals from PPG (i.e., $R_{PPG_{filt}}$ and $R_{PPG_{EMD}}$) for both experimental conditions (SB, CB), probably due to *spurious* spectral content related to autonomic system activity not present in the *true* respiration-only signal. This may be the reason of the unexpected sudden decrease of respiratory PSD in HF when extracting RESP from PPG, both with filtering and EMD approaches, mostly visible in one of the subjects (e.g., orange point in panel *a*) but generally occurring for all of them in both experimental conditions. On the contrary, PSD values computed for the first two settings (R_{RRI} and R_{PPI}) are comparable between each other and expectedly slightly increase with controlled breathing, since all the respiratory variability is centered around the respiration peak (~0.33 Hz) and spectral leakage was observed to be negligible. As regards GC measures, our results suggest that their overall behaviour is characterized by a decrease when these values are computed using respiratory time series reconstructed from PPG, especially with regard to the spectral measure. Generally, in presence of bigger databases, statistical analysis is performed to detect significant changes of the investigated measures between experimental conditions or settings. In previous works, this has allowed to characterize the possibly different behaviours of time domain measures and frequency-specific measures, which have been found to be more precise and informative than overall indices, especially when the observed processes exhibit frequency-specific oscillations [41, 71]. We could not perform statistical analysis because of the small number of samples employed, but this led to some difficulties in the interpretation of the obtained causality values in time and frequency domain. Several studies have documented that the magnitude of respiratory-related fluctuations of RRI (i.e., RSA) dramatically changes according to breathing rate [72]. Moreover, it has been demonstrated that paced breathing at ~0.25 Hz does not alter efferent vagal and sympathetic modulations in the frequency range from 0.04 to 0.15 Hz in healthy subjects [73, 74]. In this study, we found a decrease of both time and spectral measures in CB with respect to SB in 3 subjects (50%), while increased or unchanged values were detected in the remaining samples. These findings must be further investigated according to the subject-specific spontaneous breathing rate and with bigger datasets. The possibility to enroll a greater number of subjects and carry out statistical analyses represents surely one of the future extensions of this study, in order to more clearly assess the feasibility of the proposed extraction approaches of respiration from PPG. Nonetheless, our preliminary findings suggest that causality measures behave similarly if PPG is used for the detection of heart period and sampling of respiration instead of ECG. On the other hand, the filtering and EMD approaches for the extraction of respiration from PPG may be less accurate in the quantification of time domain and spectral measures, and especially of respiratory PSD. Indeed, applying a PPG bandpass filter to identify respiratory dynamics may cause misdetection of power content if other oscillatory components are present within the selected HF range or if the peak bandwidth is too large. Conversely, the EMD extraction technique is based on how well the detection of PPG peaks is performed, and this may pose a problem when the acquired waveform is noisy, due e.g. to motion artifacts.

Conclusion. In the present study, the feasibility of using the less invasive and cost-efficient PPG technique instead of the more invasive ECG for investigating cardiorespiratory interactions has been proved, since the computation of respiratory parameters and patterns of causality in time and spectral domain provided similar results when heart period is taken as the PPI sequence instead of RRI sequence. However, our findings also suggest that the use of PPG for the extraction of the respiration signal could lead to underestimates of the computed measures. This aspect should be further investigated and validated with wider databases and statistical analyses.

5.4 Linear Parametric Assessment of Cardiorespiratory Interactions During Spontaneous and Controlled Breathing

This section reports the analyses published in [75] and carried out on an historical dataset previously employed for assessing the cardiovascular control in healthy subjects [52, 76]. We performed a linear parametric analysis of cardiorespiratory interactions in bivariate time series of heart period and respiration measured in 19 healthy subjects during spontaneous breathing and controlled breathing at varying breathing frequency. The analysis is carried out computing measures of the total and causal interaction between HP and RESP variability in both time and frequency domains.

Experimental Protocol and Data Analysis. Data were acquired on 19 healthy subjects (11 females, age: 27–35 years; median = 31 years), during an experimental protocol consisting of an initial period of spontaneous breathing (SB), followed by controlled breathing sessions at 10 (C10), 15 (C15) and 20 (C20) breaths/min carried out in a random order. From the acquired electrocardiographic and respiratory flow signals, stationary heart period, HP, and respiration, RESP, time series of length $L = 255$ beats were then extracted for each breathing condition, and normalized to zero mean. Further details on signals acquisition and preprocessing can be found in [52, 76]. Each pair of HP and RESP time series was taken as a realization of a bivariate stochastic process $\mathbf{Y} = [Y_1 Y_2]$, with $Y_1 = RESP$ and $Y_2 = HP$, which was then described using a linear parametric model as in (2.4). After spectral representation of (2.4), the spectral measures of total dependence (4.25) and linear feedback (4.30) were computed, together with the instantaneous causality term (4.33). Each of these spectral measures was integrated along the whole frequency axis to obtain the corresponding time domain measures, as done for the linear feedback in (4.30). Further, assuming to have fast (i.e., within-beat) interactions from RESP to HP [67, 68], we summed the instantaneous term to the GC from RESP to HP, in order to obtain the time and frequency domain measures of extended causality $F_{RESP \to \cdot HP} = F_{RESP \to HP} + F_{RESP \cdot HP}$, $f_{RESP \to \cdot HP} = f_{RESP \to HP} + f_{RESP \cdot HP}$. Moreover, frequency-specific measures were obtained integrating the terms of the Geweke decomposition in the LF ([0.04–0.12] Hz) and HF bands, being the latter chosen for each subject as $[f_R \pm 0.04]$ Hz, with f_R the respiratory peak identified in the range [0.04–0.15] Hz. The statistical significance of the obtained results was checked with $\alpha = 5\%$ significance level. Moreover, to check the significance of the time and frequency domain TD and GC measures, for each subject $N_s = 100$ surrogate time series were generated using the $iAAFT$ algorithm, setting the significance threshold to 95% (see COUPLED DYNAMICS BETWEEN PAIRS OF NODES).

Results and Discussion. Results are shown in Fig. 5.4. The time domain measures of coupling and causality obtained as the whole-band integral of the spectral measures did not show any statistically significant variation during controlled breathing when compared with the SB condition (first row). The same findings where obtained when the spectral measures were integrated within the HF band (third row). In both cases, the number of subjects with significant coupling was very high, decreasing slightly along the direction from HP to RESP (middle column). The frequency measures integrated within the LF band were statistically significant in a lower number of subjects, especially when computed from RESP to HP (second row, right column). In this direction, a significant increase of the coupling was detected during C15 ($p = 0.03$) compared to SB. Even if not significant ($p = 0.06$), a tendency to increase during C20 was observed for the measure $F_{RESP \to \cdot HP}$ in the LF band. Our results evidence that the time domain linear measures of coupling and causality are

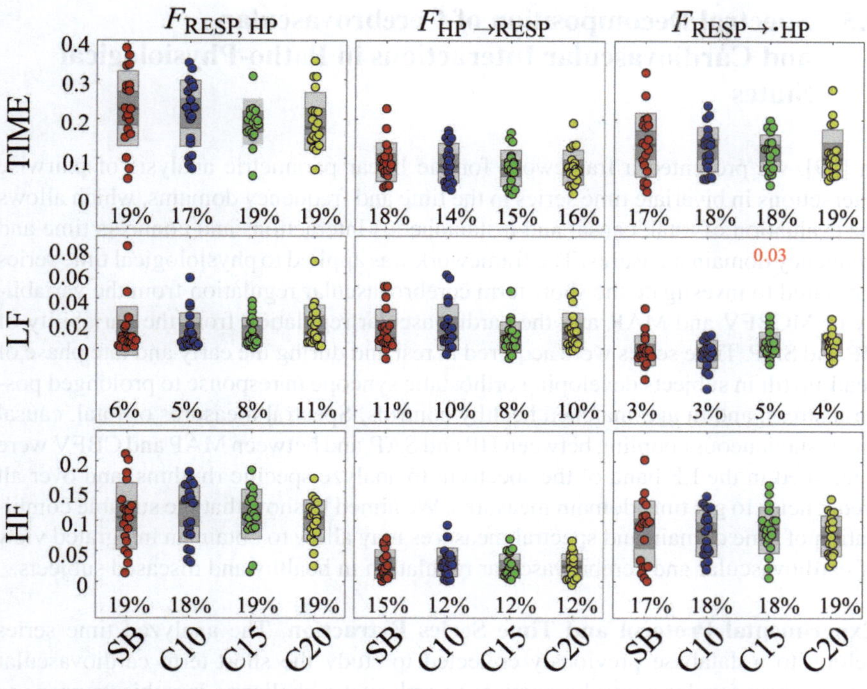

Fig. 5.4 Boxplot distributions (95% C.I. and 1 standard deviation) and individual values of time domain (top row) and spectral measures integrated in LF (middle row) and HF (bottom row) bands, in the four breathing conditions (SB, C10, C15 and C20). Left panels, total pairwise coupling; middle panels, coupling from HP to RESP; right panels, extended coupling from RESP to HP. Top red values: p-values returned by the *Wilcoxon non-parametric test* comparing the given distribution to the reference (SB); bottom values: number of subjects (out of 19) with statistically significant coupling. The figure is adapted from [75]

not significantly altered by the paced breathing condition. Frequency-specific interactions are stronger in the RSA-related HF band, wherein they are directed mostly from RESP to HP and do not change significantly across conditions. LF interactions are weaker, prevalent along the direction from HP to RESP that is usually less investigated in the literature, and appear more influenced by the paced breathing maneuver along this pathway. These results confirm, from the point of view of bivariate cardiorespiratory interactions, previous findings on the same dataset based on measures of high-order interactions (taking also into account SAP) showing that paced breathing evokes significant effects within the LF band of the frequency spectrum, but not in the HF band classically studied [77].

5.5 Spectral Decomposition of Cerebrovascular and Cardiovascular Interactions in Patho-Physiological States

In [59], we presented a framework for the linear parametric analysis of pairwise interactions in bivariate time series in the time and frequency domains, which allows the evaluation of total, causal and instantaneous interactions and connects time and frequency domain measures. The framework was applied to physiological time series measured to investigate the short-term cerebrovascular regulation from the variability of MCBFV and MAP, and the cardiovascular regulation from the variability of HP and SAP. Time series were acquired at rest and during the early and late phase of head-up tilt in subjects developing orthostatic syncope in response to prolonged postural stress, and in age-matched healthy controls. Spectral measures of total, causal and instantaneous coupling between HP and SAP, and between MAP and CBFV were integrated in the LF band of the spectrum to analyze specific rhythms, and over all frequencies to get time domain measures. We aimed to show that the suitable combination of time domain and spectral measures may allow to obtain an integrated view of cardiovascular and cerebrovascular regulation in healthy and diseased subjects.

Experimental Protocol and Time Series Extraction. The analyzed time series belong to a database previously collected to study the short-term cardiovascular and cerebrovascular control responses to orthostatic challenge in subjects prone to neurally-mediated syncope and healthy controls via the analysis of spontaneous variability of systemic variables [28, 57]. The study included 13 subjects (age: 28 ± 9 years; 5 males) with previous history of unexplained syncope (SYNC, reporting >3 syncope events in the previous 2 years) and 13 age-matched healthy subjects (nonSYNC, age: 27 ± 8 years; 5 males), enrolled at the *Neurology Division of Sacro Cuore Hospital, Negrar, Italy*. The protocol consisted of 10 min of recording in the resting supine position, followed by 60° head-up tilt test. All SYNC subjects experienced presyncope signs (i.e., a vasovagal episode characterized by hypotension and reflex bradycardia leading to partial loss of consciousness) during the tilt session; when signs were reported, the subject was returned to the resting position and a spontaneous recovery occurred. None of the nonSYNC subjects experienced presyncope symptoms during tilt. The signals analyzed in this study are the ECG (lead II), the continuous AP measured at the level of middle finger through a photopletysmographic device (Finapres, Enschede, The Netherlands), and the CBFV signal measured at the level of the middle cerebral artery by means of a transcranial doppler ultrasonographic device (Multi-Dop T, Compumedics, San Juan Capistrano, CA, USA). From these signals, cardiovascular and cerebrovascular variability time series were extracted according to the procedure described in the following. A template matching algorithm was employed to detect the QRS complexes and locate the R peaks from ECG signals, in order to calculate HP values as the temporal distance between two consecutive R peaks [28]. A manual correction procedure was followed to mitigate the effects of ectopic or isolated arrhythmic beats or missing events using linear interpolation between the closest unaffected values. The nth SAP was defined as the maximum value within the corresponding HP. A low-pass sixth-order Butterworth filter with cut-off frequency 10 Hz was applied to the CBFV signal. For the analysis of cerebrovascular variability, values of MAP and CBFV were computed respectively integrating the waveform of the sampled pressure and velocity signals within each detected diastolic pulse interval, divided by the duration of the interval itself. The beat-to-beat variability series of HP, SAP, MAP and MCBFV, herein referred respectively as H, S, M, and F, were then produced as the sequences of consecutive values collected during three stationary time windows of length $L = 250$ beats during the following physiological conditions [28, 57]: *(i)* supine rest (REST); *(ii)* early tilt (ET), starting after the onset of the head-up tilt maneuver, excluding transient change of the physiological variables; and *(iii)* late tilt (LT), starting at least 5 min after the onset of the head-up tilt maneuver for subjects, or occurring just before the pressure decrease due to presyncope (start at 16 ± 8 min after the head-up tilt) for SYNC subjects. Selection of the sequences was performed randomly in each experimental condition and repeated if non-stationarities of the mean and the variance were present. The series were visually inspected and eventually corrected through cubic spline interpolation, with corrections not exceeding the 5% of the overall length of the sequence. Further information about the experimental protocol, signal acquisition and variability series extraction can be found in [28, 57].

Data and Statistical Analysis. Standard time domain statistical parameters such as mean (μ) and variance (σ^2) were computed on the H, S, M and F time series measured for each subjects and experimental condition; the corresponding symbols and measurement units are μ_H [ms], σ_H^2 [ms^2], μ_S [mmHg], σ_S^2 [mmHg^2], μ_M [mmHg], σ_M^2 [mmHg^2], μ_F [cm s^{-1}], and σ_F^2 [cm^2 s^{-2}]. Each series was first detrended with an AR high-pass filter with zero phase (cutoff frequency 0.015 cycles/beat) [69]. Then, a bivariate model in the form of (2.4) was fitted separately on each pair of MAP and CBFV series to study cerebrovascular interactions, and on each pair of HP and SAP series to study cardiovascular interactions. In the first case, a strictly causal model with $Y_1 = F$ and $Y_2 = M$ was used, since the overlap between the time of measurement of the nth MAP sample and the nth CBFV sample did not allow unambiguous setting of instantaneous effects. In the second case, an extended model in the form of (7.1), with $\mathbf{Y} = [Y_1, Y_2]$, $Y_1 = H$ and $Y_2 = S$, $k = 0, \ldots, p$ and $A_{\mathbf{Y},0} = \begin{bmatrix} 0 & a_{Y_1Y_2,0} \\ 0 & 0 \end{bmatrix}$, was used, allowing the presence of instantaneous effects in the direction from SAP to HP, to account for fast (within-beat) baroreflex influences. For details about the construction of extended models including zero-lag effects by previous physiological knowledge, we refer the reader to Appendix B. Model identification was performed via the OLS approach, setting the model order p according to the AIC for each subject (with maximum model order equal to 8) (see Sect. 2.3.5). After AR identification, computation of time and frequency domain interaction measures of coupling and causality was performed from the estimated model parameters and spectra of the processes. Spectral analysis was performed assuming the series as uniformly sampled with the mean HP taken as the sampling period. Specifically, the

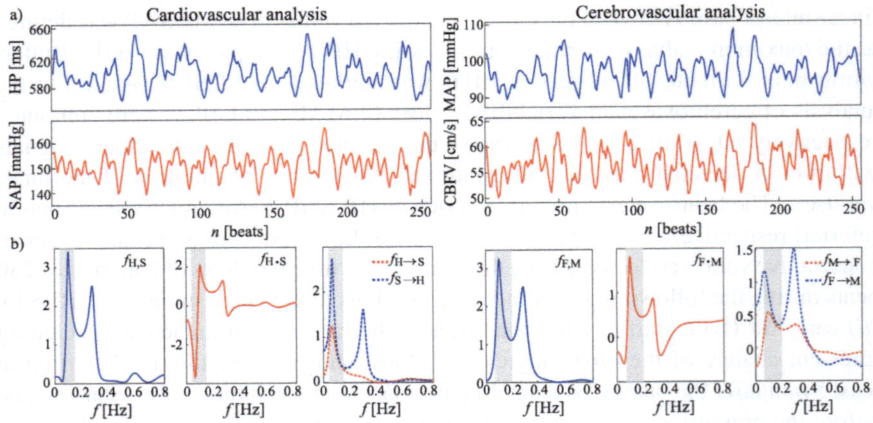

Fig. 5.5 **a** Example of HP, SAP (left), MAP and CBFV (right) time series for a representative subject in the LT condition (model order $p = 7$). **b** Example of spectral analysis of (left) *cardiovascular interactions*, and (right) *cerebrovascular interactions*. Grey areas indicate LF bands. Panels along the bottom row: (left) total linear dependence, (middle) instantaneous term, (right) linear feedback along the baroreflex ($f_{S \to H}$), the Cushing reflex ($f_{F \to M}$), the mechanical feedforward ($f_{H \to S}$) or the pressure-to-flow link ($f_{M \to F}$). The figure is adapted from [75]

spectral measures appearing in the Geweke decomposition of total dependence in (4.32) were computed; frequency-specific measures were then obtained integrating the spectral measures within predefined bands. In cardiovascular analysis, the measures were averaged within the LF band of the spectrum ([0.04–0.15] Hz) to minimize the effects of non-baroreflex mechanisms on the assessed measures and to avoid the confounding effects of respiration on SAP and HP, which are confined in the HF band [41, 78, 79]. In cerebrovascular analysis, the measures were averaged within the LF band conventionally adopted for studying MAP-CBFV spectral interactions ([0.07–0.2] Hz) [80]. An example of time series and computation of the spectral measures for a representative subject is illustrated in Fig. 5.5 (left: cardiovascular time series and measures; right: cerebrovascular time series and measures).

The distributions of the computed indices were tested for normality using the *Anderson-Darling test*. Since the hypothesis of normality was rejected for most distributions, and given the small sample size, non-parametric tests were employed. For any given group (SYNC and nonSYNC), the *non-parametric one-way Friedman test* was employed to assess the statistical significance of the differences in the median of the distributions among groups, followed, in case of rejection, by a post-hoc pairwise comparison carried out through the *paired Wilcoxon non-parametric test with Bonferroni-Holm correction for multiple comparison* to assess the differences between pairs of distributions (REST vs. ET, REST vs. LT, ET vs. LT). Additionally, the statistical significance of the differences between the two groups (SYNC vs. nonSYNC) in a given condition (i.e., REST, ET or LT) was assessed using the *unpaired Wilcoxon rank sum non-parametric test*. No statistically significant differences are detected between groups (nonSYNC vs. SYNC) for a given condition. All statistical tests were carried out with 5% significance level.

Results and Discussion: Cardiovascular Variability Analysis. For brevity, we do not report here results pertaining to the analysis of the time domain statistical parameters μ and σ of the investigated time series H and S; we refer to [59] for details. Figure 5.6A shows the results of the time domain analysis of baroreflex and feedforward interactions between HP and SAP time series, performed for the two groups in the three analyzed conditions. We remark that the time domain measures of total and causal linear dependence correspond to the equivalent frequency domain measures integrated over the whole frequency axis, and that instantaneous causality is absent because zero-lag effects are assigned to the direction $SAP \rightarrow HP$ using the extended AR model. In both groups, the total coupling decreased significantly while moving from LT to ET (panel *a*); in SYNC subjects the index was also significantly lower during LT than during REST. The decrease in total coupling was mainly determined by a lower interaction along the feedforward direction from HP to SAP, as documented by the significant decrease of the feedback $H \rightarrow S$ during LT compared to REST in both groups, and during LT compared to ET in nonSYNC subjects (panel *b*). In SYNC subjects, the linear interaction along the baroreflex direction from SAP to HP decreased significantly moving from ET to LT (panel *c*).

The results of the frequency domain analysis of cardiovascular interactions are reported in Fig. 5.6B. All spectral measures were averaged within the LF band of the spectrum, quantifying the total information shared in this band between HP and SAP ($f_{H,S}(LF)$), the information transferred along the two causal directions ($f_{H\rightarrow S}(LF)$, $f_{S\rightarrow H}(LF)$), and the information related to the mixing between the two directions ($f_{H\cdot S}(LF)$). The significant changes observed in the total coupling measure integrated over all frequencies (Fig. 5.6A, panel *a*) were observed when focusing on the LF band only for the nonSYNC subjects, for which the measure $f_{H,S}(LF)$ increased significantly moving from REST to ET, and decreased significantly moving from ET to LT; the measure did not change significantly in SYNC subjects (Fig. 5.6B, panel *a*). These trends are mainly linked to the modifications of the instantaneous term in LF, which indeed increased from REST to ET and decreased from ET to LT in nonSYNC subjects but not in SYNC subjects (Fig. 5.6B, panel *c*). The linear interaction from HP to SAP showed a tendency to decrease progressively moving from REST to ET and LT, but the decrease was statistically significant only comparing LT and REST in SYNC subjects (Fig. 5.6B, panel *b*). On the contrary, the linear interaction from SAP to HP increased markedly while moving from REST to ET in both groups, and moving from ET to LT (Fig. 5.6B, panel *d*).

The time domain analysis of baroreflex and feedforward interactions based on the considered measures of linear dependence showed that the two groups behaved rather similarly in their response to postural stress, showing a significant reduction of the total coupling between HP and SAP during late tilt (Fig. 5.6A). This result indicates that the prolongation of the orthostatic stress produces an overall weakening of the closed-loop cardiovascular regulation. The decomposition into measures of directed interaction evidenced that the drop is related to the $HP \rightarrow SAP$ direction in both

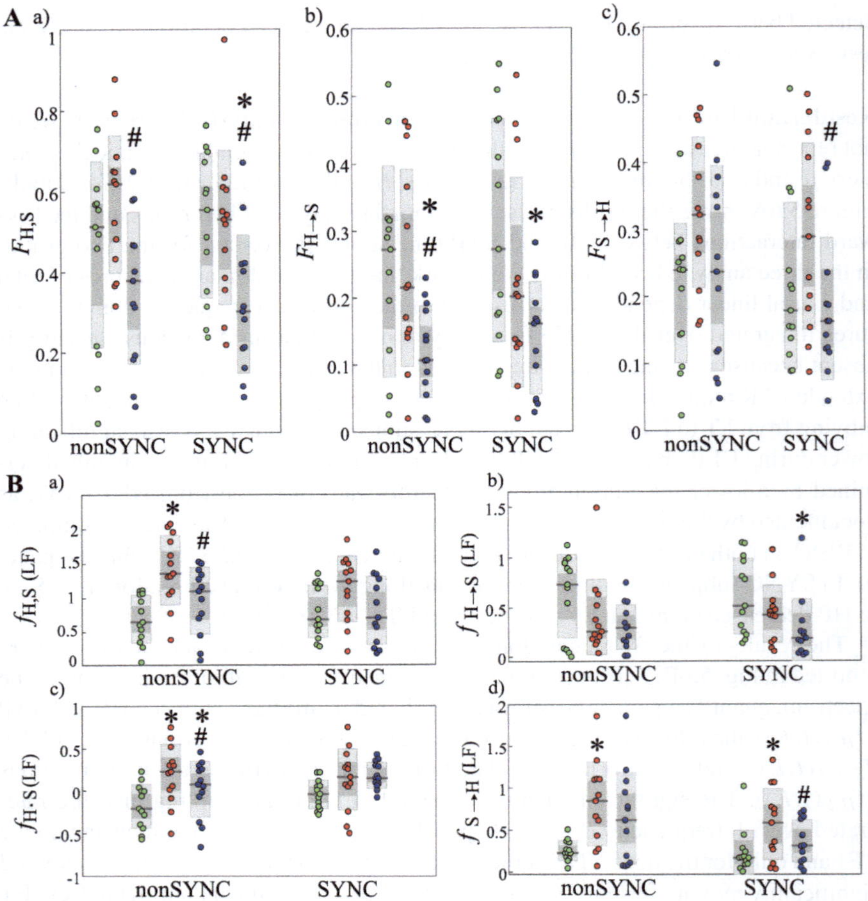

Fig. 5.6 **A** Time and **B** LF domain analysis of cardiovascular interactions. Plots depict the individual values and boxplot distributions across subjects of the **a** total linear coupling, **b** directed interaction from HP to SAP and from SAP to HP (panels **A-c**, **B-d**), and the instantaneous term (panel **B-c**). All the measures are computed at rest (REST, left bars and green circles) and during head-up tilt (ET, middle bars and red circles; LT, right bars and blue circles). Statistically significant differences: *, REST versus ET and REST versus LT; #, ET versus LT. The figure is adapted from [75]

groups, and also to the $SAP \rightarrow HP$ direction in the subjects prone to orthostatic syncope. On the other hand, the time domain analysis did not document an increase of coupling between HP and SAP moving from the resting supine position to the upright position during ET. Such an increase, which is expected from the results of a number of previous studies documenting a larger involvement of the baroreflex into the cardiovascular regulation exerted during postural stress [31, 81, 82], was observed in our work in the trends of the frequency domain measures of total coupling, mixing and directed coupling from SAP to RR computed within the LF band of the spectrum (Fig. 5.6B). In previous studies, it has been related to an increased involvement of the baroreflex into cardiovascular regulation consequent to the postural stress [31, 81, 82]. The fact that this behavior becomes evident only looking at the LF band can be explained by the role played by respiration, which affects both SAP and HP acting mostly in the HF band ([0.15–0.4] Hz) and may act as a confounder of cardiovascular interactions assessed through time domain measures accounting for the whole frequency spectrum [76, 78]. The main differentiation between the two analyzed groups consisted in the decreased interaction along the feedback direction from SAP to HP observed in syncope subjects but not in healthy controls. This result was documented both by the time domain analysis (Fig. 5.6A, panel *d*) and by the spectral analysis restricted to the LF band (Fig. 5.6B, panel *d*), and suggests an impairment of the baroreflex control as a symptom of autonomic disfunction preceding postural syncope. This interpretation is in agreement with previous studies indicating that the lack of baroreflex involvement could be one of the main mechanisms responsible for the impairment of the cardiovascular control associated with syncope development [83, 84]. A similar drop of the coupling strength along the baroreflex has been detected previously in spontaneous cardiovascular variability analyses using both nonlinear information-based techniques [57, 82] and linear approaches computing the directed coherence in the frequency domain [85]. The linear interaction from HP to SAP decreased moving from rest to tilt and with prolongation of the postural stress (panels *b* of Fig. 5.6A, B). This result suggests that feedforward effects, which comprise cardiac mechanics (e.g., the Frank–Starling law) and vascular properties (e.g., vessels compliance and peripheral resistances) [86], tend to decrease progressively its contribution to the cardiovascular regulation when the orthostatic challenge intensifies. Finally, the term reflecting the mixing between feedback and feedforward interactions, evaluated in the LF band of the spectrum, increased with the transition from rest to tilt; the variation was statistically significant only for the healthy controls (Fig. 5.6B, panel *c*). As this term measures the balance between redundant and synergistic interactions along the two causal directions, the observed increase suggests that the feedback and feedforward arms of the cardiovascular closed-loop explain a lower part of the total interactions between the LF oscillations of SAP and HP during tilt than in the resting condition.

Results and Discussion: Cerebrovascular Variability Analysis. For brevity, we do not report here results pertaining to the analysis of the time domain statistical parameters μ and σ of the investigated time series M and F; we refer to [59] for details.

The results of time domain analysis of cerebrovascular interactions are depicted in Fig. 5.7A. Besides the measures of total ($F_{M,F}$) and causal ($F_{M\to F}$, $F_{F\to M}$) linear dependence between MAP and CBFV, the figure reports also the measure of instantaneous interaction $F_{M\cdot F}$, which is nonzero in this case where the strictly causal AR model was adopted to fit the time series (panel *c*). The total dependence between the two series showed a tendency to increase with tilt (Fig. 5.7A, panel *a*); the increase was statistically significant only for the SYNC subjects during ET. This tendency was supported in SYNC subjects by the significant increase of the linear interaction

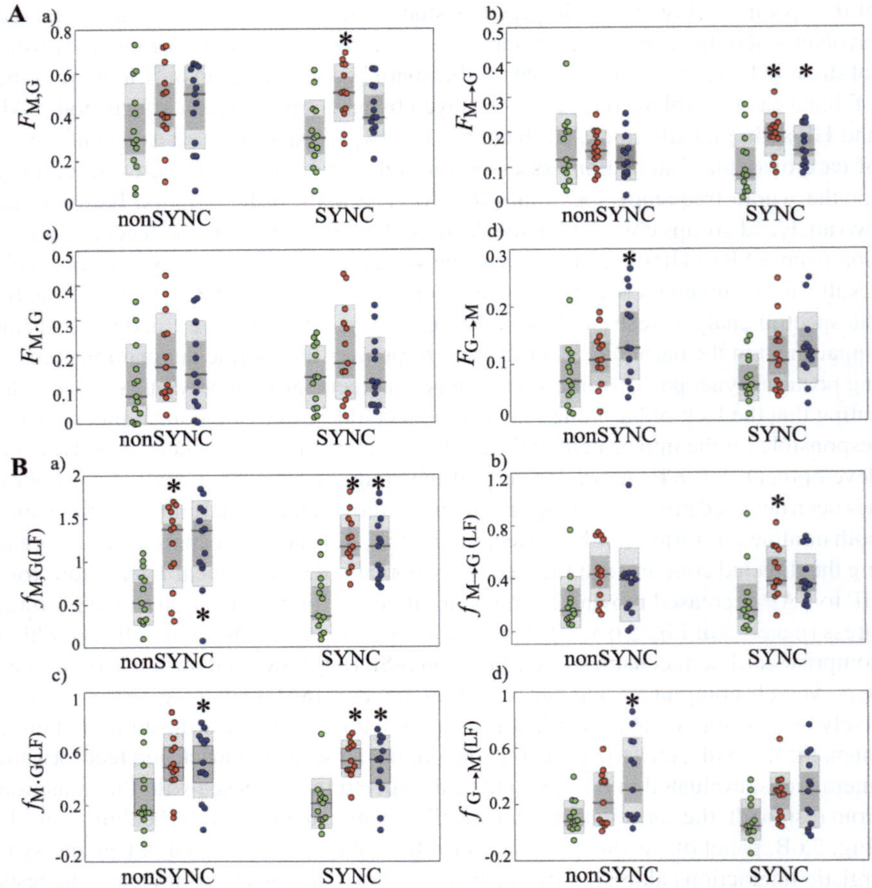

Fig. 5.7 **A** Time and **B** LF domain analysis of cerebrovascular interactions. Plots depict the individual values and boxplot distributions across subjects of the **a** total linear coupling, **b** directed interaction from MAP to CBFV and **d** from CBFV to MAP, and the **c** instantaneous term. All the measures are computed at rest (REST, left bars and green circles) and during head-up tilt (ET, middle bars and red circles; LT, right bars and blue circles). Statistically significant differences: *, REST versus ET and REST versus LT; #, ET versus LT. The figure is adapted from [75]

from MAP to CBFV during ET and LT compared to REST (panel *b*), while the linear interaction along the opposite causal direction (panel *d*) and the instantaneous interaction were substantially unaffected by tilt (panel *c*). In subjects, a tendency towards higher values of linear coupling from CBFV to MAP was observed (panel *d*), with significantly higher values of $F_{F\to M}$ during LT compared to REST; the coupling from MAP to CBFV and the instantaneous coupling did not change across conditions in this group (panel *b*). The results of the frequency domain analysis of cerebrovascular interactions are reported in Fig. 5.7B. The spectral measures were averaged within the LF band of the spectrum, quantifying the total information shared in this band between CBFV and MAP ($f_{M,F}(LF)$), the information transferred along the two causal directions ($f_{M\rightarrow F}(LF)$, $f_{F\rightarrow M}(LF)$, which can be negative in some cases), and the information related to the mixing between the two directions ($f_{F\cdot M}(LF)$). When assessed within the LF band, the total coupling (Fig. 5.7B, panel *a*) exhibits the same behaviour in the two groups, i.e., a significant increase during both ET and LT compared to REST; in the time domain, such an increase was reported only in SYNC individuals during ET (Fig. 5.7A, panel *a*). The increased cerebrovascular coupling during tilt was reflected in its most part by the measure quantifying mixing and instantaneous effects (Fig. 5.7B, panel *c*), as documented by the significantly higher values observed for $f_{F\cdot M}(LF)$ during LT in subjects and during both ET and LT in SYNC subjects. A statistically significant increase is also detected during LT for the linear interaction from CBFV to MAP in subjects (Fig. 5.7B, panel *d*), and during LT for the linear interaction from MAP to CBFV in SYNC subjects (Fig. 5.7B, panel *b*).

The time domain analysis of the pressure-to-flow interactions reported in Fig. 5.7A, panel *b* documented that, while no significant changes of the directed coupling from MAP to CBFV were detected in healthy controls, the measure increased significantly during both tilt epochs in the syncope patients. The same trends were observed when the directed interaction from MAP to CBFV was computed within the LF band ([0.07–0.2] Hz), documenting that the effect is relevant to this portion of the frequency spectrum (Fig. 5.7B, panel *b*). The increase of $MAP \rightarrow CBFV$ causal interactions in SYNC subjects suggests that the cerebrovascular autoregulation mechanisms may be impaired during tilt in this group. Indeed, cerebral autoregulation aims at maintaining MCBF relatively constant in presence of modifications of MAP [26, 27], thus attempting to preserve a certain degree of uncoupling between MCBF and MAP in a range of values of MAP as wide as possible. Accordingly, an increased causal coupling from MAP to CBFV may be interpreted as an indication of a less effective or impaired cerebrovascular autoregulation [23, 87]. This result is in line with previous findings using GC [88] and TE [57, 89] measures suggesting that the postural challenge reduces the effectiveness of the cerebrovascular control in patients prone to postural syncope. Looking at the directed interactions along the Cushing reflex, we found that the causal coupling from CBFV to MAP shows a tendency to increase progressively during postural stress; this tendency was statistically significant only in subjects after prolonged stress (Fig. 5.7A, panel *d*). The same trends were observed when the directed interaction from CBFV to MAP was computed within the LF band ([0.07–0.2] Hz), documenting that the effect is relevant to this portion of the frequency spectrum (Fig. 5.7B, panel *d*). These results may suggest that the sympathetic activation caused by prolonged postural stress is beneficial to counteract orthostatic intolerance. Indeed, since the Cushing reflex is a control mechanism enhancing sympathetic activity [90], the increase of coupling along this arm of the cerebrovascular control suggests that the prolongation of the postural stress induces a further activation of the sympathetic nervous system in subjects who do not develop syncope. Nevertheless, these findings should be confirmed in larger datasets, also because the same enhancement of the information transfer from CBFV to MAP was not clearly documented in previous studies [88, 89]. The differences may be due to the adopted methodology ([89] employed transfer entropy, and [88] assigned instantaneous causality to the causal measure) and the difference in the protocols ([89] did not distinguish the tilt epochs, and [88] did not consider healthy controls). The comparison between time domain measures accounting for whole-band interactions and frequency-specific measures computed within the LF band ([0.07–0.2] Hz) indicates that the measures of directed interaction behaved similarly (panels *b, d* of Fig. 5.7A, B), whereas the measures of total coupling and mixing exhibited different patterns in response to orthostatic stress (panels *a, c* of Fig. 5.7A, B). In fact, while the time domain measures did not evidence marked changes across conditions, the LF measures of total coupling and mixing displayed marked and statistically significant increments during the postural stress. This indicates that instantaneous effects between CBFV and MAP play an important role in cerebrovascular interactions and may confound—together with interactions occurring outside the LF band—the detection of the impact of the postural challenge on cerebral autoregulation. We remark that, according to our results, this impact is the same in SYNC and nonSYNC subjects when it is assessed in terms of measures that account for both pathways of the bidirectional interaction between MAP and CBFV; the differences between the two groups emerged considering the causal measures as discussed above.

Limitations and Conclusions. In our statistical analyses, no differences between the two analyzed groups of SYNC and nonSYNC subjects were detected. While this result may suggests that any modification of the cardiovascular or cerebrovascular control in patients prone to develop syncope is better detectable observing the response to the orthostatic challenge rather than comparing different groups, it may be a consequence of the small dataset analyzed. A main limitation of the present study concerns indeed the small number of subjects involved, which may affect the robustness of the obtained results. Future studies involving more participants should be performed to confirm the results and corroborate the interpretations of the current study. From a methodological point of view, although our work introduces an integrated framework capable of assessing total and causal interactions in bivariate time series in both time and frequency domains, it does not provide conclusive information about the treatment of instantaneous interactions in the computation of the presented measures. Here, instantaneous interactions were assigned to a specific causal direction based on physiological knowledge in the cardiovascular analysis, and were left unassigned in the absence of such knowledge in the cerebrovascular analysis. Alternative approaches have been explored, including the use of non-Gaussian modelling and independent component analysis to set the direction of instantaneous effects without prior knowledge [57, 88], and the use of disconnected AR models studied in the frequency domain to provide undirected measures of instantaneous causality and extended measures of GC [68]. Since our work confirms that instantaneous effects play a fundamental role in cardiovascular and cerebrovascular variability analysis, future studies are envisaged to compare our framework with existing approaches, to assess the agreement of in these methods in the evaluation of frequency domain interaction and shed more light on this delicate issue. For further methodological details, we refer the reader to the Appendix B.

The present work introduced a unified framework for the analysis of pairwise interactions between time series, decomposing the overall interaction into directed and instantaneous effects and considering both time and frequency domain representations. The added value of the framework stands in the tight relation between the measures defined in the time and frequency domains, which favors interpretability, and in the possibility to incorporate instantaneous influences into one causal direction or leave them isolate in the evaluation of the interactions, which brings flexibility. Exploiting these features, we contributed to describe the physiological mechanisms involved in the cardiovascular and cerebrovascular regulation in the response to a physiological stimulation (i.e., the postural stress) and to the development of an autonomic dysfunction (i.e., the pre-symptoms of postural related syncope). In perspective, the combination of time and frequency domain measures can help elucidating the mechanisms behind the oscillatory rhythms as well as the broad-band dynamics of coupled physiological variables studied in a variety of physiological conditions and diseased states. Further steps should consider the possibility to investigate the network formed by the four physiological variables as a whole, using, e.g., the OIR framework to detect patterns of HOIs among the investigated signals.

5.6 Granger Causality, Isolation and Autonomy in Closed-Loop Systems of Cerebrovascular Variables

Cerebrovascular interactions between MAP and MCBFV time series have been largely studied to investigate the cerebrovascular control and dynamic CA in a variety of physiopathological conditions [26, 27, 89]. Cerebrovascular interactions are largely determined by the so-called *pressure-to-flow link*, according to which variations of MAP drive similar changes in CBFV but also trigger CA responses whereby an homeostatic regulation of CBFV is looked for [26, 27]. In this section, we report the practical computation of the spectral measures of Granger causality, isolation and autonomy on cerebrovascular time series measured in healthy controls and subjects prone to develop postural-related syncope [28, 57], published in [55]. Here, we hypothesize that spectral indexes quantifying both the causal effects of MAP on MCBFV and the autonomous dynamics of MCBFV can identify the alteration of the physiological control mechanisms related to cerebrovascular interactions and to CA occurring with postural stress in subjects with poor orthostatic tolerance better than the more commonly used time domain indexes.

Experimental Protocol and Time Series Extraction. The analyzed time series belong to a database previously collected to study the short-term physiological regulation in subjects prone to neurally-mediated syncope and healthy controls via the analysis of spontaneous variability of systemic variables [28, 57]. The study included 13 subjects with previous history of unexplained syncope (SYNC) and 13 age-matched healthy subjects (nonSYNC). ECG, AP, and CBFV were acquired simultaneously. From these signals, the variability series of MAP and MCBFV were extracted on a beat-to-beat basis by taking the average of the AP and CBFV signals measured between the local minima occurring in the signals after each heartbeat detected from the ECG. For each subject, three sequences of 250 consecutive synchronous values of MAP and MCBFV were selected for the analysis, corresponding to the following experimental conditions: *(i)* supine rest (REST); *(ii)* early tilt (ET), starting after the onset of the head-up tilt maneuver; *(iii)* late tilt (LT), starting at least 5 min after the onset of the tilt maneuver for nonSYNC subjects, and occurring just before the pressure decrease due to presyncope for SYNC subjects. Further information about the experimental protocol, signal acquisition and time series extraction can be found in [28, 57], as well as in Sect. 5.5.

Data and Statistical Analysis. The time series extracted for each subject in the three experimental conditions were regarded as realizations of the MAP (the driver, Y_1) and MCBFV (the target, Y_2) discrete-time processes. These processes were assumed as uniformly sampled with a sampling frequency equal to the inverse of the mean HP. First, classical time domain markers such as the mean and variance of MAP ($\mu_{MAP}, \sigma^2_{MAP}$) and MCBFV ($\mu_{MCFBV}, \sigma^2_{MCFBV}$) were computed. Then, the series were pre-processed reducing the slow trends with an AR high-pass filter (zero phase; cut-off frequency 0.0156 Hz [69]) and removing the mean value. A bivariate AR model in the form of (2.4) was fitted on each pair of pre-processed series using OLS identification and setting the model order p according to the AIC (maximum scanned order $= 14$); the series and the PSD profiles were visually inspected and model orders were manually fixed where necessary, i.e. where too many or few spectral peaks were observed. After AR identification, the time domain and spectral measures of GC, GI and GA were obtained computing the parameters of the restricted models (2.8) and (2.9) from the estimated full-model parameters as described in Sect. 2.3.5.1 and then applying the derivations presented in Chap. 4. Figure 5.8a reports an example of MAP and CBFV time series, together with their estimated PSDs (panels *b*) and spectral GC (4.30), GI (4.35) and GA (4.44) profiles (panels *c, d*), measured for a representative subject. The spectral measures of GC from MAP to MCBFV, GI of MCBFV, and GA of MCBFV (respectively, $f_{MAP\rightarrow MCBFV}$, f_{MCBFV}, and a_{MCBFV}) were integrated within the two frequency bands of physiological interest for CB variability, i.e. the very-low frequency (VLF, $f \in [0.02–0.07]$ Hz) and LF ($f \in [0.07–0.2]$ Hz) [80], as well as over the whole frequency range $[0-f_s/2]$ to get the time domain values $F_{MAP\rightarrow MCBFV}$, F_{MCBFV}, and A_{MCBFV}. To test the statistical significance of the GC, GI and GA measures, a bootstrap method using explicit model equations extracted from the data was implemented, as described in AUTONOMOUS DYNAMICS IN BIVARIATE PROCESSES. A representative example is illustrated in Fig. 5.8c, d.

The distributions of the time domain markers as well as of GC, GI and GA computed across subjects for each group (SYNC and nonSYNC) were tested for normality using the *Anderson-Darling test*. Since the hypothesis of normality was rejected for most distributions, and given the small sample size, non-parametric tests

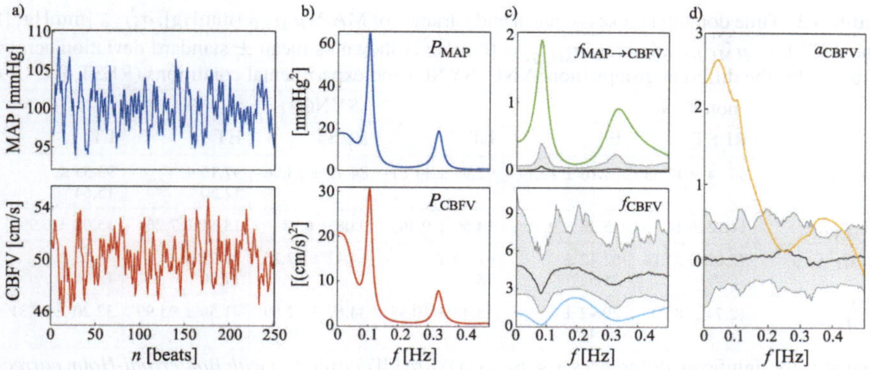

Fig. 5.8 Representative control subject in REST (model order: $p = 7$). **a** MAP and CBFV time series. **b** PSD profiles of the MAP (top) and the MCBFV series (bottom). **c** Spectral profiles of the GC from MAP to MCBFV and of the GI of MCBFV. **d** Spectral profile of the GA of MCBFV. In **c** and **d**, the distributions of the spectral GC, GI and GA measures computed from surrogate time series are depicted as shaded areas, median (black lines) and percentiles (grey lines, computed with a 5% significance level). The figure is adapted from [55]

were employed to assess the statistical significance of the differences of each index across conditions. Specifically, the *one-way Friedman test* was employed to assess the significance of the differences across conditions, followed in case of rejection by a post-hoc pairwise comparison carried out through the *paired Wilcoxon test with Bonferroni-Holm correction for multiple comparison* to assess the differences between pairs of distributions (REST vs. ET, REST vs. LT, ET vs. LT). All the statistical tests were carried out with 5% significance level.

Results and Discussion. Table 5.2 depicts the results—in terms of time domain markers (mean μ and variance σ^2) of the MAP and MCBFV series computed in the two analyzed groups during the three experimental conditions. The trends of these markers document the expected cerebrovascular response to the orthostatic stress in subjects prone to syncope and controls [89, 91]. Specifically, the average MCBFV decreased significantly during tilt in both groups as a consequence of the physiologic cerebral vasoconstriction associated with the orthostatic challenge. In the SYNC group, the drop of μ_{MCBFV} was more marked during LT and was accompanied by a significant increase, during both ET and LT compared to REST, of the average MAP, likely reflecting a progressive weakening of CA mechanisms which occurs with prolonged postural stress. The variability of the two series did not show evident trends across conditions, except for an increase of σ^2_{MAP} during ET.

Figure 5.9 reports the results of the analysis of causal, isolated and autonomous dynamics performed in the time and frequency domains as violin plots and individual values. For each distribution, the mean and interquartile range are depicted by the white circle and vertical line, respectively, while the width of the violin plot denotes probability density. All the time domain measures do not exhibit significant changes

Table 5.2 Time domain indexes of mean and variance of MAP (μ_{MAP} [mmHg], σ^2_{MAP} [mmHg^2]) and MCBFV (μ_{MCBFV} [cm/s], σ^2_{MCBFV} [(cm/s)^2]) shown as mean ± standard deviation across subjects for the different groups (nonSYNC, SYNC) and experimental conditions (REST, ET, LT)

	nonSYNC			SYNC		
	REST	ET	LT	REST	ET	LT
μ_{MAP}	98.84 ± 17.33	95.16 ± 12.17	92.94 ± 11.61	84.42 ± 13.96	97.16 ± 17.50*	93.53 ± 15.64*
σ^2_{MAP}	14.22 ± 14.79	15.38 ± 9.11	14.56 ± 9.46	9.08 ± 6.78	13.61 ± 7.26	15.08 ± 6.99
μ_{MCBFV}	72.02 ± 23.14	62.12 ± 21.52*	61.09 ± 15.72*	64.42 ± 17.25	56.25 ± 17.06*	48.12 ± 18.08*#
σ^2_{MCBFV}	12.74 ± 8.20	20.42 ± 11.42*	15.42 ± 10.38	34.67 ± 72.69	41.56 ± 95.99	32.20 ± 56.31

Statistically significant differences assessed via *paired Wilcoxon test with Bonferroni-Holm correction for multiple comparison*: *, REST versus ET, REST versus LT; #, ET versus LT

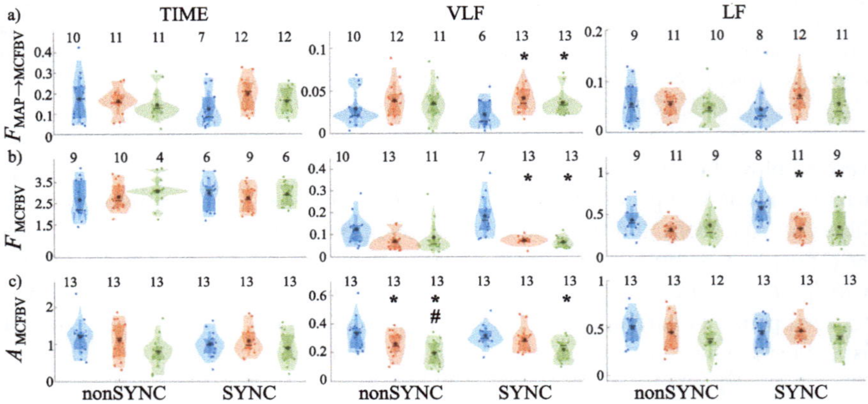

Fig. 5.9 Distributions across subjects of the GC from MAP to MCFV (**a**), the GI of MCBFV (**b**), and the GA of MCBFV (**c**) computed in the time domain (left) and integrating the spectral functions within the VLF (middle) and the LF band (right). Measures are computed at REST (blue) and during the early (ET, orange) and late (LT, green) phases of head-up tilt. Values above each distribution indicate the number of subjects for which the measure was deemed as statistically significant. *Paired Wilcoxon test with Bonferroni-Holm correction for multiple comparison*, $p < 0.05$: ∗, REST versus ET, REST versus LT; #, ET versus LT. The figure is adapted from [55]

across the three analyzed experimental conditions (Fig. 5.9, left plots). On the other hand, the evaluation of the same measures within the frequency bands of physiological interest for this application (i.e., VLF and LF) highlights some evident variations during the orthostatic stress, also differentiating the response between syncope subjects and healthy controls (Fig. 5.9, middle and right plots). These different behaviors of time domain and spectral measures evidence, for this physiological application, the need of assessing causal and autonomous dynamics in the frequency domain to capture mechanisms that remain otherwise hidden if a whole-band time domain analysis is performed.

The spectral analysis reveals a significant increase of the GC from MAP to MCBFV for the SYNC group, alongside with a significant decrease of the GI of MCBFV during both epochs of head-up tilt (ET, LT) compared to REST (Fig. 5.9a, b); the changes are observed in the VLF band for both measures, where they occur together with a marked increase in the number of subjects for which the GC and GI were statistically significant according to the surrogate data analysis, and also in the LF band for the GI measure. Methodologically, this finding confirms the simulation results showing that GC and GI provide complementary information, but also suggests that the two measures are not fully dependent on each other. Here, when assessed in specific frequency bands, the two measures describe physiological mechanisms with a different degree of discrimination: the tilt-induced enhancement of the influences of MAP on MCBFV is better captured by the GI measure. Physiologically, the presence of stronger causal interactions along the pressure-to-flow link during tilt, detected in the subjects prone to develop postural syncope but not in the healthy controls, may be indicative of a defective CA, i.e. of a reduced intrinsic ability of the cerebral vascular bed to maintain a stable perfusion despite blood pressure changes. Indeed, the increased causal coupling indicates that the variability of MCBFV is determined to a larger extent by the variability of MAP, and that the autoregulatory mechanism cannot respond fast enough to compensate for pressure changes. This interpretation agrees with that of previous studies in which a loss of CA has been associated with an increased link between AP and CBFV [23, 87]. The physiological mechanisms leading to the weakening of CA in the subjects prone to syncope are complex, and possibly include hypercapnia (i.e., augmented arterial carbon dioxide pressure) [92], which has been associated to upwards shift of the coherence between MAP and CBFV at frequencies <0.1 Hz [92], and vasoconstriction (i.e., reduction in the diameter of large vessels) [91], which can have an effect on the measured MCBFV since the Doppler ultrasound measures blood flow velocity and not absolute flow.

As regards the autonomy measure, the spectral analysis evidences a progressive reduction of the GA of MCBFV computed in the VLF band moving from REST to ET, and from ET to LT (Fig. 5.9c); the decrease is evident and statistically significant for each pairwise comparison in the nonSYNC healthy controls, while it is less marked and significant only comparing REST vs. LT in the syncope subjects. The decrease of GA with tilt indicates that the internal regulatory mechanisms of MCBFV acting in the VLF band loose progressively their strength during prolonged postural stress. As the decrease is evident particularly in the healthy controls, it seems to have a physiological rather than pathological origin; therefore, it should not regard the dynamic CA expressed in terms of interdependence between pressure and flow, which is indeed not efficiently represented by the GA measure. More likely, the decrease of GA reflects the reduced strength of exogenous effects, i.e., effects acting on MCBFV independently of MAP. Such effects might include the occurrence of hypocapnia with the orthostatic challenge in healthy subjects [93], which may have an impact on arteriolar vessel caliber, and thus on blood flow velocity. This impact, which is not observed nor quantified in our AR model, might alter the autonomous dynamics of CBFV and thus enter the computation of GA. A simulation example investigating the effects of unobserved confounders was provided in Sect. 4.2.3.3.

Limitations and Conclusion. The aim of this study was to explore, in addition to the well-known measure of Granger causality, the concepts of isolation and autonomy in coupled physiological processes, with emphasis on their frequency domain representation. In this context, the developed framework, comprising time and spectral formulations of bivariate GC, GI and GA measures, allows quantification of the concepts of causality, isolation and autonomy either considering the overall dynamics of the observed bivariate process or the oscillations at specific frequencies of physiological interest. Our experimental results document that the GI measure is complementary to GC but not trivially related to it, while GA reflects the regularity of the internal dynamics of the analyzed target process. The frequency domain formulation of GC, GI and GA is particularly useful for the analysis of dynamic processes which are rich of oscillatory content, as it allows to elicit physiological mechanisms which can be hidden in time domain due to the mixing with other spectral effects. This potential is demonstrated in our application to cerebrovascular interactions where the spectral measures highlight responses to postural stress which cannot be traced by the time domain analysis. In particular, our results suggest that GA quantifies the frequency-specific physiological response to postural stress of the slow CBFV oscillations, while GC and especially GI characterize the pathological response related to the impairment of the dynamic autoregulation of CBFV preceding the onset of postural-related syncope.

Nevertheless, this preliminary application presents some limitations. First, since the small size of analyzed group of subjects may represent an issue when one aims to generalize results to an entire population, the use of larger datasets is needed to confirm the results obtained here. Furthermore, in physiological applications where multiple complex interactions often arise, the effects of unobserved confounders are likely to occur as we have shown in Sect. 4.2.3.3. Therefore, the extension to multivariate datasets including signals possibly acting as confounders (such as, in our application, respiration and arterial carbon dioxide [92, 94]), as well as the development of multivariate extensions of the proposed measures of GC, GI and GA, are envisaged for future studies. Moreover, the inclusion of instantaneous effects in the analyzed parametric models, though not always straightforward [59, 68, 95], is recommended to provide a complete picture of causal, isolated and autonomous effects emerging in the time and frequency domains from dynamic interactions.

5.7 Cardiovascular, Cardiorespiratory and Cerebrovascular High-Order Interactions Assessed Through the O-Information Rate

Up to now, we have learnt that the representation with self-effects (e.g., for arterial compliance—Sect. 5.1) and pairwise (e.g., for cardiovascular and cerebrovascular closed-loops—Sects. 5.5 and 5.6) interactions is a powerful tool although often insufficient to provide a complete description of a complex system. It is now firmly acknowledged that many real-world systems display high-order interactions involving more than two network nodes [96]. Thus, in these systems the network behavior is integrated at different hierarchical levels and time scales. This occurs in Network Physiology, where it is important to distinguish between organ systems that interact as a pair, or as a part of a more complex structure, to produce the observed dynamics. For instance, cardiovascular interactions may arise autonomously from self-sustained mechanisms or as a result of the effects of respiration on the measured dynamics [97] (Sect. 5.5), while the interplay between cerebrovascular variables may be affected by the role of arterial carbon dioxide (p_{CO_2}) or other exogenous drivers of cerebral blood flow independently of blood pressure [92, 93] (Sect. 5.6).

In [77], a new approach to quantify both pairwise and high-order interdependencies for networks of rhythmic processes interacting across multiple time scales was proposed, defining the O-Information Rate as a new dynamic metric of HOIs (Sect. 4.3.1.1). Moreover, the possibility to assess HOIs in the frequency domain offered by the spectral expansion of the OIR (Sect. 4.3.2.1) has opened the way to the evaluation of redundant and/or synergistic interactions within specific frequency bands with physiological meaning [77]. In this context, in [98, 99] we applied the time and spectral OIR measures to networks of physiological time series, with the aim of investigating the nature of multivariate interactions underlying the communication among different physiological systems. To this end, we focused on multivariate physiological time series reflecting the dynamics of heart period, arterial pressure, breathing volume and cerebral blood flow to investigate the joint cardiovascular, cerebrovascular and respiratory regulation during postural stress [28, 57]. The application features physiological time series rich of oscillatory content, which thus lend themselves to the spectral analysis of HOIs performed by the proposed tool.

Experimental Protocol and Data Analysis. The OIR framework [77] was applied to a database of physiological time series collected from 13 young healthy subjects during supine resting (REST) and passive standing in the 60° upright position reached after head-up tilt (TILT). For each subject and condition, $Q = 5$ stationary sequences of 250 beat-to-beat values of HP (H), SAP and DAP (S, D), respiration (R), and MCBFV (F), were considered for the analysis [28, 57] (see Sect. 5.5 for details about the experimental protocol, signal acquisition and time series extraction). A VAR model (2.6) was fitted on the time series measured for each subject and condition, with model order set according to the AIC. Then, the spectral OIR (4.64) was computed for each multiplet of order $N = 3, 4, 5$ and integrated within

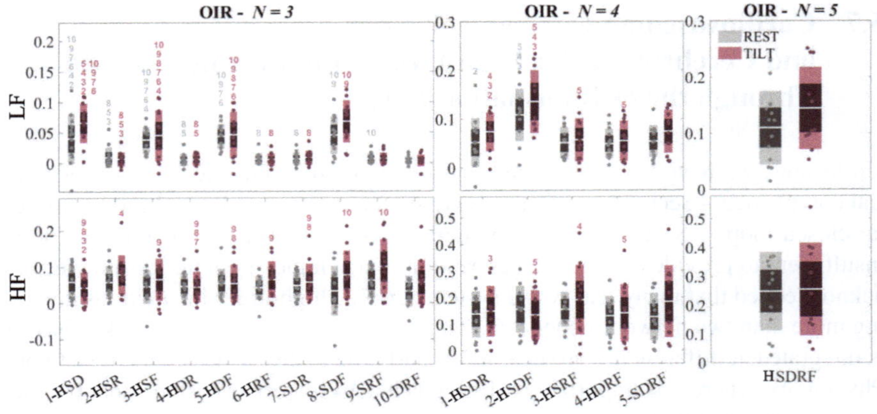

Fig. 5.10 Distribution across subjects and individual values of the spectral OIR integrated in the LF (top) and HF (bottom) band, computed at REST (grey) and during TILT (violet) for all possible multiplets of order 3,4,5 obtained grouping the time series of heart period (H), systolic pressure (S), diastolic pressure (D), respiration (R) and mean cerebral blood flow velocity (F). Numbers in the OIR(3) and OIR(4) panels indicate pairs of distributions for which the mean OIR differed significantly in a given condition (*Student t-test for paired data*, $p < 0.05$). The figure is adapted from [99]

the LF ([0.04–0.15] Hz) and HF ([0.15–0.4] Hz) bands of the spectrum to obtain frequency-specific measures of HOIs.

Results and Discussion. The results collected in Fig. 5.10 show that the OIR integrated in both LF and HF bands was positive in the large majority of subjects in both the analyzed conditions, and showed a tendency to increase with the number of series in the analyzed multiplets. This finding suggests that physiological networks probed by beat-to-beat variability series are dominated by redundancy. The result confirms similar findings observed in cardiovascular and cardiorespiratory networks [100, 101], and extends them to cerebrovascular and integrated physiological networks. Multivariate interactions were found to be stronger for HF oscillations than in the LF band, suggesting a main role of respiration, whose oscillations are typically mostly confined within the HF band [97], in driving redundant interactions in the cardiovascular and cerebrovascular networks. This result is confirmed by the observation that in the LF band the highest redundancy was displayed by multiplets including the series H, S, D and F, while significantly lower values were found in multiplets including the series R, in both experimental conditions. For these multiplets, a tendency towards an increase in redundancy (though not statistically significant) was observed moving from REST to TILT, suggesting a possible role of sympathetic activation in driving the redundancy of LF oscillations in cardiovascular and cerebral blood flow variables [100]. Conversely, results in the HF band showed some statistically significant differences between multiplets only during TILT; specifically, the multiplets containing the series R, S and F displayed the highest redundancy. No statistically significant differences between REST and TILT were detected.

Overall, these results confirm for HOIs the redundant nature of cardiovascular and cerebrovascular interactions previously reported for triplets of physiological processes [77, 97, 100], and document the relevance of separating LF and HF contributions to elicit the role of respiration on cardiovascular and cerebrovascular interactions. Moreover, the tendency of the OIR to increase with tilt was not statistically significant, suggesting that these redundant effects are preserved during postural stress. The understanding of the different and complex ways of dynamic integration of organ systems as a complex network remains one of the biggest problems in field of Network Physiology. Physiological systems exhibit complex dynamics, operate at different time scales and are regulated by multi-component mechanisms, which has been known to challenge the study of physiologic coupling and causality [102, 103]. These aspects, together with the evidence that cardiovascular and cerebrovascular interactions occur through the coupling of rhythms in different frequency bands with different physiological meaning [104], make our spectral approach eligible to probe HOIs in these networks [101]. Our results document that respiration acts as a major driver of multivariate redundant interactions in physiological networks, confirming that HOIs can have different nature for different rhythms because synergistic and redundant behaviors generally alternate in different bands of the frequency spectrum [101, 105].

5.8 Gradients of O-Information in Multi-organ Networks

Although physiological systems generally display dynamic behaviors, where the temporal correlations between different processes acquire a non negligible significance, studying the nature of the high-order node-specific dependencies between groups of variables may lead to draw meaningful conclusions on the interplay between these variables at lag zero. To this aim, in [106] we applied the OI framework (Sect. 3.2) to a database of physiological time series collected to study the effect of postural stress on cardiovascular, cerebrovascular and respiratory variability [28, 57], with the aim of investigating static HOIs among physiological signals through the use of OI gradients, calculated using the Gaussian Copula approach described in [107] to estimate entropy terms. In this section, we report the main results and attempt to make a comparison between static and dynamic (Sect. 5.7) analyses performed on the same network.

Experimental Protocol and Data Analysis. The original dataset is comprised of 13 young healthy subjects, enrolled at the *Neurology Division of Sacro Cuore Hospital, Negrar, Italy*. ECG was acquired together with AP measured at the level of middle finger through a photopletysmographic device. CBFV and respiration were measured at the level of the middle cerebral artery by means of a transcranial Doppler ultrasonographic device and through a thoracic impedance belt, respectively. From the raw signals, the physiological beat-to-beat variability series of HP, SAP, MAP, MCBFV and RESP were measured as detailed in [28, 57] and reported in Sect. 5.5 during two stationary time windows of length 250 beats in the following physiological conditions: *(i)* supine rest (REST) and *(ii)* head-up tilt test with table inclination of 60° (TILT). Prior to network analysis, each series was high-pass filtered to remove slow trends and normalized to zero mean and unit variance. Then, the first (3.12) and second (3.15) order gradients were evaluated for the physiologic network constituted by the five time series $\{HP, SAP, MAP, MCBFV, RESP\}$ (average over subjects). Bootstrap data analysis was applied to assess the statistical significance of the computed measures for each subject: gradients were considered significantly redundant (synergistic) when the 5th (95th) percentile of the bootstrap distribution was higher (lower) than zero (AN APPROACH TO THE STATISTICAL ASSESSMENT OF HIGH- ORDER INTERACTIONS).

Results and Discussion. The first and second order gradients are reported in Fig. 5.11. Looking at the first order gradients (panel *a*), we see that the heart plays a synergistic role in the resting state, whilst in the orthostatic position the system becomes dominated by redundancy with a disconnection of respiration. The analysis of the second order gradients provided similar results (panel *b*), with the cardiovascular link between HP and SAP showing a synergistic character during the supine rest, and with increasing redundant behavior of the whole network after head-up tilt.

Our results document a well known fact in physiology, i.e., that cardiovascular, cerebrovascular and respiratory interactions are highly redundant. This is also highlighted by previous dynamic analysis (Sect. 5.7, Fig. 5.10), where the first-order OIR gradient was found to be mostly redundant (left column). On the other hand, here we show an interesting and probably novel aspect of cardiovascular oscillations: the heart rate plays a synergistic role in the resting state analyzed with our static analysis; synergy could result from the fact that HRV is the common target for several neuro-autonomic mechanisms including the cardiac baroreflex and the RSA mechanism [34]. We also find that redundancy is strongly enhanced by the entrainment of cardiovascular and cerebrovascular oscillations and by sympathetic activation; in particular, in the upright position all the series are highly redundant, except for the respiration signal, which is out of the redundant circuits in tilt conditions. These results agree with the tilt-induced shift of the sympatho-vagal balance towards increased sympathetic activity and decreased parasympathetic activity [108], also previously documented via information-theoretic analyses [109]. Remarkably, here we stress the convergence of static and dynamic analyses in demonstrating the *dominant synergistic nature of respiration*. Indeed, while the OI gradients show overall low values of redundancy of groups including respiration (see, e.g., the second-order gradients in Fig. 5.11b), likely to be more prominent at higher orders, the first-order OIR gradient was found to be very low—probably not significant at all—for triplets including respiration, thus suggesting the leading synergistic character of this variable (Fig. 5.10, left column).

Overall, the redundancy showed a tendency to increase with tilt, documenting an effect of sympathetic activation on the redundant interactions among cardiovascular

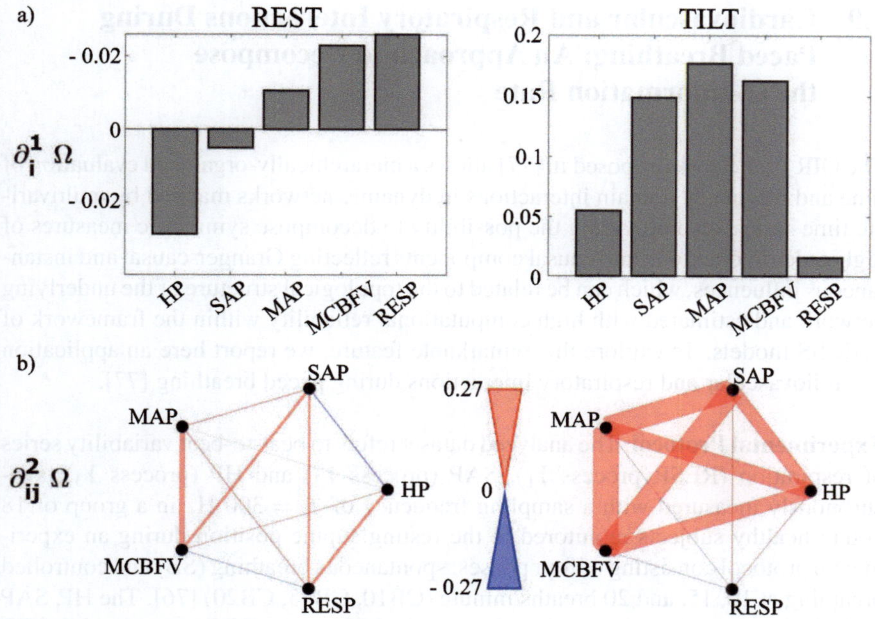

Fig. 5.11 **a** First and **b** second order gradients for the physiological system composed of the five time series $\{HP, SAP, MAP, CBFV, RESP\}$, averaged over subjects and computed in the two experimental conditions: supine resting state (REST, left) and head-up tilt (TILT, right). Colors indicate redundant (red) and synergistic (blue) characters of interaction. Width of the links indicates the strength of the gradients. The figure is adapted from [106]

and cerebrovascular oscillations [77], as also demonstrated in Sect. 5.7 (Fig. 5.10). Bootstrap data analysis [110] confirmed these findings. Specifically, as regards the first order gradients in the resting state, one out of the five showed significant synergy in 50% of subjects (HP), while in TILT significant redundancy was found for three of the five (SAP, MAP, MCBFV). Going to the second order gradients, only in the TILT condition redundancy was significant in more than 50% of subjects for the pairs HP-MAP, HP-MCBFV, SAP-MAP, SAP-MCBFV, MAP-MCBFV. This suggests an important role of the sympathetic activation led by head-up tilt in increasing redundancy in physiologic networks. Interactions involving respiratory, arterial pressure and blood flow variabilities are more shifted to redundant rather than synergistic modes of interplay; significance increases moving from the supine to the upright position, ultimately highlighting that redundancy is significantly strongly enhanced during the orthostatic stress.

5.9 Cardiovascular and Respiratory Interactions During Paced Breathing: An Approach to Decompose the O-Information Rate

The OIR framework proposed in [77] allows a hierarchically-organized evaluation of time and frequency domain interactions in dynamic networks mapped by multivariate time series, also providing the possibility to decompose symmetric measures of high-order interactions into causal components reflecting Granger-causal and instantaneous influences, which can be related to the topological structure of the underlying network and estimated with high computational reliability within the framework of VAR SS models. To explore this remarkable feature, we report here an application to cardiovascular and respiratory interactions during paced breathing [77].

Experimental Protocol. The analyzed dataset refers to beat-to-beat variability series of respiration (RESP, process Y_1), SAP (process Y_2) and HP (process Y_3), synchronously measured with a sampling frequency of $f_s = 300$ Hz in a group of 18 young healthy subjects monitored in the resting supine position during an experimental protocol consisting of four phases: spontaneous breathing (SB) and controlled breathing at 10, 15, and 20 breaths/minute (CB10, CB15, CB20) [76]. The HP, SAP and RESP time series were extracted respectively from the ECG, non-invasive ABP and nasal respiration flow as the sequences of the duration of the cardiac cycle (RRI), of the local maximum of the blood pressure signal within each detected cardiac cycle, and of the value of the respiration signal sampled at the onset of each cardiac cycle. This measurement convention implies that instantaneous influences can be described as causal effects from RESP to SAP and HP and from SAP to HP (directions $Y_1 \to Y_2$, $Y_1 \to Y_3$, $Y_2 \to Y_3$) [111]. The analysis was performed on stationary segments of the time series including 256 heartbeats, selected by visual inspection for each subject and experimental condition [76]. We note that the dataset herein analysed is the same as in [75] (Sect. 5.4), where however only the closed-loop cardiorespiratory interactions between heart rate and respiration were taken into account.

Data Analysis. Pre-processing consisted on detrending and mean removal for each time series. The VAR model (2.6) fitting the three series was identified through the OLS method (Sect. 2.3.5), selecting the order p in the range 3–14 by means of the AIC [112]. The analysis was focused on decomposing the OIR (4.51) of the three processes in OIR increments (4.52) obtained when the HP process is added to the bivariate process {RESP, SAP}. Specifically, starting from the estimated VAR parameters, we computed $\delta_{Y_1,Y_2 \to Y_3}(\bar{f})$, $\delta_{Y_3 \to Y_1,Y_2}(\bar{f})$ and $\delta_{Y_1,Y_2 \cdot Y_3}(\bar{f})$ from the terms of the spectral decomposition (4.63), then deriving $\nu_{Y_1,Y_2,Y_3}(\bar{f}) = \delta_{Y_1,Y_2;Y_3}(\bar{f})$ via (2.6, 2.7). From these spectral measures, time domain measures were obtained through integration over the whole frequency axis or within the LF range ([0.04–0.12] Hz) and the HF range (± 0.04 Hz around the peak respiratory frequency f_R). Given the possibility to ascribe instantaneous effects to specific causal directions (see above), the analysis is performed summing the information shared instantaneously between {RESP,

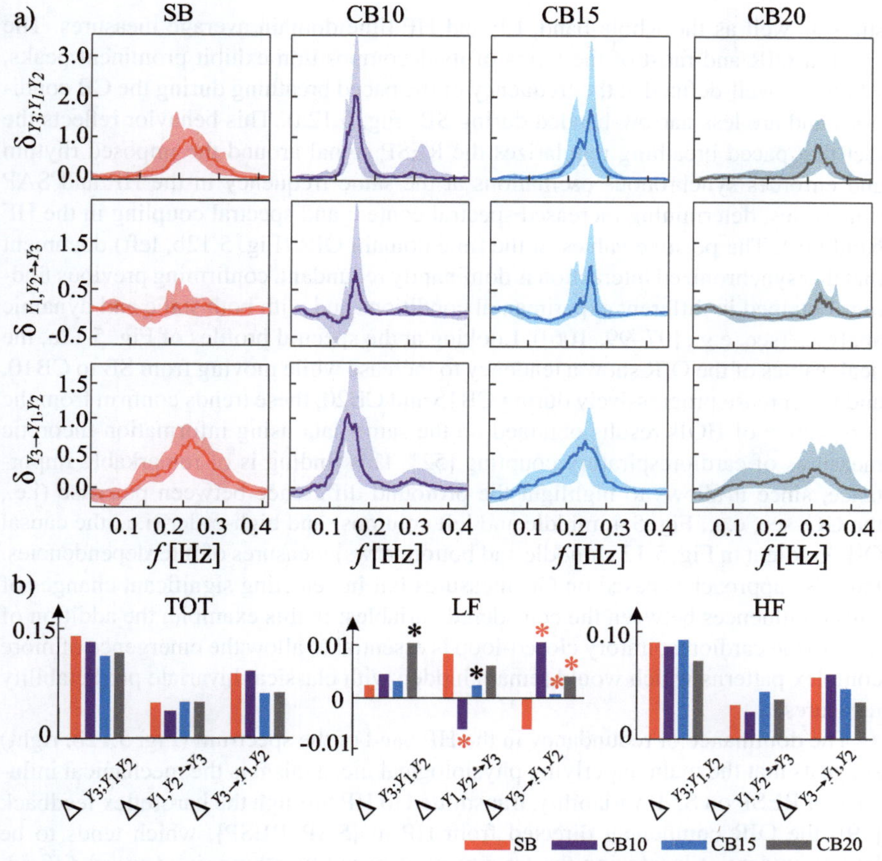

Fig. 5.12 **a** Average spectral profiles across subjects (line: median; shades: 1st–3rd quartiles) of the OIR increment obtained with the addition of HP to {SAP, RESP} (upper panels) and of its decomposition in causal terms (middle and lower panels) computed during spontaneous (SB) and controlled breathing (CB). **b** Time domain values of the mean OIR increments obtained integrating the spectral measures over the whole frequency axis (TOT), in the LF and in the HF bands; *, statistically significant differences ($p < 0.05$, CB vs. SB, *Wilcoxon signed-rank test*: black, uncorrected; red, *Bonferroni-Holm correction for multiple comparisons*). The figure is adapted from [77]

SAP} and HP to the information transferred from {RESP, SAP} to HP, i.e., computing the spectral and time domain measures $\delta_{Y_1,Y_2 \dot{\rightarrow} Y_3}(\bar{f}) = \delta_{Y_1,Y_2\rightarrow Y_3}(\bar{f}) + \delta_{Y_1,Y_2\cdot Y_3}(\bar{f})$ and $\Delta_{Y_1,Y_2 \dot{\rightarrow} Y_3} = \Delta_{Y_1,Y_2\rightarrow Y_3} + \Delta_{Y_1,Y_2\cdot Y_3}$. Spectral analysis was performed assuming the series as uniformly sampled with sampling frequency equal to the inverse of the mean HP. If the reader is interested in the full analysis of the individual measures and methodological details, we refer to the supplemental material of [77].

Results and Discussion. The results of OIR computation and decomposition are reported in Fig. 5.12, showing the grand average of the frequency domain measures as well as the whole-band, LF and HF time domain average measures. The spectral OIR and most of the terms of its decomposition exhibit prominent peaks, which are well-defined at the frequency of the paced breathing during the CB conditions and are less narrow-banded during SB (Fig. 5.12a). This behavior reflects the fact that paced breathing regularizes the RESP signal around the imposed rhythm and enforces synchronous oscillations at the same frequency in the HP and SAP time series, determining increased spectral content and spectral coupling in the HF band [76]. The positive values of the time domain OIR (Fig. 5.12b, left) document that this synchronized interaction is dominantly redundant, confirming previous findings obtained in different experimental conditions and with both static and dynamic analyses (see, e.g., [97–99, 106]). Looking at the spectral profiles of Fig. 5.12a, the peak values of the OIR show a tendency to increase while moving from SB to CB10, and to decrease progressively during CB15 and CB20; these trends confirm from the perspective of HOIs results obtained on the same data using information-theoretic measures of cardiorespiratory coupling [52]. This finding is of remarkable importance, since it allows to highlight the profound difference between pairwise (i.e., the GC; see, e.g., Fig. 5.4, middle and left columns) and high-order (i.e., the causal OIR gradient in Fig. 5.12a, middle and bottom rows) measures of interdependencies. Pairwise approaches based on GC measures fail in detecting significant changes of causal influences between the considered variables; in this example, the addition of SAP to the cardiorespiratory closed-loop is essential to allow the emergence of more complex patterns which would remain hidden with classical bivariate predictability measures.

The dominance of redundancy in the HF band of the spectrum (Fig. 5.12b, right) suggests that the main underlying physiological mechanism is the mechanical influence of RESP on SAP variability, transmitted to HP through the baroreflex feedback [79]; the OIR component directed from HP to {SAP, RESP}, which tends to be less redundant at increasing the frequency of paced breathing, is of more difficult interpretation and is likely dominated by the mechanical feedforward effects from HP to SAP [113]. The dominance of redundant mechanisms around the respiratory frequency impacts substantially the whole-band time domain OIR, which show comparable values across the analyzed conditions (Fig. 5.12b, left). On the other hand, the measures integrated within the LF band vary significantly moving from spontaneous to paced breathing (Fig. 5.12b, middle): the information transfer from {SAP, RESP} to HP becomes mostly synergistic during CB10, and during CB15 and CB20 returns progressively to the redundant values observed at SB; the information transfer along the direction HP→ {SAP, RESP} is prevalently synergistic at rest and shifts to redundant values during CB. The shift to synergy observed at CB10 for $\Delta_{Y_1,Y_2 \rightarrow Y_3}$ suggests that, when the respiratory activity slows down and tends to overlap with the Mayer waves typically observed in SAP and HP [114], the baroreflex (SAP→HP) and respiratory sinus arrhythmia (RESP→HP) mechanisms operate independently in determining the variability of heart rate.

5.10 Partial Information Rate Decomposition in Physiological Networks

As pointed out in Sect. 4.3.1.1, pairwise and fully conditioned TE analyses may encounter challenges in the presence of synergy or redundancy in time series data; indeed, pairwise causality fails to reveal synergistic effects while fully conditioned causality may fail to reveal redundant effects [115]. Moreover, high-order metrics such as the OIR framework do not allow to dissect separately mechanisms of redundancy and synergy occurring in complex networks of multiple nodes, since they do not put in evidence multiplets of variables which are both redundant and synergistic with equal strength. Conversely, we have shown how the PIRD is a powerful tool which separately evaluates redundant and synergistic patterns of interactions among groups of four variables in the context of a dynamic analysis. To show the promising features of the method, we report here the practical application of the PIRD framework (Sects. 4.3.1.1 and 4.3.2.1) to a database including 13 subjects at rest and during prolonged postural stress leading to pre-syncope [28, 57]. The application was first presented in [116], and then in [117].

Experimental Protocol and Data Analysis. Beat-to-beat variability series of HP (H), MAP (M), MCBFV (F) and RESP (R) were extracted during three stationary epochs of 250 beats in the supine rest (REST), early tilt (ET), and late tilt (LT) phases. Series were filtered to reduce slow trends and normalized to zero mean. Further details about the protocol, signal acquisition and time series extraction can be found in [28, 57] and are reported in Sect. 5.5. The physiological network comprising the F, M, H and R time series was investigated to decompose the rate of information dynamically shared between the target ($Y = F$) and a set of source processes ($\mathbf{X} = \{H, M, R\}$, with $H = X_1$, $M = X_2$, $R = X_3$).

Each series was first detrended with an AR high-pass filter with zero phase (cutoff frequency 0.015 cycles/beat) [69]. Then, a VAR model in the form of (2.6) was fitted to the four time series; model identification was performed via the OLS approach, setting the model order p according to the AIC for each subject (with maximum model order equal to 14). After VAR identification, computation of time and frequency domain interaction measures of mutual, unique, redundant and synergistic information rates was performed from the estimated model parameters and spectra of the processes. Spectral analysis was performed assuming the series as uniformly sampled with the mean HP taken as the sampling period ($f_s = \frac{1}{<H_n>}$). Specifically, the spectral MIRs between the target and groups of sources were computed; then, according to coarse-graining PIRD, the unique, redundant and synergistic information rates calculated from the spectral redundancy rate in (4.68) were computed through integration within the low frequency (LF, [0.04–0.15] Hz) and high frequency (HF, [0.15–0.4] Hz) bands of the spectrum to analyze specific rhythms with physiological meaning, as well as through integration over all frequencies, to get overall time-domain measures.

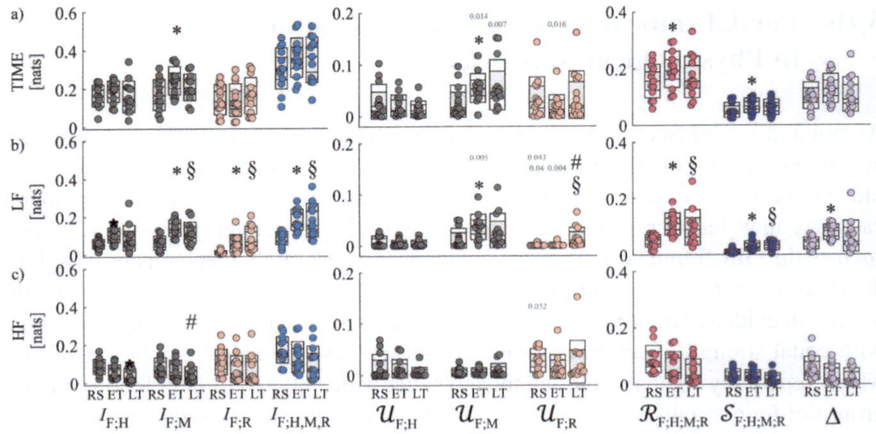

Fig. 5.13 Coarse-grained PIRD applied to the physiological network of mean cerebral blood flow velocity (F), mean arterial pressure (M), heart period (H) and respiration (R) assessed in syncope patients. The mutual, unique, redundant and synergistic information rates shared between the target F and the sources $\{H, M, R\}$, as well as the redundancy-synergy balance Δ, are shown as boxplot distributions and individual values in the REST (left, RS for brevity), ET (middle) and LT (right) conditions along **a** the whole frequency axis (TIME), **b** the LF and **c** the HF bands of the spectrum. *Wilcoxon signed rank test for paired data*, $p < 0.05$: $*$, REST versus ET; §, REST versus LT; #, ET versus LT. The figure is adapted from [117]

Given the small size of the surveyed population, non-parametric statistical tests were applied to assess statistically significant differences between indexes evaluated in the three phases of the experimental protocol, i.e., REST, ET and LT conditions. Specifically, the *Wilcoxon signed rank test for paired data* was applied on the MIRs shared between the target and each source, as well as on the PIRD terms (i.e., the joint MIR shared between the target and all the sources, the unique, redundant and synergistic information rates), evaluated in the time domain and along the LF and HF bands of the spectrum during the three phases of the protocol. Further, the statistical differences among pairs of unique information rate measures, as well as between redundancy and synergy, were tested in the time domain, LF and HF bands for each experimental condition via *Wilcoxon signed rank test for paired data* with *Bonferroni-Holm correction for multiple comparisons*. For all the statistical tests, the significance level was set to 0.05.

Results and Discussion. Results are shown in Fig. 5.13 as boxplot distributions and individual values of the PIRD measures computed in the REST, ET and LT conditions. The MIRs ($I_{F;H}, I_{F;M}, I_{F;R}, I_{F;H,M,R}$) and unique information rates ($\mathcal{U}_{F;H}, \mathcal{U}_{F;M}, \mathcal{U}_{F;R}$) assessed in the time domain (panels *a*) and within the LF (panels *b*) and HF (panels *c*) bands of the spectrum are depicted in the left and middle columns, respectively, while the redundant and synergistic contributions and their balance (Δ) are shown in the right column. The statistical differences among pairs of unique information rates (middle panels) were tested via Wilcoxon signed rank test for paired data ($p < 0.05$) with Bonferroni-Holm correction for multiple comparisons: significant p-values are shown above the boxplot distributions with colors indicating the distribution to be compared with (i.e., black if compared with $\mathcal{U}_{F;H}$, grey if compared with $\mathcal{U}_{F;M}$).

The transition to ET induced an increase of $I_{F;M}$ and $\mathcal{U}_{F;M}$ mainly visible in the LF band of the spectrum (Fig. 5.13a, b, left and middle columns), in line with previous observations indicating that the increase of the information transfer from M to F is related to altered CA in syncope patients [55, 59, 89]. The LT-induced decrease of $I_{F;M}$ in HF is probably related to the modulating effect of respiration [28], and thus is not detected by the unique contribution of arterial pressure ($\mathcal{U}_{F;M}$, panel **c**, middle column). As regards the interactions between cerebral blood flow and respiration, we found that the MIR $I_{F;R}$ increases during the early phase of tilt in LF probably due to confounding effects of H, M, while REST and ET values of $\mathcal{U}_{F;R}$ are very close to zero for most of subjects (Fig. 5.13b, left and middle columns), suggesting that the unique information shared between R and F may be negligible during this phase of the protocol. This result is further confirmed by the significantly lower values of $\mathcal{U}_{F;R}$ than $\mathcal{U}_{F;H}$ and $\mathcal{U}_{F;M}$ in the REST and ET phases. However, a significant increase of the unique information rate shared between the respiratory and the cerebral system in the LF band is observed in LT, thus shedding light on the remarkable role of respiration in influencing the dynamics of CBFV suddenly before the occurrence of syncope [28, 94]. It is worth noting that the tilt-induced significant increase of cerebro-vascular and cerebro-respiratory interactions in the LF band is well visible as significant increases of the joint MIR $I_{F;H,M,R}$ (Fig. 5.13b, left column), which however is not a *source-specific* measure and thus cannot distinct between the pathways of information flow.

A major result is related to the interactions between cerebral and cardiac processes, which indeed do not change in response to the orthostatic challenge within the investigated network ($I_{F;H}$, $\mathcal{U}_{F;H}$, black distributions of left and middle panels), thus implying invariance of these relationships with the postural stress. Nonetheless, besides being significantly lower than $\mathcal{U}_{F;M}$ during tilt as suggested by p-values < 0.05 (Fig. 5.13a, middle column), a result that emerges better in the LF band (Fig. 5.13b, middle column), the role played by heart rate in influencing CBFV variability in this group of subjects cannot be marked as negligible throughout the experimental protocol outlined in the study. Instead, we remark here the prevalence of the cerebro-vascular interactions between cerebral blood flow and arterial pressure. Overall, what emerges from the analysis of mutual and unique information rates is that the latter is able to capture direct mechanisms of interaction which can be masked by the presence of other unobserved variables in the MIR measure: looking at the unique contributions, only the M–F interactions seem to be significantly affected by the early postural stress in the LF band, while respiration may play a role in shaping CBFV dynamics in the late phase of tilt.

In addition, the significant increase of the redundant and synergistic information rates assessed in the LF band moving from REST to ET and LT, with redundancy always significantly higher than synergy as documented by statistical tests, suggests that the orthostatic stress is responsible for the emergence of predominantly redundant patterns of interaction between cardiovascular and cardiorespiratory processes sharing information with the cerebral flow velocity taken as the target process. The significant increase of Δ in the early phase of tilt (Fig. 5.13b, right panel) remarks the importance of LF oscillations within the network and the LF-dependent nature of the interactions among the investigated signals.

Overall, these results confirm the redundant nature of cardiovascular and cerebrovascular interactions previously reported for similar triplets of physiological processes [77, 97, 100], and document the relevance of separating LF and HF contributions to elicit the different roles of heart rate, arterial pressure and respiration on cardiovascular and cerebrovascular interactions. Moreover, the significant increase of redundancy with tilt suggests that these redundant effects are enhanced during postural stress, likely as a consequence of sympathetic activation and vagal withdrawal [100]. Therefore, the evidence that cardiovascular and cerebrovascular interactions occur through the coupling of rhythms in different frequency bands with different physiological meaning [104], make the proposed spectral PIRD eligible to probe high-order interactions in these networks [101].

5.11 High-Order Links in Cardiovascular and Respiratory Networks

The OIR and PIRD framework allow a comprehensive description of node- and network-specific dependencies among the nodes of the observed network, thus characterizing the role of single nodes (OIR gradient) and whole groups (OIR) in the redundant/synergistic informational circuits, as well as decomposing the pairwise information shared into unique, redundant and synergistic components (PIRD). Further, the characterization of the functional links sustaining physiological networks is essential to provide an overall and complete description of the observed network. Link-specific analysis in the context of a HONA can be carried on through the novel approach of the B-Index Rate, which combines the assessment of high-order interactions with statistical inference, brings to new comprehensive assessment of physiological interactions and complements existing strategies for the classification of patho-physiological states. To show its potential, we report here the application of the B-Index Rate measure to physiological networks probed measuring simultaneously the spontaneous variability of several cardiac and vascular parameters during different physiological states [118].

Experimental Protocol and Data Analysis. We analyzed a network with five nodes performing a dynamic analysis of the continuous random processes descriptive of the heart period, systolic and diastolic pressure, cardiac output and peripheral resistance measured on a beat-to-beat basis [118]. Analysis is performed on a group of young healthy volunteers, recruited at the *Jessenius Faculty of Medicine, Comenius Univer-*

sity, Martin, Slovakia, where ethics approval and informed consent were obtained. Subjects were monitored according to an experimental protocol consisting of recording cardiovascular signals for 15 min in the resting supine position (REST), followed by further 8 min of recordings in the upright position (TILT) reached after tilting the subjects to 45° the motor-driven bed table to evoke mild orthostatic stress. Details about the protocol can be found in [119]. The study involved 39 volunteers (22 women, age 19.4 ± 2.3 years), in whom the ECG (CardioFax ECG-9620, NihonKohden, Tokyo, Japan) and the continuous finger ABP collected non-invasively by the photoplethysmographic volume-clamp method (Finometer Pro, FMS, Netherlands) were measured simultaneously with the ICG (CardioScreen 2000, Medis, Germany). Signals were digitized with 1 kHz sampling rate. The analyzed beat-to-beat time series were the HP, SAP, DAP, CO and TPR obtained from the ECG, ABP and ICG signals as follows: HP_n is the duration of the current RRI; SAP_n is the maximum ABP value measured within HP_n; DAP_n is the minimum ABP value measured between the occurrence times of SAP_n and SAP_{n+1}; $CO_n = 60 \cdot \frac{SV_n}{HP_{n-1}}$, where the stroke volume is computed as $SV_n = \beta \cdot Z_n'^{max} \cdot LVET_n$, being $LVET_n$ the left ventricular ejection time, $Z_n'^{max}$ the maximum of the time-derivative of the impedance signal taken within HP_n, and β a correcting factor accounting for thorax volume and base impedance; and $TPR_n = \frac{MAP_n}{CO_n}$, where MAP_n is the mean ABP measured between the occurrence times of DAP_{n-1} and DAP_n. This measurement convention is typically adopted in computational physiology [113, 120] and is illustrated in Fig. 5.14a. For each subject and condition, stationary realizations of $L = 300$ points of the multivariate process $\{HP, SAP, DAP, CO, TPR\}$ were obtained as described above (an example is reported in Fig. 5.14b). Then, the dynamic analysis of the network with $Q = 5$ nodes was performed estimating MIR and cMIR through the regression-based method described in Sect. 4.3.1.2, implemented setting the VAR model (2.6) order through the AIC and using $q = 20$ lags to identify the restricted VAR models (2.7), assessing the significance of MIR and cMIR through the use of $N_s = 100$ *iAAFT* surrogates (significance $\alpha = 0.05$), and finally computing the B-index rate from the thresholded MIR and cMIR (we refer to LINK- SPECIFIC ANALYSIS OF HOIS for technical details about surrogate data analysis).

Results and Discussion. The dynamic formulation of the proposed framework was exploited to explore in detail the interactions among several cardiovascular variables related to cardiac rhythm and contractility, as well as to the regulation of AP. The results in Fig. 5.15 indicate that four of the nodes of this network, i.e., those mapping the variability of TPR, CO, DAP and SAP, form a fully connected subnetwork which is stable in the two analyzed conditions. This finding is documented by the statistically significant values of both MIR and cMIR, as well as of their balance, consistently observed at rest and during tilt for the links $TPR - CO, TPR - SAP$, $TPR - DP$, $CO - SAP$ and $CO - DAP$ (Fig. 5.15a, b). In particular, the link between TPR and CO is very strong and exhibits the highest values of both MIR and cMIR, as a consequence of the inverse relation existing between the two processes. Notably, for this link cMIR is consistently higher than MIR, resulting in significantly

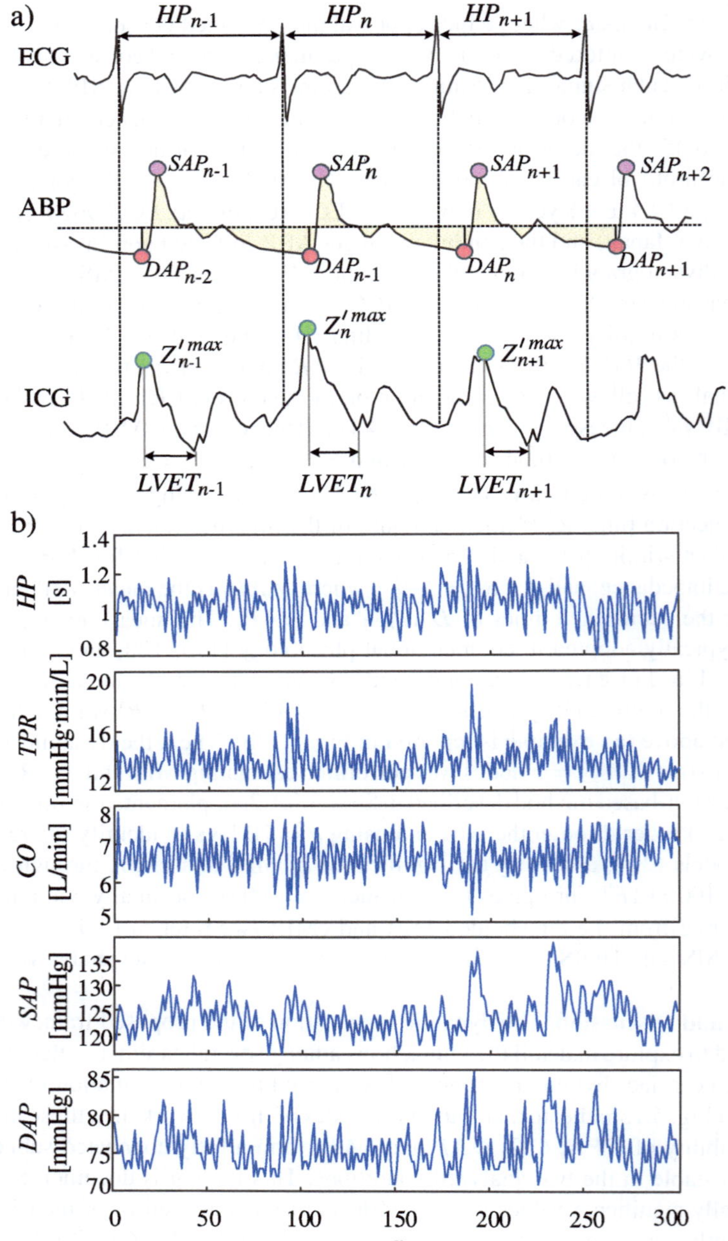

Fig. 5.14 Bio-signal processing and measurement convention for the extraction of the time series used in the dynamic analysis of cardiovascular networks. The HP, TPR, CO, SAP and DAP representative time series measured for a subject monitored in the resting supine position and shown in **b** are extracted from the acquired ECG, ABP and ICG signals shown in **a**. The figure is adapted from [118]

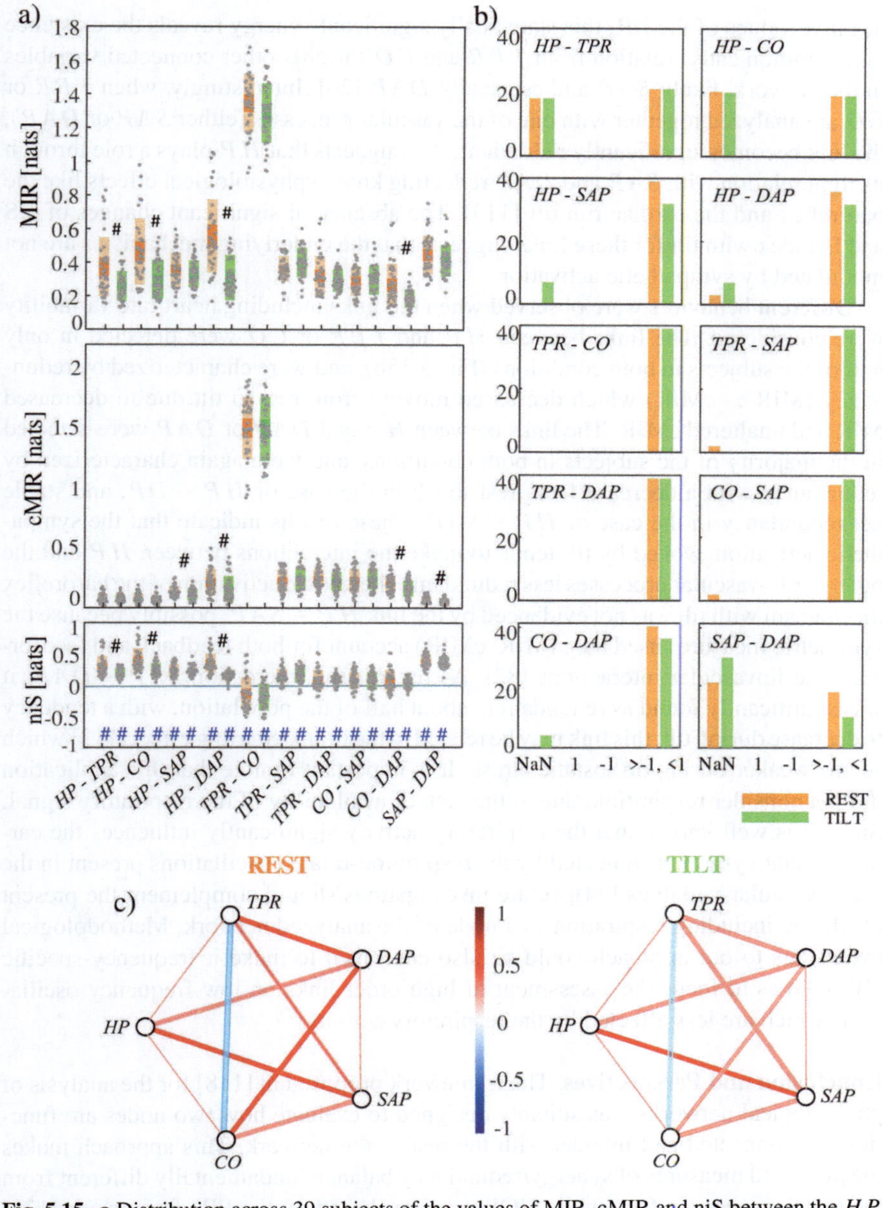

Fig. 5.15 **a** Distribution across 39 subjects of the values of MIR, cMIR and niS between the HP, SAP, DAP, CO and TPR time series measured in the supine position (REST) and during postural stress (TILT); black and blue hash symbols denote statistically significant differences between REST and TILT, and between MI and cMI ($p<0.05$, *paired Wilcoxon test*). **b** Number of subjects for which both MIR and cMIR were not statistically significant (B-index=NaN), only MIR was significant ($B = 1$), only cMIR was significant ($B = -1$), or both MIR and cMIR were significant ($|B| < 1$) in the two conditions. **c** Reconstructed network structure; link thickness: number of subjects with significant MIR and cMIR, link color maps: average B-index rate. The figure is adapted from [118]

negative values of the IIR; this statistically significant synergy reveals the existence of a common cause relation from TPR and CO towards other connected variables in the network, likely SAP and especially DAP [20]. Interestingly, when TPR or CO are analyzed together with one of the vascular processes (either SAP or DAP), the link becomes significantly redundant; this suggests that HP plays a role through its tight relation with SAP and DAP reflecting known physiological effects like the baroreflex and the cardiac run-off [113]. The absence of significant changes of niS and B-index with tilt for these links suggests that the underlying mechanisms are not modulated by sympathetic activation.

Different behaviors were observed when the links including heart rate variability were considered. The links between HP and TPR or CO were detected in only half of the subjects in both conditions (Fig. 5.15b), and were characterized by redundancy (MIR > cMIR) which decreased moving from rest to tilt due to decreased MIR and unaltered cMIR. The links between HP and DAP or DAP were detected in the majority of the subjects in both conditions, and were again characterized by redundancy with a decrease from rest to tilt in the case of $HP - DP$, and stable net redundancy in the case of $HP - SAP$. These results indicate that the sympathetic activation evoked by tilt tends to make the interactions between HP and the other cardiovascular processes less redundant. The known activation of the baroreflex mechanism with tilt was not evidenced by the link $HP - SAP$, possibly because the symmetric measures used here (MIR, cMIR) account for both feedback and feedforward cardiovascular interactions [82]. As regards the interaction $SAP - DAP$, it was significantly found as redundant in about half of the population, with a tendency to decrease during tilt; this link may be related to the Frank-Starling effect [113] which seems weaker during orthostatic stress. It is important to note that this application did not consider respiration, due to the lack of availability of the respiratory signal. Since it is well known that the respiratory activity significantly influences the cardiovascular system, as reflected by the respiration-related oscillations present in the cardiovascular variables [34], future investigations should complement the present results by including respiration as a node of the analyzed network. Methodological extensions to our approach could be also envisaged to make it frequency-specific [101], so as to focus the assessment of high-order links on low-frequency oscillations which are less affected by the respiratory activity.

Conclusion and Perspectives. The framework proposed in [118] for the analysis of physiological networks was suitably designed to evaluate how two nodes are functionally connected and interact with the rest of the network. This approach makes the proposed measures of synergy/redundancy balance fundamentally different from the existing ones. In fact, while HOI measures based on the PID framework [121] concentrate on one network node and relate its activity to that of two or more other nodes, and the recently proposed OI-based measures [77, 122] concentrate on the whole network analyzed collectively, the niS measure puts the focus on each specific link, thus allowing to represent high-order effects as networks. Moreover, the normalization leading to the B-index rate and the associated analysis for statistical significance allow to prune the indirect links determined by cascade/common drive or common target relations, thus making it possible to exploit HOIs for inferring the structure of the analyzed functional network. These methodological advantages are particularly relevant in the analysis of cardiovascular networks, which have been up to now analyzed without accounting for high-order interactions [4, 113, 119] or considering them exclusively at the level of the whole network [77, 79, 97, 101].

The application to physiological networks showed how the proposed framework can elicit mechanisms of cardiovascular regulation investigated at rest and in response to postural stress. The use of the proposed dynamic measures of synergy/redundancy balance and link strength can favor a deeper investigation of physiology as well as the discovery of new clinical markers. These measures can also be exploited to complement existing analyses for empowering the automatic classification of pathophysiological states.

5.12 Summary

In the field of Network Physiology, the dynamic activity of diverse physiological systems has been widely investigated using dynamic measures of complexity and causality in different experimental conditions and patho-physiological states. Our applications to cardiovascular, cardiorespiratory and cerebrovascular closed-loop systems evidenced well-known behaviors including *(i)* the LF-specific increase of the causal interaction along the baroreflex in healthy subjects and the drop of cardiovascular interactions in patients prone to develop postural-related syncope, *(ii)* the increase of the information transferred along the pressure-to-flow link with the postural stress, suggesting a progressive weakening of cerebral autoregulation in syncope subjects, *(iii)* the tilt-induced decrease of the rate of information shared between cardiac and respiratory variables, reflecting a decrease of the strength of cardiorespiratory interactions due to sympathetic activation and vagal withdrawal, and *(iv)* the emergence of a significant causal interaction directed from respiration to the heart within the LF band of the frequency spectrum due to paced breathing at 0.25 Hz. Nevertheless, we moved a bit forward demonstrating that univariate and bivariate analyses, in spite of their broad applicability and computational ease, are often insufficient to provide a complete description of the complex physiological mechanisms involved in the maintenance of the homeostatic balance. Indeed, the application of our new measure of linear self-predictability to short-term arterial compliance data, as well as the utilization of measures of coupling and causality for investigating cardiorespiratory, cardiovascular and cerebrovascular closed-loop systems, suggested that, in physiological contexts where multiple complex interactions often arise, the effects of unobserved confounders are likely to occur. Therefore, the extension to multivariate datasets including signals possibly acting as external drivers (such as, respiration and arterial carbon dioxide influencing respectively cardiac and cerebral dynamics) are envisaged.

More sophisticated analyses exploiting hierarchical interaction measures [77, 123] have highlighted the redundant nature of cardiovascular, cardiorespiratory and cerebrovascular interactions, and documented the relevance of separating LF and HF contributions to elicit the role of respiration on cardiovascular and cerebrovascular variables. Indeed, our results document that respiration acts as a major driver of multivariate redundant interactions in physiological networks, confirming that HOIs can have different nature for different rhythms because synergistic and redundant behaviors generally alternate in different bands of the frequency spectrum. Overall, our applications have proved the significance of computing frequency-specific hierarchical interaction measures in the case of physiological variables rich of oscillatory components with different frequencies and shape, suggesting that the changes of these measures in the time domain may be confined to specific bands of the spectrum. Indeed, the added value of the framework proposed in [77] and then in [123] stands in the tight relation between the measures defined in the time and frequency domains, which favors interpretability [33, 124]. Exploiting this feature, we contributed to describe the physiological mechanisms involved in the cardiovascular, cardiorespiratory and cerebrovascular regulations in the response to a physiological stimulation (e.g., the postural stress or controlled breathing) and to the development of autonomic dysfunctions (e.g., the pre-symptoms of postural related syncope). In perspective, the combination of time and frequency domain measures can help elucidating the mechanisms behind the oscillatory rhythms as well as the broad-band dynamics of coupled physiological variables studied in a variety of physiological conditions and diseased states. Moreover, the inclusion of instantaneous effects in the analyzed parametric models, though not always straightforward [59, 68, 95], is recommended to provide a complete picture of causal, isolated and autonomous effects emerging in the time and frequency domains from dynamic interactions.

References

1. Plamen Ch Ivanov. "The new field of network physiology: building the human physiolome". In: *Frontiers in Network Physiology* 1 (2021), p. 711778.
2. Klaus Lehnertz, Timo Bröhl, and Thorsten Rings. "The human organism as an integrated interaction network: Recent conceptual and methodological challenges". In: *Frontiers in Physiology* 11 (2020), p. 598694.
3. Thomas E Dick et al. "Cardiorespiratory coupling: common rhythms in cardiac, sympathetic, and respiratory activities". In: *Progress in brain research* 209 (2014), pp. 191–205.
4. Steffen Schulz et al. "Cardiovascular and cardiorespiratory coupling analyses: a review". In: *Philosophical Transactions of the Royal Society A: Mathematical, Physical and Engineering Sciences* 371.1997 (2013), p. 20120191.
5. Andreas Voss et al. "Short-term heart rate variability—influence of gender and age in healthy subjects". In: *PloS one* 10.3 (2015), e0118308.
6. Junichiro Hayano and Fumihiko Yasuma. "Hypothesis: respiratory sinus arrhythmia is an intrinsic resting function of cardiopulmonary system". In: *Cardiovascular research* 58.1 (2003), pp. 1–9.
7. Fred Shaffer, Rollin McCraty, and Christopher L Zerr. "A healthy heart is not a metronome: an integrative review of the heart's anatomy and heart rate variability". In: *Frontiers in psychology* 5 (2014), p. 1040.

8. Jack M Gorman and Richard P Sloan. "Heart rate variability in depressive and anxiety disorders". In: *American heart journal* 140.4 (2000), S77–S83.
9. Thomas Benichou et al. "Heart rate variability in type 2 diabetes mellitus: A systematic review and meta-analysis". In: *PloS one* 13.4 (2018), e0195166.
10. Massimo Pagani and Daniela Lucini. "Autonomic dysregulation in essential hypertension: insight from heart rate and arterial pressure variability". In: *Autonomic Neuroscience* 90.1-2 (2001), pp. 76–82.
11. Francesco Buccelletti et al. "Heart rate variability and myocardial infarction: systematic literature review and metanalysis." In: *European Review for Medical & Pharmacological Sciences* 13.4 (2009).
12. Maja Elstad et al. "Cardiorespiratory interactions in humans and animals: rhythms for life". In: *American Journal of Physiology-Heart and Circulatory Physiology* 315.1 (2018), H6–H17.
13. Massimo Piepoli et al. "Origin of respiratory sinus arrhythmia in conscious humans: an important role for arterial carotid baroreceptors". In: *Circulation* 95.7 (1997), pp. 1813–1821.
14. Dwain L Eckberg. "Point: counterpoint: respiratory sinus arrhythmia is due to a central mechanism vs. respiratory sinus arrhythmia is due to the baroreflex mechanism". In: *Journal of applied physiology* (2009).
15. Alberto Porta et al. "Model-based assessment of baroreflex and cardiopulmonary couplings during graded head-up tilt". In: *Computers in biology and medicine* 42.3 (2012), pp. 298–305.
16. J Krohova et al. "Information domain analysis of respiratory sinus arrhythmia mechanisms". In: *Physiological Research* 67 (2018), S611–S618.
17. Gary G Berntson, John T Cacioppo, and Karen S Quigley. "Respiratory sinus arrhythmia: Autonomic origins, physiological mechanisms, and psychophysiological implications". In: *Psychophysiology* 30.2 (1993), pp. 183–196.
18. Gustav G Belz. "Elastic properties and Windkessel function of the human aorta". In: *Cardiovascular drugs and therapy* 9 (1995), pp. 73–83.
19. Vasco Sequeira and Jolanda van der Velden. "Historical perspective on heart function: the Frank–Starling Law". In: *Biophysical reviews* 7 (2015), pp. 421–447.
20. Dávid Švec et al. "Short-Term Arterial Compliance Changes in the Context of Systolic Blood Pressure Influence". In: *Physiological Research* 70.Suppl 3 (2021), S339.
21. Isabella Tan et al. "Heart rate dependency of large artery stiffness". In: *Hypertension* 68.1 (2016), pp. 236–242.
22. Laura Sparacino et al. "A method to assess linear self-predictability of physiologic processes in the frequency domain: application to beat-to-beat variability of arterial compliance". In: *Frontiers in Network Physiology* 4 (2024), p. 1346424.
23. Rong Zhang et al. "Transfer function analysis of dynamic cerebral autoregulation in humans". In: *American Journal of Physiology-Heart and Circulatory Physiology* 274.1 (1998), H233–H241.
24. Patrice Brassard et al. "Losing the dogmatic view of cerebral autoregulation". In: *Physiological reports* 9.15 (2021), e14982.
25. Niels A Lassen. "Cerebral blood flow and oxygen consumption in man". In: *Physiological reviews* 39.2 (1959), pp. 183–238.
26. Rune Aaslid et al. "Cerebral autoregulation dynamics in humans." In: *Stroke* 20.1 (1989), pp. 45–52.
27. OB Paulson, S Strandgaard, and L Edvinsson. "Cerebral autoregulation." In: *Cerebrovascular and brain metabolism reviews* 2.2 (1990), pp. 161–192.
28. Vlasta Bari et al. "Nonlinear effects of respiration on the crosstalk between cardiovascular and cerebrovascular control systems". In: *Philosophical Transactions of the Royal Society A: Mathematical, Physical and Engineering Sciences* 374.2067 (2016), p. 20150179.
29. Deborah Anne Burton, Keith Stokes, and George M Hall. "Physiological effects of exercise". In: *Continuing Education in Anaesthesia, Critical Care & Pain* 4.6 (2004), pp. 185–188.
30. Gary G Berntson et al. "Autonomic cardiac control. III. Psychological stress and cardiac response in autonomic space as revealed by pharmacological blockades". In: *Psychophysiology* 31.6 (1994), pp. 599–608.

31. William H Cooke et al. "Human responses to upright tilt: a window on central autonomic integration". In: *The Journal of physiology* 517.2 (1999), pp. 617–628.
32. John F Geweke. "Measures of conditional linear dependence and feedback between time series". In: *Journal of the American Statistical Association* 79.388 (1984), pp. 907–915.
33. Daniel Chicharro. "On the spectral formulation of Granger causality". In: *Biological cybernetics* 105 (2011), pp. 331–347.
34. Michael A Cohen and J Andrew Taylor. "Short-term cardiovascular oscillations in man: measuring and modelling the physiologies". In: *The Journal of physiology* 542.3 (2002), pp. 669–683.
35. Fred Shaffer and Jay P Ginsberg. "An overview of heart rate variability metrics and norms". In: *Frontiers in public health* 5 (2017), p. 258.
36. Nicolaas Westerhof et al. *Snapshots of hemodynamics: an aid for clinical research and graduate education*. Springer, 2018.
37. Dávid Švec and Michal Javorka. "Noninvasive Arterial Compliance Estimation". In: *Physiological Research* 70.Suppl4 (2021), S483–S494.
38. Julio A Chirinos. "Arterial stiffness: basic concepts and measurement techniques". In: *Journal of cardiovascular translational research* 5 (2012), pp. 243–255.
39. Alan J Bank et al. "Direct effects of smooth muscle relaxation and contraction on in vivo human brachial artery elastic properties". In: *Circulation research* 77.5 (1995), pp. 1008–1016.
40. Tatsuya Arai et al. "Preliminary application of a novel algorithm to monitor changes in preflight total peripheral resistance for prediction of post-flight orthostatic intolerance in astronauts". In: *Acta Astronautica* 68.7-8 (2011), pp. 770–777.
41. Riccardo Pernice et al. "Comparison of frequency domain measures based on spectral decomposition for spontaneous baroreflex sensitivity assessment after Acute Myocardial Infarction". In: *Biomedical Signal Processing and Control* 68 (2021), p. 102680.
42. Masahiro Hasegawa and Simon Rodbard. "Effect of posture on arterial pressures, timing of the arterial sounds and pulse wave velocities in the extremities". In: *Cardiology* 64.2 (1979), pp. 122–132.
43. Auke MT Huijben et al. "Aortic augmentation index and pulse wave velocity in response to head-up tilting: effect of autonomic failure". In: *Journal of Hypertension* 30.2 (2012), pp. 307–314.
44. Massimo Nardone, Anthony V Incognito, and Philip J Millar. "Evidence for pressure-independent sympathetic modulation of central pulse wave velocity". In: *Journal of the American Heart Association* 7.3 (2018), e007971.
45. Victoria L Cooper and Roger Hainsworth. "Carotid baroreceptor reflexes in humans during orthostatic stress". In: *Experimental Physiology* 86.5 (2001), pp. 677–681.
46. Jun Sugawara et al. "Effects of transient change in carotid arterial stiffness on arterial baroreflex during mild orthostatic stimulation". In: *Artery Research* 6.3 (2012), pp. 130–135.
47. Maja Elstad and Lars Walløe. "Heart rate variability and stroke volume variability to detect central hypovolemia during spontaneous breathing and supported ventilation in young, healthy volunteers". In: *Physiological measurement* 36.4 (2015), p. 671.
48. Barbora Czippelova et al. "Arterial stiffness and endothelial function in young obese patients-vascular resistance matters". In: *Journal of Atherosclerosis and Thrombosis* (2019), p. 47530.
49. Michal Javorka et al. "Role of respiration in the cardiovascular response to orthostatic and mental stress". In: *American Journal of Physiology-Regulatory, Integrative and Comparative Physiology* 314.6 (2018), R761–R769.
50. Maja Elstad et al. "Low-frequency fluctuations in heart rate, cardiac output and mean arterial pressure in humans: what are the physiological relationships?" In: *Journal of hypertension* 29.7 (2011), pp. 1327–1336.
51. Troy E Brown et al. "Important influence of respiration on human RR interval power spectra is largely ignored". In: *Journal of applied physiology* 75.5 (1993), pp. 2310–2317.
52. Alberto Porta et al. "Information domain analysis of cardiovascular variability signals: evaluation of regularity, synchronisation and co-ordination". In: *Medical and Biological Engineering and Computing* 38 (2000), pp. 180–188.

53. Martina Valente et al. "Cardiovascular and respiratory variability during orthostatic and mental stress: A comparison of entropy estimators". In: *2017 39th Annual International Conference of the IEEE Engineering in Medicine and Biology Society (EMBC)*. IEEE. 2017, pp. 3481–3484.
54. Maja Elstad et al. "Respiratory sinus arrhythmia: opposite effects on systolic and mean arterial pressure in supine humans". In: *The Journal of physiology* 536.1 (2001), pp. 251–259.
55. Laura Sparacino et al. "A method to assess granger causality, isolation and autonomy in the time and frequency domains: theory and application to cerebrovascular variability". In: *IEEE Transactions on Biomedical Engineering* (2023).
56. Chiara Barà et al. "Direct causality measures unravel complex networks of cardiovascular oscillations and their modifications with postural stress". In: *bioRxiv* (2025), pp. 2025–03.
57. Luca Faes et al. "Investigating the mechanisms of cardiovascular and cerebrovascular regulation in orthostatic syncope through an information decomposition strategy". In: *Autonomic Neuroscience* 178.1-2 (2013), pp. 76–82.
58. Laura Sparacino et al. "Statistical Approaches to Characterize Functional Connectivity in Brain and Physiologic Networks on a Single-Subject Basis". In: *2023 45th Annual International Conference of the IEEE Engineering in Medicine & Biology Society (EMBC)*. IEEE. 2023, pp. 1–4.
59. Riccardo Pernice et al. "Spectral decomposition of cerebrovascular and cardiovascular interactions in patients prone to postural syncope and healthy controls". In: *Autonomic Neuroscience* (2022), p. 103021.
60. Luca Faes, Giandomenico Nollo, and Alberto Porta. "Non-uniform multivariate embedding to assess the information transfer in cardiovascular and cardiorespiratory variability series". In: *Computers in biology and medicine* 42.3 (2012), pp. 290–297.
61. Gabriele Volpes et al. "A portable multisensor system to assess cardiorespiratory interactions through photoplethysmography". In: *2022 IEEE International Symposium on Medical Measurements and Applications (MeMeA)*. IEEE. 2022, pp. 1–6.
62. Riccardo Pernice et al. "A portable system for multiple parameters monitoring: towards assessment of health conditions and stress level in the automotive field". In: *2019 AEIT International Conference of Electrical and Electronic Technologies for Automotive (AEIT AUTOMOTIVE)*. IEEE. 2019, pp. 1–6.
63. Riccardo Pernice et al. "Low invasive multisensor acquisition system for real-time monitoring of cardiovascular and respiratory parameters". In: *2020 IEEE 20th mediterranean electrotechnical conference (MELECON)*. IEEE. 2020, pp. 306–310.
64. Jiapu Pan and Willis J Tompkins. "A real-time QRS detection algorithm". In: *IEEE transactions on biomedical engineering* 3 (1985), pp. 230–236.
65. Alessandra Fusco et al. "On how to extract breathing rate from PPG signal using wearable devices". In: *2015 IEEE Biomedical Circuits and Systems Conference (BioCAS)*. IEEE. 2015, pp. 1–4.
66. Peter H Charlton et al. "Extraction of respiratory signals from the electrocardiogram and photoplethysmogram: technical and physiological determinants". In: *Physiological measurement* 38.5 (2017), p. 669.
67. Luca Faes et al. "Estimating the decomposition of predictive information in multivariate systems". In: *Physical Review E* 91.3 (2015), p. 032904.
68. D Nuzzi et al. "Extending the spectral decomposition of Granger causality to include instantaneous influences: application to the control mechanisms of heart rate variability". In: *Philosophical Transactions of the Royal Society A* 379.2212 (2021), p. 20200263.
69. Giandomenico Nollo et al. "Synchronization index for quantifying nonlinear causal coupling between RR interval and systolic arterial pressure after myocardial infarction". In: *Computers in Cardiology 2000. Vol. 27 (Cat. 00CH37163)*. IEEE. 2000, pp. 143–146.
70. STEFANO Strano et al. "Respiratory sinus arrhythmia and cardiovascular neural regulation in athletes." In: *Medicine and Science in Sports and Exercise* 30.2 (1998), pp. 215–219.
71. Laura Sparacino et al. "Causal and Non-Causal Frequency Domain Assessment of Spontaneous Baroreflex Sensitivity after Myocardial Infarction". In: *2020 11th Conference of the European Study Group on Cardiovascular Oscillations (ESGCO)*. IEEE. 2020, pp. 1–2.

72. J Philip Saul et al. “Transfer function analysis of the circulation: unique insights into cardiovascular regulation”. In: *American Journal of Physiology-Heart and Circulatory Physiology* 261.4 (1991), H1231–H1245.
73. John E Sanderson et al. “Impact of changes in respiratory frequency and posture on power spectral analysis of heart rate and systolic blood pressure variability in normal subjects and patients with heart failure”. In: *Clinical science* 91.1 (1996), pp. 35–43.
74. Gian Domenico Pinna et al. “Effect of paced breathing on ventilatory and cardiovascular variability parameters during short-term investigations of autonomic function”. In: *American Journal of Physiology-Heart and Circulatory Physiology* 290.1 (2006), H424–H433.
75. Riccardo Pernice et al. “Assessment of Cardiorespiratory Interactions During Spontaneous and Controlled Breathing: Linear Parametric Analysis”. In: *2022 12th Conference of the European Study Group on Cardiovascular Oscillations (ESGCO)*. IEEE. 2022, pp. 1–2.
76. Alberto Porta et al. “Accounting for respiration is necessary to reliably infer Granger causality from cardiovascular variability series”. In: *IEEE transactions on biomedical engineering* 59.3 (2011), pp. 832–841.
77. Luca Faes et al. “A new framework for the time-and frequency-domain assessment of high-order interactions in networks of random processes”. In: *IEEE Transactions on Signal Processing* 70 (2022), pp. 5766–5777.
78. Alberto Porta et al. “Quantifying the strength of the linear causal coupling in closed loop interacting cardiovascular variability signals”. In: *Biological cybernetics* 86 (2002), pp. 241–251.
79. Jana Krohova et al. “Multiscale information decomposition dissects control mechanisms of heart rate variability at rest and during physiological stress”. In: *Entropy* 21.5 (2019), p. 526.
80. Jurgen AHR Claassen et al. “Transfer function analysis of dynamic cerebral autoregulation: a white paper from the International Cerebral Autoregulation Research Network”. In: *Journal of Cerebral Blood Flow & Metabolism* 36.4 (2016), pp. 665–680.
81. Alberto Porta et al. “Causal relationships between heart period and systolic arterial pressure during graded head-up tilt”. In: *American Journal of Physiology-Regulatory, Integrative and Comparative Physiology* 300.2 (2011), R378–R386.
82. Luca Faes, Giandomenico Nollo, and Alberto Porta. “Mechanisms of causal interaction between short-term RR interval and systolic arterial pressure oscillations during orthostatic challenge”. In: *Journal of applied physiology* 114.12 (2013), pp. 1657–1667.
83. Rogelio Mosqueda-Garcia et al. “Sympathetic and baroreceptor reflex function in neurally mediated syncope evoked by tilt.” In: *The Journal of clinical investigation* 99.11 (1997), pp. 2736–2744.
84. Giandomenico Nollo et al. “Assessing causality in normal and impaired short-term cardiovascular regulation via nonlinear prediction methods”. In: *Philosophical Transactions of the Royal Society A: Mathematical, Physical and Engineering Sciences* 367.1892 (2009), pp. 1423–1440.
85. Luca Faes et al. “Causal cross-spectral analysis of heart rate and blood pressure variability for describing the impairment of the cardiovascular control in neurally mediated syncope”. In: *IEEE transactions on biomedical engineering* 53.1 (2005), pp. 65–73.
86. Alberto Porta and Luca Faes. *Assessing causality in brain dynamics and cardiovascular control*. 2013.
87. RB Panerai et al. “Frequency-domain analysis of cerebral autoregulation from spontaneous fluctuations in arterial blood pressure”. In: *Medical and Biological Engineering and Computing* 36.3 (1998), pp. 315–322.
88. L Schiatti et al. “Extended Granger causality: a new tool to identify the structure of physiological networks”. In: *Physiological measurement* 36.4 (2015), p. 827.
89. Vlasta Bari et al. “Cerebrovascular and cardiovascular variability interactions investigated through conditional joint transfer entropy in subjects prone to postural syncope”. In: *Physiological Measurement* 38.5 (2017), p. 976.
90. Saqib Saleem et al. “Is the Cushing mechanism a dynamic blood pressure-stabilizing system? Insights from Granger causality analysis of spontaneous blood pressure and cerebral

blood flow". In: *American Journal of Physiology-Regulatory, Integrative and Comparative Physiology* 315.3 (2018), R484–R495.
91. BP Grubb et al. "Cerebral vasoconstriction during head-upright tilt-induced vasovagal syncope. A paradoxic and unexpected response." In: *Circulation* 84.3 (1991), pp. 1157–1164.
92. RB Panerai et al. "Effect of CO2 on dynamic cerebral autoregulation measurement". In: *Physiological measurement* 20.3 (1999), p. 265.
93. Simone Cencetti, Gabriele Bandinelli, and Alfonso Lagi. "Effect of PCO2 changes induced by head-upright tilt on transcranial Doppler recordings". In: *Stroke* 28.6 (1997), pp. 1195–1197.
94. Cesare Porta et al. "Influence of respiratory instability during neurocardiogenic presyncope on cerebrovascular and cardiovascular dynamics". In: *Heart* 94.11 (2008), pp. 1433–1439.
95. Luiz A Baccalá and Koichi Sameshima. "Frequency domain repercussions of instantaneous Granger causality". In: *Entropy* 23.8 (2021), p. 1037.
96. Federico Battiston et al. "Networks beyond pairwise interactions: Structure and dynamics". In: *Physics reports* 874 (2020), pp. 1–92.
97. Luca Faes et al. "Information decomposition in multivariate systems: definitions, implementation and application to cardiovascular networks". In: *Entropy* 19.1 (2016), p. 5.
98. Luca Faes et al. "Quantifying High-Order Interactions in Cardiovascular and Cerebrovascular Networks". In: *2022 12th Conference of the European Study Group on Cardiovascular Oscillations (ESGCO)*. IEEE. 2022, pp. 1–2.
99. Laura Sparacino et al. "Quantifying High-Order Interactions in Complex Physiological Networks: A Frequency-Specific Approach". In: *International Conference on Complex Networks and Their Applications*. Springer. 2022, pp. 301–309.
100. Alberto Porta et al. "Quantifying net synergy/redundancy of spontaneous variability regulation via predictability and transfer entropy decomposition frameworks". In: *IEEE Transactions on Biomedical Engineering* 64.11 (2017), pp. 2628–2638.
101. Luca Faes et al. "Information decomposition in the frequency domain: A new framework to study cardiovascular and cardiorespiratory oscillations". In: *Philosophical Transactions of the Royal Society A* 379.2212 (2021), p. 20200250.
102. Plamen Ch Ivanov and Ronny P Bartsch. "Network physiology: mapping interactions between networks of physiologic networks". In: *Networks of Networks: the last Frontier of Complexity*. Springer, 2014, pp. 203–222.
103. Amir Bashan et al. "Network physiology reveals relations between network topology and physiological function". In: *Nature communications* 3.1 (2012), p. 702.
104. Alberto Porta and Luca Faes. "Wiener–Granger causality in network physiology with applications to cardiovascular control and neuroscience". In: *Proceedings of the IEEE* 104.2 (2015), pp. 282–309.
105. Yuri Antonacci et al. "Measuring high-order interactions in rhythmic processes through multivariate spectral information decomposition". In: *IEEE Access* 9 (2021), pp. 149486–149505.
106. Tomas Scagliarini et al. "Gradients of O-information highlight synergy and redundancy in physiological applications". In: *Frontiers in Network Physiology* 3 (2024), p. 1335808.
107. Robin AA Ince. "Measuring multivariate redundant information with pointwise common change in surprisal". In: *Entropy* 19.7 (2017), p. 318.
108. Nicola Montano et al. "Power spectrum analysis of heart rate variability to assess the changes in sympathovagal balance during graded orthostatic tilt." In: *Circulation* 90.4 (1994), pp. 1826–1831.
109. Luca Faes, Giandomenico Nollo, and Alberto Porta. "Information domain approach to the investigation of cardio-vascular, cardio-pulmonary, and vasculo-pulmonary causal couplings". In: *Frontiers in physiology* 2 (2011), p. 80.
110. Dimitris N Politis. "The impact of bootstrap methods on time series analysis". In: *Statistical Science* (2003), pp. 219–230.
111. Luca Faes et al. "A framework for assessing frequency domain causality in physiological time series with instantaneous effects". In: *Philosophical Transactions of the Royal Society A: Mathematical, Physical and Engineering Sciences* 371.1997 (2013), p. 20110618.

112. Luca Faes, Silvia Erla, and Giandomenico Nollo. “Measuring connectivity in linear multivariate processes: definitions, interpretation, and practical analysis”. In: *Computational and mathematical methods in medicine* 2012 (2012).
113. Michal Javorka et al. “Basic cardiovascular variability signals: mutual directed interactions explored in the information domain”. In: *Physiological Measurement* 38.5 (2017), p. 877.
114. Claude Julien. “The enigma of Mayer waves: facts and models”. In: *Cardiovascular research* 70.1 (2006), pp. 12–21.
115. Sebastiano Stramaglia et al. “Disentangling high-order effects in the transfer entropy”. In: *Physical Review Research* 6.3 (2024), p. L032007.
116. Laura Sparacino, Yuri Antonacci, and Luca Faes. “Partial Information Rate Decomposition in Physiological Networks”. In: *2024 13th Conference of the European Study Group on Cardiovascular Oscillations (ESGCO)*. IEEE. 2024, pp. 1–2.
117. Laura Sparacino et al. “Decomposing Multivariate Information Rates in Networks of Random Processes”. In: *arXiv preprint* arXiv:2502.04555 (2025). https://arxiv.org/abs/2502.04555.
118. Gorana Mijatovic et al. “Assessing High-Order Links in Cardiovascular and Respiratory Networks via Static and Dynamic Information Measures”. In: *IEEE Open Journal of Engineering in Medicine and Biology* (2024).
119. Jana Krohova et al. “Vascular resistance arm of the baroreflex: methodology and comparison with the cardiac chronotropic arm”. In: *Journal of Applied Physiology* 128.5 (2020), pp. 1310–1320.
120. Donald P Bernstein. “Continuous noninvasive real-time monitoring of stroke volume and cardiac output by thoracic electrical bioimpedance.” In: *Critical care medicine* 14.10 (1986), pp. 898–901.
121. Paul L Williams and Randall D Beer. “Nonnegative decomposition of multivariate information”. In: *arXiv preprint* arXiv:1004.2515 (2010).
122. Fernando E Rosas et al. “Quantifying high-order interdependencies via multivariate extensions of the mutual information”. In: *Physical Review E* 100.3 (2019), p. 032305.
123. Laura Sparacino et al. “Measuring hierarchically-organized interactions in dynamic networks through spectral entropy rates: theory, estimation, and illustrative application to physiological networks”. In: *Neurocomputing* 630 (2025), p. 129675.
124. John Geweke. “Measurement of linear dependence and feedback between multiple time series”. In: *Journal of the American statistical association* 77.378 (1982), pp. 304–313.

Chapter 6
Applications to Brain Networks

The human brain has always fascinated researchers and neuroscientists. Its complexity lies in the combined spatial- and temporal-evolving activities that different functionally and structurally interconnected regions explicate over a three-dimensional space. These networks exhibit distinct patterns of activity in the resting state or during task execution, but also interact with each other in various spatio-temporal modalities, being connected both by anatomical tracts and by functional associations [1, 2]. In fact, to understand the mechanisms of perception, attention, and learning, as well as to manage neurological and mental diseases such as epilepsy, neurodegeneration, and depression, it is necessary to map the patterns of neural activation and connectivity that are both spatially distributed and temporally dynamic. Specifically, *functional connectivity* was defined as the temporal dependence between spatially remote neurophysiological events [3], and has been widely exploited to describe the relationship between the neuronal activation patterns of anatomically separated brain regions, reflecting the level of functional communication between these regions. The analysis of the complex interactions between brain areas has been shaping the research field of *connectomics* [4, 5]. The effort to map the *human connectome*, which consists of brain networks, their structural connections and functional interactions [4], has given life to a number of different approaches, each with its own specifications and interpretations [6–9]. Some of these methods, such as covariance structural equation modeling [10] and the dynamic causal modeling [11, 12], are based on the definition of an underlying structural and functional model of brain interactions. Conversely, some others, such as Granger causality [13] and transfer entropy [14], as well as directed coherence [15, 16], partial directed coherence [17, 18] and directed transfer function [19], are data-driven and based on the statistical analysis of multivariate time series. Interestingly, while non-linear model-free and linear model-based approaches are apparently unrelated as they look at different aspects of multivariate dynamics, they become clearly connected if some assumptions, like the Gaussianity of the joint probability distribution of the variables drawn from the data [20, 21], are met. Under these assumptions, connectivity measures such as Granger causality

L. Sparacino, *Spectral Information Dynamics in Network Neuroscience and Physiology*,
Studies in Computational Intelligence 1235,
https://doi.org/10.1007/978-3-032-05416-6_6

and transfer entropy, as well as coherence [22] and mutual information rate [23, 24], can be mathematically related to each other; this equivalence leaves the basis for a model-based frequency-specific interpretation of inherently model-free information-theoretic measures [25]. Furthermore, emerging trends, such as the development of high-order interaction measures, are coming up in the neurosciences to respond to the need of providing more exhaustive descriptions of brain network interactions. These measures allow to deal with multivariate representations of complex systems [26–28], showing their potential in disentangling physiological mechanisms involving more than two units or subsystems [29].

A number of brain mapping modalities have been used in the last decades to investigate the human connectome. The most known technique used so far in this context is the neuroimaging technique of *functional magnetic resonance imaging* (fMRI), which allows to quantify hemodynamic changes (i.e., spontaneous blood oxygen level-dependent—BOLD—signal fluctuations) following the activation of specific brain areas by examining resting-state networks (RSNs) in the resting or relaxed state in response to neuronal activity [1, 30–33], as well as to explore brain functional connectivity associated with both healthy and neuro-pathologic functions [33–35]. However, fMRI lacks in time resolution and therefore cannot be entrusted with detecting short-living events, which can instead be investigated by *electroencephalography* (EEG), a low-cost non-invasive imaging technique allowing to study the dynamic relations between the activity of cortical brain regions and providing different information with respect to fMRI [36]. Specifically, in spite it features lower spatial resolution than fMRI, EEG provides quicker, affordable, and accessible insights about brain function. It measures voltage from the scalp and can sample data at the order of kHz, meaning that it can provide data on how on the response of a large population of pyramidal cells with the same orientation changes over the course of milliseconds [37, 38]. fMRI, on the other hand, being a hemodynamic response that reflects changes in blood oxygenation as neurons engage in a process called the BOLD signal, evolves over the order of seconds. As a result, a compromise must be made between temporal resolution when using EEG or incredibly high spatial resolution with fMRI.

This chapter represents an overall review of exemplary applications to static and dynamic neural networks probed by fMRI data (Sect. 6.1) and EEG signals (Sect. 6.2), respectively. Limitations of the proposed methodologies and further advancements in the context of Network Neuroscience will be thoroughly discussed.

6.1 Static Networks of Random Variables: Applications to fMRI Data

Resting-state fMRI connectivity has been generally assessed via *static measures of pairwise coupling* calculated over the course of an entire scan session. These derivations are based on the null hypothesis of i.i.d. Gaussian variables that only takes into account the observed (static) spatial correlations and ignores temporal features. Approaches exploring pairwise connectivity patterns, such as mutual information [24], are easily applicable, require little computational effort and offer a straightforward interpretation of the findings. Although highly effective, these methods are inherently restricted by the constructional requirement that every interaction must be between two elements. However, there is mounting evidence that such measures cannot fully capture the interplay among the multiple units of a complex system [28, 39]. Consequently, recognizing and modeling *high-order functional structures*, characterized by statistical interactions involving more than two network units, has become a crucial and evolving area of Network Neuroscience [39, 40], where HOIs have been suggested as fundamental components of complexity and functional integration in brain networks [41], and proposed to be linked to emergent mental phenomena and consciousness [42]. Nevertheless, in spite of their promising significance, the investigation of HOIs in the brain is a relatively unexplored domain. Given that these interactions are not typically accessible through the well-established pairwise measures of functional connectivity network analysis, their study has been often limited by the lack of formal tools as well as by the involvement of inherent computational and combinatorial challenges. While many different information-theoretic metrics have been proposed throughout the years, all attempting to capture the information shared by triplets of random variables or processes [43–46], a recent work [26] suggests the potential use of information theory for identifying HOIs in multivariate systems, as well as for distinguishing between qualitatively distinct modes of information sharing, i.e. redundancy and synergy. Synergy is a potentially intriguing phenomenon as it reflects the ability of the human brain to generate new information by combining the interplay of anatomically distinct but functionally connected brain areas. A measurement of this quantity in the context of HOIs comes from the O-Information [26], which provides an overall evaluation of whether a system is dominated by redundancy or synergy.

Taking these premises into account, in this section we review exemplary applications of the proposed static measures of pairwise (i.e., the MI) and high-order (i.e., the OI, its gradient and the B-index) connectivity defined in Chap. 3 to resting-state fMRI data acquired in both normal and neuro-pathologic states. Overall, we remark that future developments should consider transient behaviors [47], enhance the identification of topological and causal structures [48, 49], and go beyond the use of the first order gradient (i.e., the OI increment defined in (3.12)). Expanding the study of increments of information across orders, as well as of their assessment through surrogate and bootstrap approaches, would allow to further unveil synergistic structures and to investigate their role in brain complex networks of multiple interacting nodes [29]. Furthermore, the exploration of dynamic forms of pairwise and high-order connectivity [27, 28], which account for temporal correlations in the detection of brain functional couplings, may spark great interest among neuro-scientists and assume high relevance in the field of fMRI data analysis.

6.1.1 Pairwise and High-Order Brain Functional Connectivity Signatures on a Single-Subject Basis

Keeping up with the shift towards personalized neuroscience essentially requires the derivation of meaningful insights from individual brain signal recordings, by analyzing descriptive indexes of physio-pathological states through statistical methods that prioritize subject-specific differences under varying experimental conditions. While such validation is typically performed at the level of subject groups, in clinical practice, where the goal is to optimize the individual treatment plan and look into the effects of interventions on a single patient [33, 35, 50], statistical analyses should be focused on subject-specific differences between different states. To address this, methods to determine the accuracy (confidence limits) of individual estimates of the considered indexes are necessary. This is especially important as the accuracy of estimates may vary over time and depend on factors such as the individual patho-physiologic state. The absence of confidence limits or error bounds on estimates may lead to biased clinical decisions, making it imperative to ensure a reliable assessment of the patient's underlying condition.

Resting-State Patterns of Mutual Information and O-Information. Within this framework, in [51] we proposed a methodology for assessing the value of *single-subject fingerprints of brain functional connectivity*, assessed both by standard pairwise (i.e., the MI) and novel high-order measures (i.e., the OI). The significance and variations across different conditions of functional pairwise and HOIs between groups of brain signals were statistically verified on an individual level through the utilization of surrogate and bootstrap data analyses. The approach is illustrated on exemplary single-subject recordings of resting-state fMRI (rest-fMRI) signals acquired in a pediatric patient treated at *IRCCS–ISMETT (Scientific Institute for Research, Hospitalization and Healthcare—Mediterranean Institute for Transplantation and Advanced Specialized Therapies), Palermo, Italy*. The patient featured a cavernous transformation of the portal vein, an obstruction also known as portal cavernoma, a common cause of portal hypertension in children. Even in cases where liver function appears normal, this disease can result in the development of hepatic encephalopathy (HE) [52] due to the presence of portal-systemic shunts, which cause an increase in plasma ammonia levels and toxic brain catabolites deposition in the globi pallidi. HE is a serious condition that can have a profound impact on the patient's ability to perform daily tasks, causing psychomotor sluggishness, attention deficits and a decline in fine motor performance. Although HE is currently diagnosed using psychometric and electrophysiological examinations, the administration and interpretation of psychometric tests can be influenced by a number of variables, including but not limited to age, educational attainment, and the potential impact of learning effects. HE may go undiagnosed if these variables are disregarded. To overcome this issue, we investigated the potential of rest-fMRI with BOLD echo-planar imaging (EPI) technique to assess brain functional connectivity, in order to detect cognitive impairment related to the presence of HE in the analysed pediatric patient.

Moreover, the possibility of cognitive improvement following surgical correction of the disease using Meso-Rex surgery, as described in [53], is also investigated.

Data Acquisition and Pre-Processing. The patient, an 8-year-old boy, was admitted with cognitive impairment characterized by psychomotor sluggishness and a decline in fine motor performance, attention deficits and a profound reduction in the ability to perform daily tasks. Standard liver function tests and hematologic markers were determined by obtaining and analyzing blood samples from veins using conventional methods. A measurement of venous ammonia confirmed the presence of ammonia. The patient underwent Doppler ultrasonography, magnetic resonance (MR) imaging, and MR angiography to diagnose and assess portal cavernoma, collaterals, and spontaneous shunts. Specifically, baseline MR imaging (MRI) examinations were performed on a 3T MRI scanner (Discovery 750w, General Electric Medical System) utilizing a 32-channel head coil, during PRE, i.e. before the surgical correction of the portal cavernoma by means of Meso-Rex surgery, and during two follow-up phases, i.e. after 1 month (POST1) and 12 months (POST12) the surgical treatment. The subject was positioned in the scanner with his head comfortably restrained by foam padding to reduce head movement. Earplugs were used to reduce the noise of the scanner. During the resting-state scan, the subject was instructed to keep his eyes closed, remain as motionless as possible, and clear his head of any particular thoughts. A standard multi-parametric MRI protocol was carried out with fast spin-echo T1-weighted and T2-weighted MR images, fluid-attenuated inversion recovery, T2*-weighted gradient-recalled-echo, susceptibility-weighted imaging, and standard 3-direction diffusion-weighted imaging. Isotropic T1-weighted volumetric imaging (3D-SPGR or MPRAGE) was acquired as anatomical reference images for rest-fMRI with BOLD EPI technique, which was then performed to assess spontaneous neuronal activity in the RSNs and evaluate brain network connectivity [32, 35, 50, 54].

The volume of T1-weighted morphological data and functional slices, obtained respectively through MR and BOLD imaging, was appropriately pre-processed following a series of steps. First, morphological scans were pre-processed by correcting motion artifacts. The original data volume was transformed and normalized to the standard EPI template in Montreal Neurological Institute (MNI) atlas[1] and restored to $3 \times 3 \times 3\text{mm}^3$. The resulting images were spatially smoothed with an 8-mm full width at half-maximum Gaussian kernel. Non-brain tissues were removed from the scans and segmentation of brain tissues was performed. Atlas-based cortical parcellation was obtained, and seed selection was carried out using Brodmann areas[2] after transforming the coordinates from the MNI atlas into the Talairach atlas.[3] Then, confounds, i.e., noise variables representing fluctuations of non-neuronal origin such as residual physiological effects derived from subject motion, were estimated. These confounding effects were minimized by performing the so-called *denoising*

[1] https://brainmap.org/training/BrettTransform.html.

[2] https://www.brainm.com/software/pubs/dg/BA_10-20_ROI_Talairach/functions.htm.

[3] https://brainmap.org/training/BrettTransform.html.

procedure. To this aim, the CONN toolbox[4] was used, which is an open-source MATLAB/SPM-based cross-platform software.[5] The *CompCor* function in CONN was used for spatial and temporal pre-processing to minimize the impact of motion and physiological noise factors, as well as to define and remove confounds in the BOLD signals. Regression of first-order derivative terms for the whole brain, ventricular, and white matter signals was also included in the correlation pre-processing to reduce the influence of spurious variance on neuronal activity.

Resting State Networks Identification. The assessment of brain functional connectivity for this patient was obtained for a given number of RSNs, selected through a seed-based correlation approach. A seed region of interest (ROI) was first identified, and then the linear correlation of the seed region with all the other voxels of the entire brain was computed making use of statistical analysis [1, 55, 56]. Among the commonly known and analysed 36 RSNs [57], this procedure, for whose details we refer the reader to [33, 56], allowed the identification of 8 resting-state networks with the best signal-to-noise ratio, following the fMRI image denoising and realignment steps. The chosen networks were then used to evaluate subject-specific cognitive fingerprints at baseline and after disease correction, and to show any significant improvement in the individual functional connectivity after surgery. All ROIs encompassing the 8 selected RSNs (Default Mode—DM, SensoriMotor—SM, Visual—VS, Salience—SAL, Dorsal Attention—DA, FrontoParietal—FP, Language—L, Cerebellar—CB) were imported into the CONN Toolbox, then used to perform seed-based extraction of $M = 32$ BOLD fMRI time series as sequences of $L = 200$ consecutive synchronous values, $\mathbf{v} = \{\mathbf{v}_1, \ldots, \mathbf{v}_M\}$, considered as a realization of the network $\mathbf{V} = \{V_1, \ldots, V_M\}$ describing the neural dynamics.

Data and Statistical Analysis. Linear models in the form of (2.1) were fitted on each pair of BOLD time series $\mathbf{x} = \mathbf{v}_i$ and $\mathbf{y} = \mathbf{v}_j$ $(i, j = 1, \ldots, M, i \neq j)$, pre-processed by removing the mean value and scaled to have unitary standard deviation, for which the time-domain MI (3.7) was then obtained as a measure of pairwise functional connectivity. In each experimental condition, the existence of every pairwise link was evaluated applying surrogate data analysis and assessing the significance of the estimated MI using *iAAFT* surrogates, as detailed in COUPLED DYNAMICS BETWEEN PAIRS OF NODES. Furthermore, the OI measure ($\Omega(\mathbf{v}^N)$, r.h.s. of (3.12)) was computed for a predefined number of multiplets from order $N = 3$ to order $N = 8$. Specifically, among all the possible combinations of order 3 derived from the $M = 32$ time series, 56 triplets were selected randomly from different RSNs. These triplets were then used as roots for building 40 multiplets of order 4, where the additional time series was chosen randomly within the remaining RSNs. The procedure was iterated for higher orders, eventually obtaining 30 multiplets of order 5, 20 multiplets of order 6, 5 multiplets of order 7 and 1 multiplet of order 8. For each order and multiplet, the significance of the estimated OI and OI increments ($\Delta(OI)$)

[4] https://web.conn-toolbox.org/.

[5] https://www.nitrc.org/projects/conn.

was assessed by applying the bootstrap method as detailed in AN APPROACH TO THE STATISTICAL ASSESSMENT OF HIGH- ORDER INTERACTIONS. Specifically, we set $b_s = 50$ for the generation of bootstrap fMRI time series of length $L = 200$. Moreover, the significance of the differences δ between the MI/OI values measured in two conditions (PRE vs. POST1, PRE vs. POST12, POST1 vs. POST12) was assessed comparing the MI/OI distributions obtained through the block bootstrap method. Specifically, when the MI and OI measures are computed on a single-subject basis in two different experimental conditions, the bootstrap distributions can be employed to assess the significance of the difference between the two conditions through a statistical test. To this end, the bootstrap data generation procedure is executed in both the analyzed experimental conditions, and the *parametric Student t-test for unpaired data* is then employed to assess the statistical significance of the difference between pairs of bootstrap distributions for a given measure. We refer to STATISTICAL SIGNIFICANCE OF THE DIFFERENCE BETWEEN EXPERIMENTAL CONDITIONS for technical details. In this work, a significance level $\alpha = 0.05$ was used both to compute confidence intervals of the surrogate and bootstrap distributions as well as to perform statistical tests.

Results and Discussion. Results of the analysis relevant to the application of the proposed methodology to the clinical case are reported in Fig. 6.1, showing the subject-specific maps of brain functional connectivity in the three conditions (Fig. 6.1a), the distributions of the OI values for all the orders ($N = 3, \ldots, 8$), where each order comprises a given number of analysed multiplets (Fig. 6.1b; in each box, the central black mark indicates the mean, and the bottom and top edges of the box indicate the 25th and 75th percentiles; red, blue and grey circles indicate positive-redundant, negative-synergistic, and non-significant OI values, respectively), and the values of the OI increment computed as in (3.12), for each target $\mathbf{v}_j$ within the multiplet $\mathbf{v}^N$ at order N (Fig. 6.1c; red, blue and grey squares indicate positive-redundant, negative-synergistic, and non-significant $\Delta(OI)$, respectively, brought by that target to the whole multiplet, for a given order). Values of MI, OI and $\Delta(OI)$ are expressed in nats, i.e., natural units. The pre-surgery phase is characterized by a relatively sparse functional connectivity network (Fig. 6.1a), with 41% of the MI values detected as statistically significant by surrogate data analysis. Compared to this phase, the immediate post-surgery period shows a weakening of functional brain connectivity, as evidenced by the lower number of statistically significant MI values between pairs of BOLD series determined by the surrogate data approach (Fig. 6.1a, PRE vs. POST1). Indeed, the global density of the network, i.e., the number of significant connections, decreases from 41% in PRE to 25% in POST1. The number of significant connections increased markedly twelve months after surgery (Fig. 6.1a, POST12, 52% of significant connections), suggesting that the proposed surgery correction of the portosystemic shunt worked in recovering brain functional connectivity in this patient. Interestingly, the local densities, i.e., the number of significant functional connections within (and between) RSNs, are characterized by a drop in the immediate post-surgery period (Table 6.1, POST1) followed by an increase 12 months after the treatment (Table 6.1, POST12), for almost all the (pairs of) RSNs. This suggests

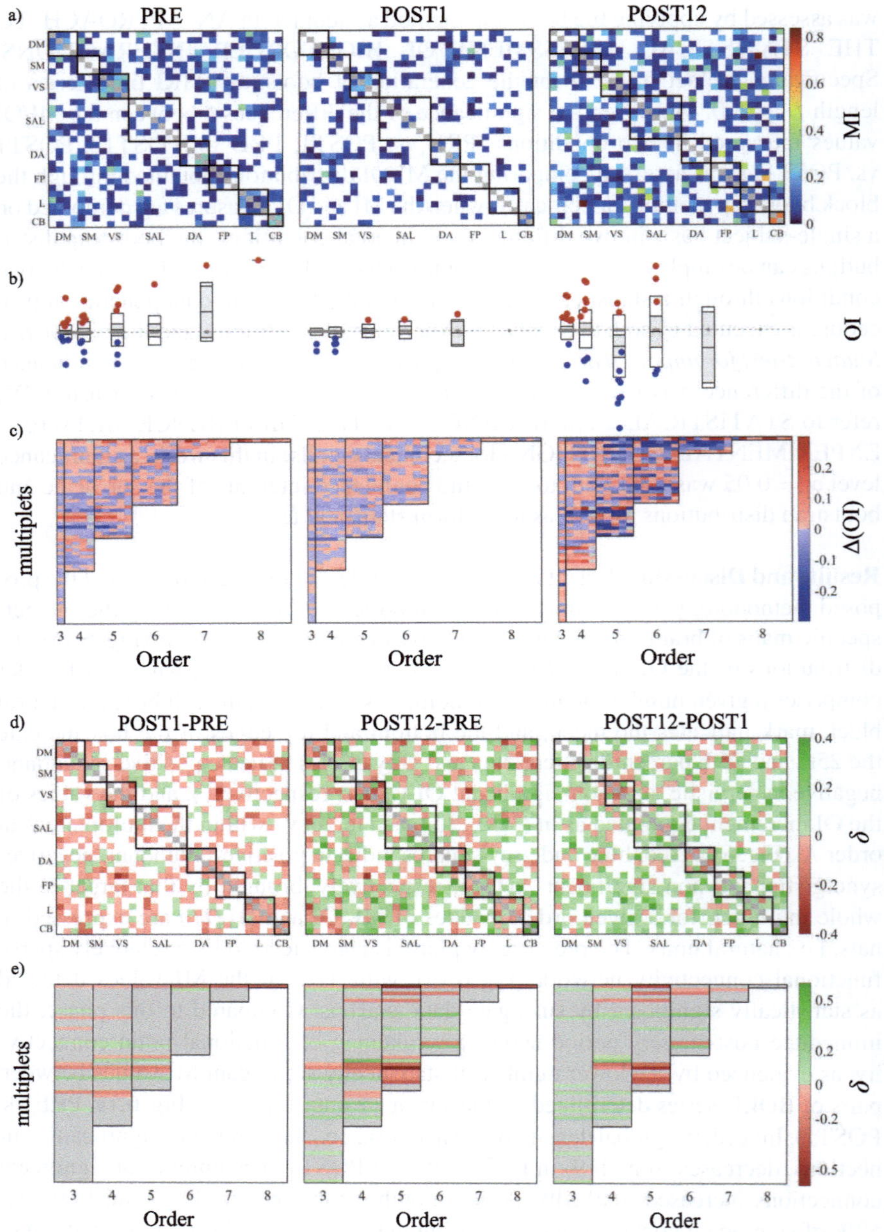

Fig. 6.1 **a** MI-weighed significant connections among pairs of variables (white squares: non-significant connections; black squares: time series belonging to the same RSN). **b** Boxplot distributions of the OI for all the multiplets from order 3 to 8 (dashed grey horizontal line: zero level). **c** $\Delta(OI)$ values computed for each target inside the multiplets at all orders (separated by black vertical lines). Networks of the differences between **d** MI and **e** OI values (white squares/grey rectangles: non-significant differences). The figure is adapted from [51]

Table 6.1 Local density within and between resting-state networks, expressed in % of significant connections, before the treatment (PRE, left), one month (POST1, middle) and twelve months (POST12, right) after the treatment

PRE

	DM	SM	VS	SAL	DA	FP	L	CB
DM	50	67	25	54	31	38	69	38
SM		0	58	52	67	33	58	33
VS			100	14	25	69	31	25
SAL				33	29	36	39	21
DA					33	25	31	38
FP						50	13	25
L							83	63
CB								100

POST1

	DM	SM	VS	SAL	DA	FP	L	CB
DM	16	25	25	18	31	38	31	13
SM		33	33	33	58	33	25	17
VS			67	25	25	6	19	25
SAL				14	29	18	25	7
DA					50	13	0	0
FP						17	19	25
L							50	13
CB								100

POST12

	DM	SM	VS	SAL	DA	FP	L	CB
DM	50	58	50	46	38	44	44	63
SM		67	42	71	67	58	67	33
VS			50	43	31	63	63	75
SAL				62	32	50	57	64
DA					33	63	50	38
FP						50	50	38
L							67	50
CB								100

that the weakening and reduction of the number of links within the network is not localized to a specific brain area but spread over the whole cortex.

The decrease of functional connectivity one month after the treatment and its increase twelve months after, observed with the pairwise estimator of MI, are translated into consequent decreases and increases of the OI values, respectively. Specifically, while all the three phases are characterized by the presence of a great number of non-significant connections, the strength and number of these high-order links is decreased one month after and again increased twelve months after the treatment (Fig. 6.1b; e.g., in case of $N = 3$, the significance rate goes from 16% in PRE to 3.6% and 16% in POST1 and POST12, respectively). This finding confirms the main result coming from MI analysis, i.e., that the surgery correction of the portosystemic shunt worked in recovering brain connectivity in this patient. Moreover, in the last experimental phase, the number of synergistic interactions predominates over redundancy, suggesting that the recovered brain can display synergy as an emergent behavior, as well as that synergistic interactions may serve to integrate and complement redundant sub-networks in recovered physiological conditions. These findings are confirmed by the decrease of $\Delta(OI)$ values in POST1, and their subsequent increase in POST12 (Fig. 6.1c, POST1 vs. POST12), characterized by a tendency towards synergy (dark blue rectangles).

Panels *d*, *e* of Fig. 6.1 show the maps of the pairwise differences between the MI (panel *d*) and OI (panel *e*) values computed in two different experimental conditions. The impairment of brain connectivity in POST1 and its recovery in POST12 is confirmed for this patient looking at the variations of the MI across conditions established by the bootstrap technique (Fig. 6.1d). This analysis documents indeed that the functional connectivity is decreased one month after surgery (red squares in POST1-PRE) but is markedly increased one year after (green squares in POST12-PRE and POST12-POST1), suggesting an improvement of cognitive functions for this patient. The utilization of the bootstrap technique for the detection of OI variations across conditions confirmed the results shown in Fig. 6.1b, c. In detail, the high number of non-significant differences, indicated by grey rectangles in Fig. 6.1e, reflects the presence of non-significant OI values in the three conditions (Fig. 6.1b, grey circles) with only a few significant connections left. Looking at the differences between the experimental conditions, the decrease of OI values toward synergy twelve months after the treatment is documented by the predominance of red rectangles in Fig. 6.1e, POST12-PRE and POST12-POST1. This reduction is localized to specific multiplets, suggesting that the recovery of high-order interactions is specific to certain areas of the pediatric brain.

Our preliminary results agree with previous findings [58], obtained by applying multivariate information metrics to fMRI data and documenting the presence of copious and widely distributed synergistic subsystems across the entire cerebral cortex. In our application to fMRI data, we randomly selected nodes from different RSNs to build high-order structures comprising between 3 and 8 regions, and showed that synergistic subsets are ubiquitous, arising at higher orders systematically across the cortex. Specifically, while redundant interactions dominate at larger subset sizes, especially during the pre-operative and the immediate post-operative phases, the late post-operative phase is characterized by the appearance of a previously hidden repertoire of synergistic ensembles, as also demonstrated by the bootstrap data analysis applied to detect subject-specific differences between conditions. In detail, these randomly sampled assemblies expressing synergy were found to involve nodes from the DM, SAL and FP networks in the pre-operative phase, together with the VS, L and CB networks twelve months after the surgical correction, when the number of synergistic pathways of interaction was definitely increased. The application of multivariate information measures demonstrates that high-order synergies represent a kind of *shadow structure* emerging from resting-state brain activity and missed by bivariate functional connectivity approaches, which indeed reveal redundancy-dominated correlations and do not provide an overall map of the statistical structure of the network [58–60]. Given the novelty of our findings, the significance of these synergistic dependencies remains almost entirely unknown, although the clinical importance of studying and comprehending these intriguing patterns persists unaltered.

Conclusion. We proposed here a subject-specific statistical evaluation of functional network connectivity analysis in the peculiar case of pediatric portal cavernoma. Our mathematical framework, based on the combined use of pairwise and high-order functional connectivity measures, allowed to display subject-specific features of brain connectivity in this patient before and after the surgical correction of the portosystemic shunt. Moreover, the utilization of surrogate and bootstrap data analyses was essential to statistically validate the functional connectivity maps obtained before surgery and during the follow-up phases (after 1 month and 12 months the surgical treatment), as well as the differences between pairs of them. This has great clinical relevance for single-subject investigations and treatment planning, particularly when it is necessary to study the effects of clinical diseases on single individuals and the subject-specific responses to personalized diagnosis and care. The statistical assessment of intra-subject connectivity network changes over time could be interpreted as evidence of statistically significant increases/decreases of functional connectivity related to an event, i.e., the surgical procedure to remove the shunt in our clinical case. Specifically, the overall increase of the number and strength of functional pairwise and synergistic connections after the surgical treatment, resulting from our analyses, was confirmed by clinical findings: during the follow-up phases, the patient recovered well from HE, as evidenced by the improvement of his cognitive functions, the recovery from psychomotor sluggishness and attention deficits and the subsequent return to school, which he had dropped out before the treatment. Therefore, the proposed statistical approaches can successfully help scientists and clinicians to identify significant pairwise but especially high-order brain functional connectivity signatures on a single-subject basis in different physiological and diseased conditions.

While this approach should be tested on a larger number of individuals to validate the clinical findings, it still revealed clinically and physiologically plausible patterns of brain pairwise connectivity in the reported application. The possibility to investigate brain connectivity and its post-treatment functional developments at a high-order level was essential to fully capture the complexity and modalities of the recovery. The results here obtained, albeit in a preliminary fashion, support the need of investigating the complex behaviour of brain structures and their emergent synergistic patterns. We assert that high-order interactions in the brain represent a vast and under-explored space that, now accessible with tools of multivariate information theory, may offer novel scientific insights even in today clinical practice.

High-order links assessed via the B-index. The rest-fMRI data acquired in the pediatric patient affected by hepatic cavernoma was further utilized to assess the presence and significance of high-order links within the investigated network in the three different experimental conditions (PRE, POST1, POST12) [61]. To this aim, the B-index measure was calculated based on MI and CMI under the assumption of Gaussian data (3.28). Indeed, while node—(Δ(OI)) and network-specific (OI) measures of HOIs are extremely useful in shaping the roles that nodes and groups have in information sharing within the investigated network, link-specific measures can help elucidating the network structure in terms of the redundant/synergistic nature of the structural network connections.

Experimental Protocol and Data Analysis. The analysis was performed for each experimental condition by computing the MI between the two selected time series $\{\mathbf{x}, \mathbf{y}\}$, and the CMI between them conditioned to the rest of the system $\mathbf{z} = \mathbf{v} \setminus \{\mathbf{x}, \mathbf{y}\}$; their statistical significance was assessed with surrogate data generated by independently shuffling in random order the sequences relevant to each variable, so as to make the surrogate variables independent while preserving their marginal distributions (THE DYNAMICS OF THE SINGLE NODE); the B-index was evaluated as in (3.28) following the procedure described in Sect. 3.2. The application was proposed in [61].

Results and Discussion. Results of the analysis are reported in Fig. 6.2, showing the subject-specific maps of brain connectivity (Fig. 6.2, above) and the reconstructed network structures (Fig. 6.2, below), in the three experimental conditions. As also reported in [51], compared to the pre-surgery phase (panel *a*, PRE), the immediate

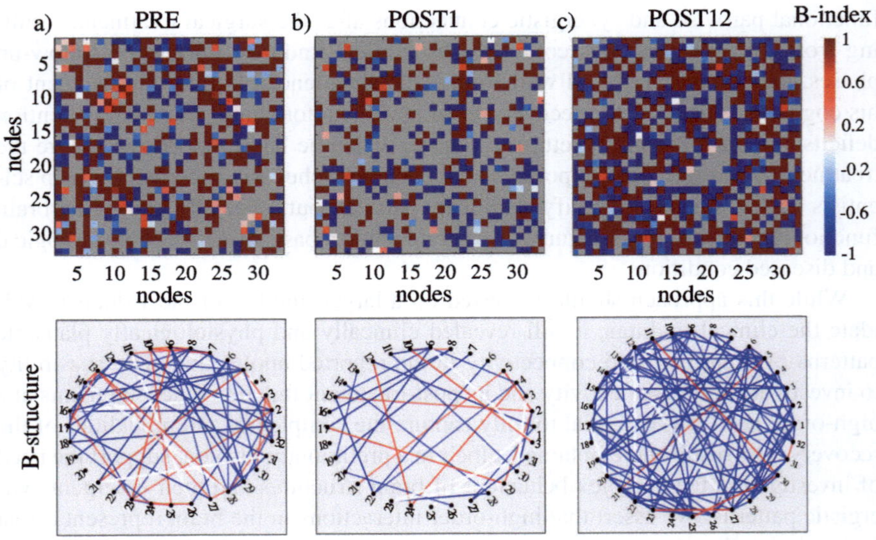

Fig. 6.2 **a** B-index estimates and **b** the reconstructed network structures computed across the resting-state networks of 32 BOLD fMRI series representing the pre-surgery state (PRE), 6 months post-surgery (POST1) and 12 months post-surgery (POST12) due to a hepatic cavernoma issue in a pediatric patient. The figure is adapted from [61]

post-surgery period (panel *b*, POST1) is characterized by a weakening of brain connectivity, as demonstrated by lower density of both redundant and synergistic links in the B structure. Interestingly, twelve months after the treatment (panel *c*, POST12), the brain network is mostly synergistic, with increased number of true connections demonstrated by an augmented B density (the same can be inferred from Fig. 6.1b, c). In this patient, the proposed surgery correction of the portosystemic shunt worked in recovering brain cognition, as evidenced by stronger brain connectivity and an increase of synergistic interactions between the nodes of the investigated resting-state networks. Overall, the utilization of the B-index to quantify and characterize high-order links supplied results in agreement with the analysis of the OI and the first-order OI gradient detailed in the above paragraph, with the remarkable peculiarity of providing network reconstruction in terms of structural links.

6.1.2 Gradients of O-Information Highlight Synergistic and Redundant Informational Circuits

The study of high-order dependencies in complex systems has recently led to the introduction of statistical synergy, a novel quantity corresponding to a form of emergence in which patterns at large scales are not traceable from lower scales. As a consequence, several works in the last years dealt with the synergy and its counterpart, the redundancy, e.g., the O-Information [26]. In spite of its growing use, this metric does not provide insight about the role played by low-order scales in the formation of high order effects. To fill this gap, the framework for the computation of the OI has been recently expanded introducing the so-called gradients of this metric, which measure the irreducible contribution of a variable (or a group of variables) to the high order informational circuits of a system [29]. In [62], we reviewed the theory behind the OI and its gradients (Sect. 3.2) and presented the potential of these concepts in the fields of Network Physiology (Sect. 5.8) and Network Neuroscience, showing an application relevant to brain networks probed by fMRI. Gradients of O-Information were calculated using the Gaussian Copula approach described in [63] to estimate entropy terms.

Experimental Data. We considered the data from the Human Connectome Project [64] corresponding to 1083 healthy subjects whose organization of networks in the human cerebrum was explored using rest-fMRI [65]. The fMRI data acquisitions were performed on a Siemens 3T Skyra scanner at Washington University. In order to construct a best-estimate parcellation of the human cerebral cortex to serve as a reference for future studies, a clustering algorithm was used to parcellate the cerebral cortex into networks of functionally coupled regions. Parcellations were examined for a coarse solution that organized the cortex into 7 networks as well as a finer solution that identified 17 networks. The estimated networks were found to be consistent across the discovery and replication data samples and were confirmed by region-based functional connectivity MRI analyses. Here we consider the parcellation in 7 clusters, each corresponding to the following connectivity networks: Default, Control, Limbic, Visual, Somatosensor, Ventral Attention, and Dorsal Attention. For each subject, we analyze the corresponding $M = 7$ fMRI time series. The significance of the detected HOIs is assessed using the statistics of subjects: gradients are considered significantly redundant (synergistic) when the 5th (95th) percentile of the bootstrap distribution is higher (lower) than zero (AN APPROACH TO THE STATISTICAL ASSESSMENT OF HIGH- ORDER INTERACTIONS).

Results and Discussion. In bottom panel of Fig. 6.3 we depict the first-order gradient computed for the seven intrinsic connectivity networks. Except for the Default, all the regions are significantly redundant, hence suggesting that the DMN it is the region for whom the balance synergy-redundancy is less leaning towards redundancy. Going to the second order gradients (middle panel of Fig. 6.3), four pairs of regions are significantly redundant: somatomotor—dorsal attention, somatomotor—visual, dorsal attention—ventral attention, dorsal attention—visual. Concerning third order gradients, we find a significantly redundant triplet, default—control—dorsal attention, and a significantly synergistic triplet: ventral attention—somatomotor—visual (see top panel of Fig. 6.3). These results evidence peculiar intrinsic connectivity networks contributing to redundancy and synergy in the large-scale organization of the overall fMRI network, and confirm that gradients of increasing order tend to highlight less redundant/more synergistic interactions.

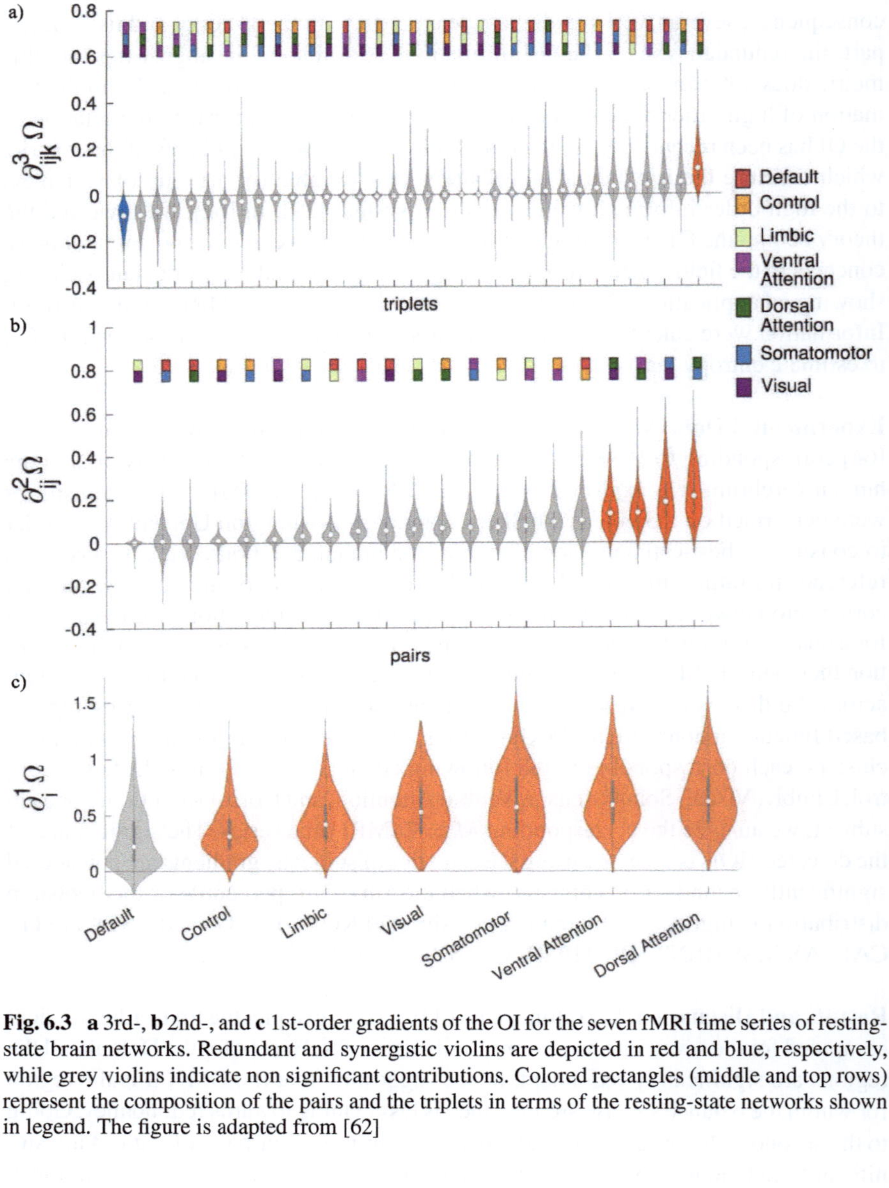

Fig. 6.3 **a** 3rd-, **b** 2nd-, and **c** 1st-order gradients of the OI for the seven fMRI time series of resting-state brain networks. Redundant and synergistic violins are depicted in red and blue, respectively, while grey violins indicate non significant contributions. Colored rectangles (middle and top rows) represent the composition of the pairs and the triplets in terms of the resting-state networks shown in legend. The figure is adapted from [62]

In [60], the synergistic and redundant districts of the resting brain have been explored, and it has been found that redundant interactions are especially prominent in the primary sensory, primary motor and insular cortices, corresponding to the brain's somatomotor and salience subnetworks. In contrast, regions with higher relative importance for synergy have been affiliated with the DMN and fronto-parietal executive control subnetworks. We note that in [58] an analogous synergy-redundancy gradient as in [60] has been found using partial entropy decomposition. It is worth mentioning that in [60] dynamical synergy and redundancy (from the double redundancy lattice, [66]), have been explored for each pair of the 232 regions of the augmented Schaefer atlas. In agreement with [60], we find that the default network has the minimum first order gradient, i.e., it is the less redundant; moreover a major redundant role is played by the somatosensor network. However our results refer to a different spatial scale and, coming from a static analysis, are not expected to be fully reproducing with the results in [60]: notice that the emergence of a synergistic circuit made of visual, somatomotor and ventral attention has not been observed in previous studies. We remark that in a recent paper it has been observed that ventral attention and motor network connectivity are relevant to functional impairment after right brain stroke [67]; moreover higher functional connectivity of ventral attention and visual network has been found to play a role to maintain cognitive performance in white matter hyperintensity [68]. These findings render even more interesting our results, i.e., these three networks belonging to a synergistic informational circuit in the resting brain.

Conclusion. Summarizing, gradients of OI constitute a promising tool to analyze many-body effects in complex systems, with the advantage of providing a description of high-order phenomena which can be tuned and can even be at the level of single variables or pairs. The application here described show the effectiveness of this approach for multivariate neural data. In the big-data setting, evaluation of gradients of OI remains an heavy computational burden, indeed for n variables even first-order gradients require the estimation of entropy terms of order up to n: further work will be devoted to develop approximate approaches for the evaluation of gradients so as to make it feasible also for a large number of variables.

6.2 Dynamic Networks of Random Processes: Applications to EEG Data

Exploited in a wide range of clinical and research applications [69–72], EEG has allowed to identify the spatio-temporal patterns of neuronal electric activity over the scalp with huge feasibility, thanks to advances in the technologies for its acquisition, such as the development of high-density EEG systems [73, 74] and the combination with other imaging modalities, robotics or neurostimulation [75–78]. On the whole, acquiring EEG signals is a challenging task and requires tricks to face some delicate steps such as positioning of the electrodes on the scalp or setting the more appropriate sampling frequency [79–85]. Furthermore, failure to properly perform the early stages of EEG pre-processing (e.g., resampling, channel re-referencing, data filtering, artifact rejection) can decrease the signal-to-noise ratio and introduce unwanted artifacts into the data. Indeed, due to the lack of standardization of data preparation, it is crucial to pay attention to this delicate aspect of EEG analysis, as it can impact subsequent steps of the evaluation of connectivity among brain networks [8, 79, 86]. Moreover, it is well-recognized that the scalp EEG signals do not directly indicate the location of the active neuronal populations in the brain [87, 88]. Causality and connectivity measurements applied on the scalp EEG do not allow interpretation of the interacting brain sources, since the channel sites cannot be seen as approximations of the anatomical locations of sources and then spurious connectivity can be detected between sensors on the scalp [89–93]. To overcome this issue, EEG source imaging has been widely applied over the past years to localize the anatomical sources (source space) of a given scalp measurement (sensor space) [94–98].

In this section, we remark the usefulness of EEG data in the context of Network Neuroscience and review exemplary applications of the dynamic measures of pairwise (e.g., the MIR) and high-order (e.g., the OIR) connectivity defined in Chap. 4 to EEG signals. Further details on data-driven approaches for the study of EEG-based brain connectivity, as well as on the most common and tricky pitfalls occurring during acquisition and pre-processing of electrophysiological signals, can be found in [99].

6.2.1 *Human Brain Activity During Motor Execution Assessed via the OIR Framework*

The OIR framework proposed in [28] allows a hierarchically-organized evaluation of time and frequency domain pairwise and high-order interactions in dynamic networks mapped by multivariate time series, being part of the comprehensive and unifying framework recently proposed in [100]. In this section, we show the potential of our framework in the context of EEG signals acquired in healthy subjects performing a motor execution task; the application was published in [101]. We remark that the study of human brain activity during *motor execution* is very important in clinical contexts and in neuroscience. In fact, motor actions commonly derive from the involvement of several areas in the brain causing excitatory and inhibitory coupling among different regions in the two hemispheres [102]. Deepening the investigation of changes in motor-related brain regions according to movement states would allow to elucidate the motor control mechanism of the human brain.

Experimental Protocol and Data Analysis. The data is relevant to 20 healthy subjects randomly chosen from a database of 109 participants [103, 104]—note that in [100] we used the same database selecting only one exemplary subject (see Sect. 6.2.4). We analyzed a resting-state condition in which the participants were

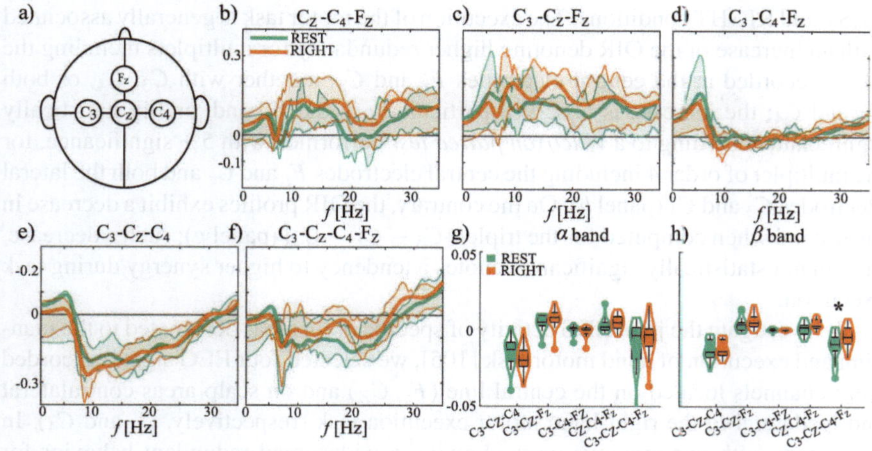

Fig. 6.4 **a** EEG electrodes montage highlighting the positions of the four selected electrodes. **b–f** Spectral profiles (mean—bold line—and 25th–75th percentiles—shades—over 20 subjects and 15 trials per subject) of the OIR computed for each multiplet during the two phases of the protocol (REST, RIGHT). **g–h** Violin plots of the distribution across participants and trials of the OIR values computed by integrating the spectral OIR within the α and β band during REST and RIGHT. *Wilcoxon test for paired data* ($p < 0.05$): *, REST versus RIGHT. The figure is adapted from [101]

relaxed (REST) and a condition in which they were asked to open and close the right fist cyclically (RIGHT). The raw signals were firstly detrended, then filtered and finally epoched to extract 15 trials of 4 s each for each participant and condition. We performed the analysis on the signals recorded by the four electrodes depicted in Fig. 6.4a, i.e. $\mathbf{X}_1 = C_3, \mathbf{X}_2 = C_z, \mathbf{X}_3 = C_4, \mathbf{X}_4 = F_z$. For each subject and trial, a VAR model (2.6) was identified from the four selected time series setting the model order according to the BIC. Then, the estimated VAR parameters were used to compute the spectral OIR (4.62) for each multiplet of order $N = 3, 4$. Finally, values indicative of HOIs occurring for the α and β brain rhythms were obtained by integrating the measures over the relevant frequency ranges (i.e., $\alpha = [8-12]$ Hz, $\beta = [16-26]$ Hz).

Results and Discussion. Figure 6.4b–f reports the grand-average over participants and trials of the frequency profiles obtained for each multiplet separately for the REST and RIGHT conditions in the frequency range [2–35] Hz. The trends reveal the prevalence of positive values of the OIR, denoting redundant interactions, for the triplets including the signal recorded at the electrode F_z (panels *b, c, d*), while the triplet $[C_3 - C_z - C_4]$ displays negative OIR values related to synergy within the α and β bands (panel *e*); synergistic interactions are detected, although weaker, also for the multiplet of order 4 including all the analyzed electrodes (panel *f*). Panels *g*, *h* depict the distributions across participants and trials of the spectral OIR integrated over the α and β frequency bands, computed for each multiplet separately for the

REST and RIGHT conditions. The execution of the motor task is generally associated with an increase of the OIR denoting higher redundancy for multiplets including the EEGs recorded at the central electrodes F_z and C_z, together with C_3, C_4, or both C_3 and C_4; the increase is observed particularly in the β band, and is statistically significant, according to a *Wilcoxon paired test* performed with 5% significance, for the multiplet of order 4 including the central electrodes F_z and C_z and both the lateral electrodes C_3 and C_4 (panel *h*). On the contrary, the OIR profiles exhibit a decrease in the α band when computed for the triplet $[C_3 - C_z - C_4]$ (panel *g*); such a decrease, though not statistically significant, denotes a tendency to higher synergy during task execution.

To investigate the joint EEG activity of specific brain areas designated to the planning and execution of hand motor task [105], we selected four EEG signals recorded from channels located on the central line (F_z, C_Z) and on scalp areas contralateral and ipsilateral to the right-hand motor execution task (respectively, C_3 and C_4). In agreement with our previous work showing a widespread redundant behavior for this network of EEG interactions [106], we document the prevalence of redundancy for HOIs. However, we also show that combinations of signals measured from both central electrodes and electrodes located in ipsi- and contra-lateral locations give rise to synergistic HOIs that reflect the emergence of interaction mechanisms not retrievable from a pairwise analysis. From a physiological point of view, these functional mechanisms involve spatial locations (C_3 and C_4) and emerge in frequency bands (α and especially β) which are linked to the well-known phenomenon of event-related desynchronisation occurring during motor execution and imagery [107]. Nevertheless, we stress the preliminary nature of our results, which need confirmation on larger datasets [104], also after adopting methodological improvements (e.g., the consideration of the statistical significant of the detected OIR values and increments [27]) and addressing common issues of brain connectivity analyses (e.g., those related to the effects of volume conduction on pairwise and high-order connectivity measures [93, 108].

6.2.2 Neural Interactions From ECoG Signals in the Anesthetized Macaque Monkey

The OIR framework proposed in [28] and applied to exemplary EEG signals acquired in healthy subjects performing a motor execution task (see previous section) is herein discussed in the context of representative monkey electrocorticographic (ECoG) signals [28]. *Electrocorticography* is a type of intracranial EEG that uses electrodes placed directly on the exposed surface of the brain to record electrical activity from the cerebral cortex, in contrast to conventional EEG electrodes that monitor this activity from the scalp. ECoG signals are recorded through an invasive procedure requiring surgical opening of the scalp and skull; they are composed of synchronized postsynaptic potentials occurring primarily in cortical pyramidal cells, and thus conducted through several layers of the cerebral cortex before reaching recording electrodes placed just below the skull. However, to reach the scalp electrodes of a conventional EEG, electrical signals must also be conducted through the skull, where potentials rapidly attenuate due to the low conductivity of bone. For this reason, the spatiotemporal resolution and signal quality of ECoG are much higher than EEG. This technique has been widely used to diagnose and treat neurological disorders in clinical settings for several decades, despite the invasiveness. Recently, it has been applied in neuroscience to explore brain functions and connectivity, brain-computer interfaces, and brain-machine interfaces (see, e.g., [109–111]).

Experimental Protocol and Data Analysis. The analyzed dataset, consisting of monkey ECoG signals downloaded from the public server,[6] was recorded with a sampling frequency of 1000 Hz in one macaque monkey using 128 electrodes, placed in pairs with an inter-electrode distance of 5 mm to cover the frontal, parietal, temporal and occipital lobes of the left hemisphere [112]. We considered two five-minutes recording sessions during which the blindfolded monkey was seated in a primate chair with tied hands, first in a resting state (REST) and then after injection of a sedative inducing anesthesia (ANES). From the 128 electrodes, a subset of 20 was selected as depicted in Fig. 6.5a to cover, considering ten bipolar ECoG signals obtained taking the differential activity between close electrodes, the following five brain regions of the DMN: *(i)* pre-frontal cortex ($\mathbf{X}_1 = [Y_1, Y_2]$), *(ii)* parietal cortex ($\mathbf{X}_2 = [Y_3, Y_4]$), *(iii)* temporal cortex ($\mathbf{X}_3 = [Y_5, Y_6]$), *(iv)* low visual cortex ($\mathbf{X}_4 = [Y_7, Y_8]$), and *(v)* high visual cortex ($\mathbf{X}_5 = [Y_9, Y_{10}]$). The ten bipolar signals were band-pass filtered between 0.5 and 200 Hz, downsampled to $f_s = 250$ Hz, epoched to extract $\sim$160 trials lasting 2 s for each condition, and finally normalized to zero mean and unit variance within each trial. Then, a VAR model (2.6) was fitted on the $Q = 10$ signals of each trial using least squares identification and setting the model order according to the BIC [22]. From the VAR parameters, the analysis of high-order interactions was performed for the $M = 5$ blocks computing the spectral OIR for all multiplets of order $N = 3, 4, 5$. Time-domain OIR values $\Omega_{\mathbf{X}^N}$ (4.51) were then obtained integrating the spectral measures $\nu_{\mathbf{X}^N}(\bar{f})$ (4.64) within the δ ([0.2−3] Hz), θ ([4−7] Hz), α ([8−12] Hz), β ([12−30] Hz) and γ ([31−70] Hz) frequency bands, as well as cumulatively between 0 70 Hz.

Results and Discussion. The results of OIR computation are reported in Fig. 6.5b, showing the grand average of the spectral OIR for five multiplets selected as the most representative of the analyzed interactions, together with the time-domain OIR obtained through whole-band and band-specific integration. The positive values of the OIR functions and of the integrated measures, observed for all multiplets in both conditions and increasing with the order of the multiplet, indicate that the analyzed system is dominated by redundancy. Moreover, the redundancy level is modulated by the experimental condition to an extent that depends on the analyzed multiplet and spectral band. Indeed, considering the multiplets of order 3 and 4 which involve

[6] neurotycho.org.

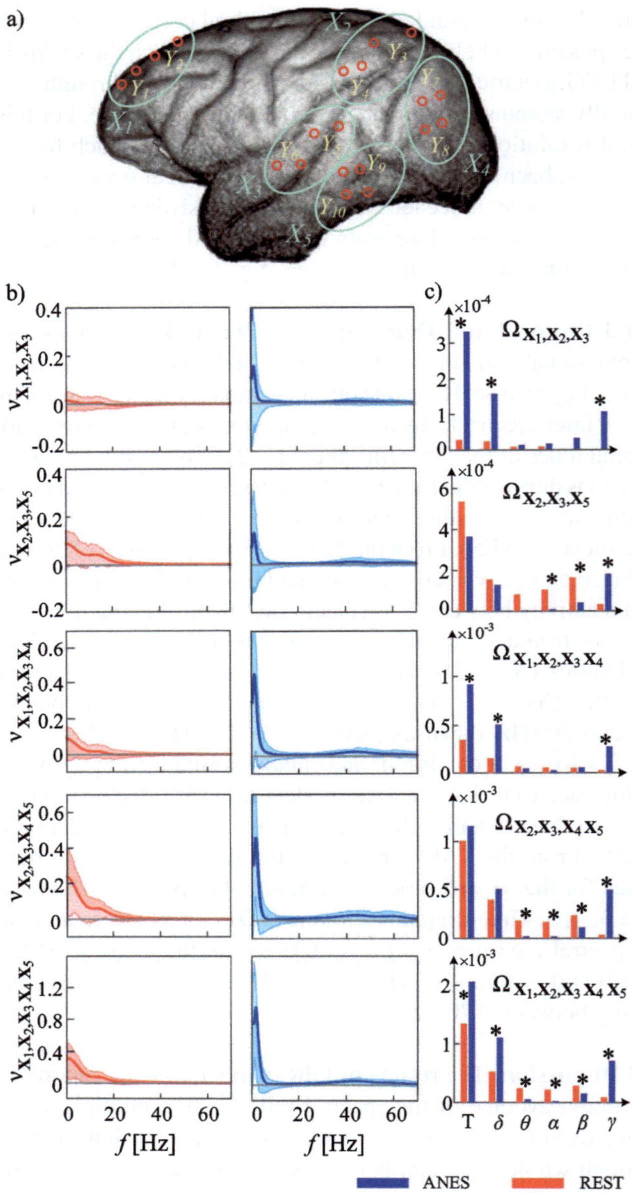

Fig. 6.5 ECoG electrode montage. Average spectral profiles across trials (line: median; shades: 1st–3rd quartiles) of the OIR computed for five representative multiplets during relaxation (REST) and anesthesia (ANES). Average time-domain OIR obtained by integrating the spectral measures over the whole frequency axis (T) or within the δ, θ, α, β and γ bands; asterisks denote statistically significant difference between REST and ANES (*Wilcoxon signed-rank test with Bonferroni correction for multiple comparisons*, $p < 0.05$). The figure is adapted from [28]

the prefrontal cortex $\mathbf{X}_1$ (first and third row of panels in Fig. 6.5b), a significant increase of the OIR is observed while moving from REST to ANES; such increase is driven by the rise of a peak in the OIR at $\sim$2 Hz (δ band) together with an increased contribution within the γ band. On the other hand, the multiplets formed by signals from the parietal, temporal and visual cortices (second and fourth row of panels in Fig. 6.5b) display a drop of redundancy in the α and β bands during ANES compared to REST. These two opposite behaviors are summarized by the OIR encompassing all five regions (fifth row of panels in Fig. 6.5b), which during ANES displays significantly higher levels of redundancy in the δ and γ bands (and in the whole band), and significantly lower redundancy in the θ, α, and β bands.

Our results indicate that the activity relevant to the α and β rhythms observed during the relaxed awake state disappears during anesthesia, leaving place to dominant interactions within the δ and γ bands. The redundancy observed at REST for the α waves is significant for the multiplets involving signals from the visual cortex, in agreement with the knowledge that these waves can be predominantly recorded from the occipital lobes during wakeful relaxation with closed eyes [113]. On the other hand, the higher redundancy reported in the δ band can be related to the slow wave oscillations ([0.1−4] Hz) typically observed under anesthesia [114]. Moreover, the fact that higher δ redundancy is observed only for multiplets including frontal cortex signals supports the knowledge that the slow oscillations are a manifestation of a coupling between the anterior and posterior axes of the brain [115]. Anesthesia evokes also an increase of redundancy related to γ oscillations, which are associated with different cognitive functions [116]. Overall, these results agree with those in [112] and support the integration theory according to which the conscious state is generated by highly integrated neural interactions that disappear in the unconscious state [117]. A recent study comparing resting wakefulness with propofol-induced anaesthesia in human fMRI data has shown how the anterior-posterior disconnection occurring during anesthesia is associated with a decrease of *integrated information* within the DMN in the left hemisphere [118]. Importantly, the concepts of integration information and that of redundancy are interrelated, as explained in [119] where it is highlighted that a drop of integrated information corresponds to an increase of redundancy. Thus, our results support the theory of an anterior-posterior disconnection during anesthesia, which in our case can be ascribed to the significant increase of the OIR documented when the frontal cortex is considered in the analyzed multiplet.

6.2.3 Rehabilitation Modulates High-Order Interactions Among Large-Scale Brain Networks in Subacute Stroke

Nowadays, several neurodegenerative diseases, such as healthy brain aging [120] and dementia or Alzheimer's disease [121], and other pathologies are known to cause brain networks disruptions. For example, the effects of *cerebrovascular stroke* on brain functional connectivity are widely investigated, due to the high incidence of stroke in the worldwide population and to the impact of resultant impairments on the quality of life of stroke survivors [122]. High-order functional connectivity approaches have not been yet applied to investigate stroke effects and the associated neuro-plasticity changes involved in recovery. To the best of our knowledge, findings reported in the literature about cerebrovascular stroke were exclusively based on metrics of pairwise interaction between signals, either from fMRI or EEG techniques, in brain regions belonging to different resting-state sub-networks [123]. In particular, great attention has been given to the evaluation of recovery-related changes within the motor network (MN) [124], since more than a half of stroke survivors experience motor deficits [125]. However, the brain is a complex system, and it has been demonstrated that disruption of connections between large-scale RSNs may as well contribute to different types of post-stroke impairments [126]. A few studies have investigated both within and between large-scale RSNs connectivity patterns in stroke populations [126–128]. In these works, altered functional connectivity has been found among different brain areas including not only sensory and sensorimotor cortices, but also high-order cognitive control networks such as the default mode network, the executive control network (ECN), and the dorsal attention network [128, 129]. Particularly, the DMN, ECN, and MN appear to be involved in connectivity disruption in stroke patients and may play a relevant role in cerebral reorganization associated with functional recovery. Yet, none of the above-mentioned studies have considered jointly the concurrent interactions between these three networks.

In [130], we proposed to investigate changes ensuing from *post-stroke rehabilitation* in the interactions among the DMN, ECN and MN, by applying high-order metrics and thus going beyond the traditional pairwise analysis. In detail, we analyzed changes in cortical connectivity estimated from resting-state EEG signals in a cohort of patients in the subacute post-stroke stage who followed a period of physical rehabilitation. Considering the high temporal resolution of EEG signals, we exploited the OIR framework developed in [28] to perform a causal and spectral high-order connectivity analysis of the temporal dynamics among large-scale brain networks.

Experimental Protocol and Clinical Assessment. Eighteen post-stroke patients participated in this study, which was approved by the local Ethics Committee *Comitato Etico Provinciale dell'Insubria* and conducted in compliance with the *Declaration of Helsinki*. Participants (7 females and 11 males, aged 67 ± 10 years) were all enrolled in the subacute stage after a single unilateral ischemic stroke and met the inclusion criteria of occurrence of the acute event less than 30 d after the first evaluation. All subjects were right-handed, had no other reported concomitant orthopedic or rheumatologic diseases, and had no global or comprehension aphasia. Each patient followed a physical rehabilitation treatment for both upper and lower limbs, tailored according to the individual residual motor capacity. More details about the dataset can be found in a previous work on the topic [131].

Subjects were evaluated by clinicians both from a clinical and an electrophysiological point of view at two time-points: at the admission to the rehabilitation center (T0, on average after 12 ± 5 days from the stroke event) and at the end of the treatment (T1, on average after 55 ± 11 days from the stroke event). The upper limb performance was evaluated with the upper extremities Fugl-Meyer Assessment (FMA), which includes a motor scale with scores ranging from 0 (hemiplegia) to 66 (normal motor performance) [132]. Likewise, the level of walking ability was assessed by the Functional Ambulation Category (FAC), a gait assessment scale that distinguishes between 6 levels of walking ability (0, not able to walk; 5, independent walk) based on the amount of physical support required [133]. The variations $\text{FMA}^{T1-T0} = \text{FMA}^{T1} - \text{FMA}^{T0}$ and $\text{FAC}^{T1-T0} = \text{FAC}^{T1} - \text{FAC}^{T0}$ were considered as primary clinical outcomes of subject's motor recovery.

EEG Acquisition, Pre-Processing and Source Reconstruction. At both T0 and T1, 5 min of eye-closed resting-state EEG recordings were collected with a Neuroscan system (Compumedics Neuroscan, Compumedics, NC, USA). 64 Ag/AgCl electrodes were placed on the scalp according to the International 10/20 standard system with the reference electrode placed between Fz and Cz positions and the ground electrode positioned anterior to Fz. Continuous data were acquired at a sampling rate 1000 Hz. EEG signals were pre-processed offline using the open source EEGLab signal processing Toolbox [134]. Data were down-sampled 128 Hz and band-pass filtered between 0.5 and 45 Hz to remove slow drifts and high-frequency components. Flat (zero amplitude for more than 5 s) and *bad* (noisy for more than 90% of the acquisition) channels were removed and the Artifacts Subspace Reconstruction (ASR) algorithm [135] was applied to all retained channels with a cut-off parameter $k = 20$. Independent Component Analysis (ICA) was then applied to the ASR-cleaned EEG exploiting the RUNICA Infomax algorithm [136], and the ICLabel automated classifier [137] was employed to guide the manual selection of non-brain artifactual components, e.g., eye, heart, muscle, line noise. The originally removed channels were reconstructed by the interpolation of the neighbor signals and the cleaned EEG data were re-referenced to a common average value. For all the acquisitions, the first 30 s were discarded to account for a period of settling before reaching the proper resting-state condition and the subsequent 1 min was used for brain sources estimation. We employed the exact low resolution brain electromagnetic tomography (eLORETA) approach, which allows the exact localization of the brain current source densities distribution implementing a discrete, linear, weighted minimum norm inverse solution [138]. Activity of brain sources was reconstructed only in the cortical grey matter (6239 isotropic voxels with 5 mm spatial resolution) in the MNI152 space.

Regions of Interest and Time Series Extraction. For the inter-network connectivity analysis, we defined 16 ROIs within the three large-scale networks. In Table 6.2, centroid coordinates in the MNI152 space and number of voxels of 6 ROIs within DMN, 5 ROIs belonging to ECN, and 5 ROIs within MN are reported according to previous MRI and EEG studies of RSNs identification in the healthy and stroke populations [139–141]. All voxels within 8 mm of radius from a seed were considered as belonging to the specific ROI, with the constraint of non-overlapping regions. In case of overlap, the voxels were assigned to the closest centroid regions. Sixteen time

Table 6.2 Regions of interest

RSN[a]	Region	Coordinates in mm (X, Y, Z)	Number of voxels
DMN	PCC/precuneus	0, −52, 27	14
	Medial prefrontal	−1, 54, 27	13
	L lateral parietal	−46, −66, 30	12
	R lateral parietal	49, −63, 33	8
	L middle temporal	−61, −24, −9	15
	R middle temporal	58, −24, −9	12
ECN	Dorsal medial PFC	0, 24, 46	15
	L anterior PFC	−44, 45, 0	7
	R anterior PFC	44, 45, 0	6
	L inferior parietal	−50, −51, 45	13
	R inferior parietal	50, −51, 45	16
MN	L M1	−33, −20, 52	12
	R M1	−36, −18, 52	13
	L PreMotor	−34, −1, 56	13
	R PreMotor	35, 0, 55	13
	SMA	0, −4, 65	10

Cerebral areas grouped by large-scale networks. Centroids coordinates (X, Y, Z) are reported in the MNI coordinate system. [a]RSN = Resting-State Network, DMN = Default Mode Network, ECN = Executive Controls Network, MN = Motor Network, PCC = Posterior Cingulate Cortex, PFC = pre-frontal cortex, M1 = primary motor cortex, SMA = Supplementary Motor Area, L = left, R = right

series of 7680 samples (1 min) each were thus obtained, by averaging the magnitude of sources activity among all voxels belonging to each ROI. For further analysis, the ROI signals were epoched in shorter windows of 10 s, hence obtaining 6 windows of 1280 samples for each acquisition.

Data and Statistical Analysis. We performed an inter-network resting-state connectivity analysis, considering the three subsets of ROIs, grouped as reported in Table 6.2, as vector targets constituting a three-node RSN. High-order analysis was conducted in a linear parametric framework based on linear AR modeling of multiple time series, under the assumption of WSS and of jointly Gaussian stochastic processes. The analysis follows the OIR framework introduced in [28]. Specifically, all measures of interactions were computed identifying the VAR model (2.6) through the OLS method. The AIC was used to select the optimum order p for each epoch of the acquired resting-state signals. The spectral OIR (4.64) with the causal decomposition terms of its gradient, i.e., $\delta_{\mathbf{X}_i \to \mathbf{X}^N_{-i}}(\bar{f})$ and $\delta_{\mathbf{X}^N_{-i} \to \mathbf{X}_i}(\bar{f})$ of (4.63), were calculated among the three RSNs separately for each of the six signals' epochs, at T0 and T1. Then, measures were integrated over the range of the three frequencies of interest, i.e., theta (θ), alpha (α), and beta (β), thus obtaining the equivalent metrics in time-domain at the specific oscillation. Delta and gamma bands were excluded from this analysis since they may be affected by noise and spurious signals. To account for interindividual variability, frequency bands ranges were defined according to the Individual Alpha Frequency (IAF) criteria [142] as follows: $\theta = [IAF - 6Hz, IAF - 2.5Hz], \alpha = [IAF - 2Hz, IAF + 2Hz]$, and $\beta = [IAF + 2.5Hz, IAF + 20Hz]$.

The six epoch values were finally averaged to obtain one value per acquisition per subject. Considering the small sample size, we employed non-parametric statistical tests to assess the rehabilitation effect in terms of inter-networks connectivity changes. T0 and T1 values of all connectivity metrics were compared by a *non-parametric one-tailed Wilcoxon's test for paired samples*. To further investigate the correlations of inter-network connectivity changes with the functional outcomes, *non-parametric Spearman's correlations* between the variation of high-order (Ω^{T1-T0}) connectivity metrics with FMA^{T1-T0} and FAC^{T1-T0} scores were computed. The statistical significance level was set at 0.05 for all tests. Because of the small sample size, in this work we did not correct the p-values from our statistical analysis for family-wise error rate. This could increase the chance of type I error, but we aimed to avoid missing any potential significance in our exploratory analysis. Therefore, we highlighted the stronger significance values ($p < 0.01$) in the results section.

Results. In Fig. 6.6a, we report the population distributions of global OIR $\Omega_{DMN,ECN,MN}$ in the three frequency bands. We observe overall positive values (expressed in *nats*) both at T0 ($\Omega(\theta) = 0.04 \pm 0.04$; $\Omega(\alpha) = 0.08 \pm 0.05$; $\Omega(\beta) = 0.17 \pm 0.17$) and T1 ($\Omega(\theta) = 0.03 \pm 0.06$; $\Omega(\alpha) = 0.07 \pm 0.06$; $\Omega(\beta) = 0.16 \pm 0.21$), indicating the prevalence of redundancy in the interactions among the three RSNs. On average, decreasing values can be noticed at T1, especially at θ, in which we found a significant T1 – T0 difference ($p = 0.049$), and α frequencies. Although OIR is still positive, on average, this trend suggests a relative shift toward less redundant and more synergistic values in the overall balance described by the OIR metric after the rehabilitation. The average decrease of $\Omega_{DMN,ECN,MN}$ at T1 is observed also in the causal decomposition of the spectral OIR, considering each of the three RSNs in turn as target process (Fig. 6.6b).

In particular, considering the statistically significant T1 – T0 differences, the increasing synergy/decreasing redundancy ($\Omega^{(T1-T0)}_{DMN,ECN,MN} < 0$) in the overall balance behavior of the system appears to be mostly explained by the MN acting as driver in two main cases: by its-own towards the rest of the system in θ and α ($p(\theta) = 0.005$ and $p(\alpha) = 0.009$), and in combination with ECN as a driver towards DMN at all the considered frequency bands. Only in the α band, we found a mild ($p(\alpha) = 0.037$) significant reduction of causality for ECN considered as a target of the vector process $\{DMN, MN\}$. Conversely, the ECN seems to play a role in the recovery of motor functions, especially of the upper limb. As shown in Fig. 6.7, we found a negative correlation between Ω^{T1-T0} and FMA^{T1-T0} when ECN acts as a driver towards the process $\{DMN, MN\}$ in both θ and α frequencies, and when ECN is driver in pairs with DMN towards MN in θ. Hence, increase in synergy driven by ECN appears to correlate with upper limb motor recovery.

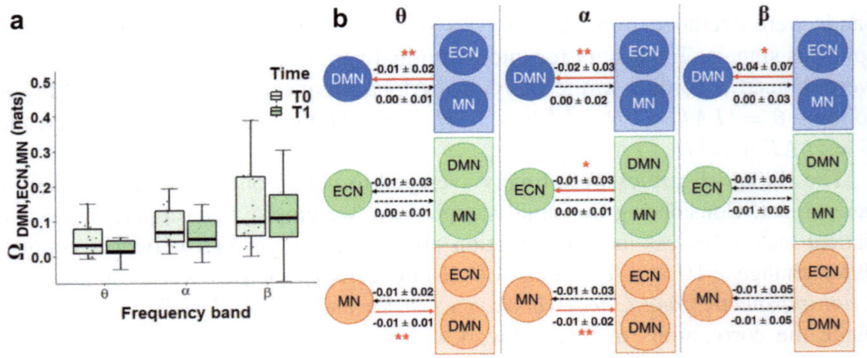

Fig. 6.6 **a** Distributions of the OIR gradient integrated within the θ, α and β bands (boxplot lengths: interquartile range, IQR; horizontal line: median value; external whiskers: data within $\pm 1.5 \times IQR$). **b** Causal decomposition of the spectral OIR gradient along the θ, α and β bands. Average differences Ω^{T1-T0} are reported as mean $\pm$ std when DMN (top row), ECN (middle row), and MN (bottom row) act as single targets with respect to the rest of the system. Red arrows: significant T1 – T0 differences. Statistical test T1 – T0: $*$, p-value < 0.05, $**$, p-value < 0.01. Average differences of 0.00: $\Omega^{T1-T0} < 10^{-2}$. The figure is adapted from [130]

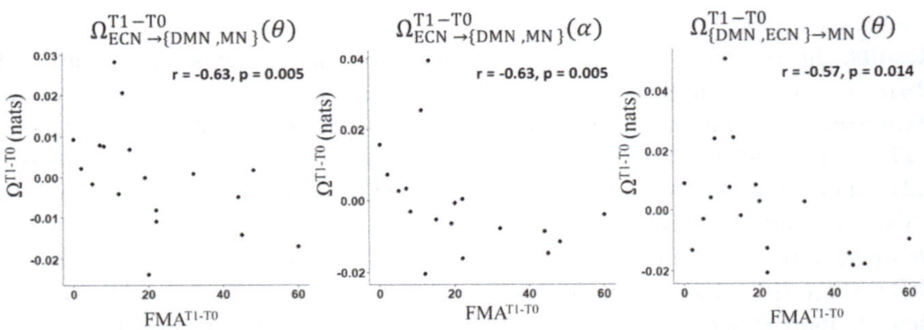

Fig. 6.7 Significant Spearman's correlations between OIR variations T1 – T0 and FMA^{T1-T0}. r, correlation coefficient, p, p-value. The figure is adapted from [130]

Discussion. The main goal of this work was to investigate HOIs among three RSNs, i.e., DMN, ECN and MN, to identify their changes after rehabilitation of subacute stroke patients. MN processes sensory input, and it is primary responsible for the execution of motor tasks [143]. DMN and ECN are high-level functional brain networks; DMN is a network which controls the interoceptive monitoring and self-referential processes [144], whereas ECN focuses on the control and execution of externally directed activities [145], including motor behavior. The relation among these three networks appears particularly of interest in post-stroke rehabilitation assessment. In fact, motor functionality recovery was found positively associated with interactions between areas belonging to MN and either DMN or ECN networks [128]. All the previous studies only investigated pairwise interactions between these networks or the correlation of intra-network connectivity with functional impairment in different domains [146], while there is a lack in the characterization of their HOIs. Moreover, it is unclear whether the three interacting networks act synergistically or redundantly in stroke patients. For this reason, alongside the pairwise connectivity descriptors, we provided also a high-order characterization of these interactions using OIR, to investigate the causal processes underlying the redundant/synergic behavior among the three RSNs.

In our analysis, we found a prevalence of redundancy (positive OIR values) in the interactions among the RSNs examined at all the frequency bands of interest, both before and after the rehabilitation. In addition, in our longitudinal evaluation of subacute stroke recovery, we observed a relative increase of shift toward less redundant and more synergistic values among DMN, ECN, and MN. Looking at the causal decomposition of the OIR, we found that this increase in net synergy appeared prevalently when MN acts alone as driver towards the other two networks, or when ECN and MN jointly drive DMN. More specifically, we found a significant increase of causal influence from T0 to T1, especially in α and θ frequencies. In the literature, the description of high-order brain networks characteristics is very limited. Up to date, very few works investigated the synergistic/redundant behavior within RSNs in healthy subjects, and, to our knowledge, no one has previously investigated a post-stroke population. In [106], the functional interactions among different cortical areas during movement execution were explored in normal subjects. In that work, dynamic EEG measures of high-order connectivity at the sensor level in the frequency domain highlighted the presence of redundancy among MN and frontal areas. This result agrees with our findings. It should be noted that the concept of redundancy and synergy has been previously applied. Luppi et al. [60] provided a first description of neuronal profiles for synergy and redundancy of different RSNs to analyze resting-state functional MRI data from 100 healthy participants of the Human Connectome Project. They found that redundancy tends to prevail within RSNs, in particular in MN, visual and salience networks, while synergistic interactions are stronger between RSNs, especially between DMN and frontoparietal regions belonging to ECN. However, these results are not directly comparable with ours since different conceptual definitions of redundancy and synergy were used. Indeed, they considered a pairwise approach, in which target and driver variables are defined based on their present and past states, evolving jointly over time. Future research is encouraged to assess separate redundant and synergistic contributions before and after rehabilitation, to confirm and complement the results of the present work.

In our study we found some interesting correlations between connectivity metrics and clinical motor recovery scales. In particular, the increased interactions and synergy, when ECN acts as a driver towards the other networks, are significantly correlated to the recovery of upper limb motor function, measured as FMA score. This result may suggest the involvement of ECN in motor functionality recovery, as previously hypothesized [127, 147]. However, this hypothesis should be further investigated and supported by studies in larger populations.

Limitations and Future Developments. Besides the three RSNs here investigated, other RSNs changes have been found to be involved in stroke recovery, such as the dorsal and ventral attention or language networks [146] as well as the auditory and the visual networks [122, 127]. Being based on linear parametric modeling, OIR requires a standard technique for the identification of the VAR model, such as the OLS or the algorithm for the solution of the YW equations. However, to avoid the increasing of bias and variance of estimation, which may result in ill-posed regression problems, the ratio between the amount of data samples available and the number of regression coefficients to be estimated should be at least equal to 10 to guarantee the accuracy of the estimation procedure [148, 149]. For this reason, we focused our analysis on three of the main RSNs known to be affected by stroke, without excessively increasing the number of time series to be fitted. In our case, we had 1280 samples available for each time series, and we could not fit more than the 16 selected ROIs, otherwise the goodness of the estimated autoregressive parameters would not be guaranteed. As for the causal interpretation of the individual network contribution to the inter-network increase in synergy, we should point out that the driver role of MN and ECN cannot be stated with total confidence. In fact, in our data, the instantaneous causality term which appears in the r.h.s. of (4.63) is not null; thus, we cannot affirm that the causal terms totally explain the redundancy/synergy balance. Indeed, the strict causality hypothesis [25] should be fulfilled to account exclusively for the causal terms. However, the instantaneous synchronization is intrinsic in EEG techniques, and, even though mitigated by source activity reconstruction, we could not completely overcome this issue. In future, this aspect should be further investigated. Besides, the OIR approach proves to be powerful for the investigation of HOIs in connectivity adding spectral and directional description. This method could be interestingly applied for the simultaneous investigation of a larger number of networks, but still, at the cost of lower-spatial resolution to obtain a lower number of cortical areas for each large-scale network.

Conclusion. The use of HOI metrics in the study of inter network connectivity in stroke patients could be a powerful tool for better understanding complex multiple relationships among RSNs, giving a particular focus on their redundant/synergistic behavior. Adopting this perspective, in our study we found a predominance of redundant interactions among the RSNs both before and after the rehabilitation in subacute stroke. After rehabilitation, we observed an increased shift toward less redundant and more synergistic behavior, mainly related to the joint effect of MN and ECN on DMN. Finally, the correlation analysis with clinical outcomes, suggests that ECN may be a relevant player in motor functionality improvement.

6.2.4 Brain Interactions at Different Orders: Exemplary Single-Subject Analysis

In [100], we introduced a coherent framework integrating several information dynamics approaches to quantify single-node, pairwise and high-order interactions in network systems. A *hierarchical organization* of interactions of different order was established using measures of entropy rate, mutual information rate and O-information rate to quantify the dynamics of individual nodes, the links between pairs of nodes, and the redundant/synergistic hyperlinks in groups of nodes. Flexibility and scalability of the proposed framework are guaranteed by the utilization of information-theoretic measures defined for scalar or vector processes, in both time and frequency domains in a way such that the two representations are tightly connected in a straightforward way, i.e., by satisfaction of the spectral integration property. Network interactions are categorized hierarchically depending on the number of nodes involved in the computation of each interaction measure: *entropy rate* describes the predictable information within a node, *mutual information rate* describes the information dynamically shared between two nodes, interaction information and *O-information rates* describe the information shared among three or more nodes through concepts of redundancy and synergy, thus opening the way to a deeper investigation of HOIs. Remarkably, there we highlighted the possibility to retrieve single-node, pairwise and high-order activities directly from the spectral representation of the VAR model (2.6) in the frequency domain, thus switching from the concept of multivariate analysis typical of previous studies on the topic (those employing, e.g., measures of conditional GC and thus focusing on the activity of two nodes of the network even when the other nodes are taken into account, such as in [150–152]) to that of *hierarchical HONA*, which allows to go beyond the framework of pairwise interactions such that the two concepts of redundancy and synergy acquire a leading role and an overall quantification of the collective interaction among groups of nodes is provided.

In this section, we report the practical application of the proposed framework to EEG signals relevant to one exemplary healthy subject performing a motor execution task [100]—note that in [101] we used the same database selecting a group of 20 representative subjects (see Sect. 6.2.1). Further, we remark that the time and frequency domain OIR measure applied to EEG and ECoG signals in the previous sections of this chapter is an essential part of our new framework, allowing the assessment of node—(through the OIR gradient) and network-specific dependencies in the context of a HONA analysis.

Experimental Protocol and Data Analysis. The dataset[7] comprises 64 EEG electrodes referenced to both mastoids (international 10–20 system, $f_s = 160$ Hz) [103, 104]. The subject was asked to open and close the right fist cyclically until a target on the right side of a screen disappeared. The raw signals were firstly detrended,

[7] https://physionet.org/content/eegmmidb.

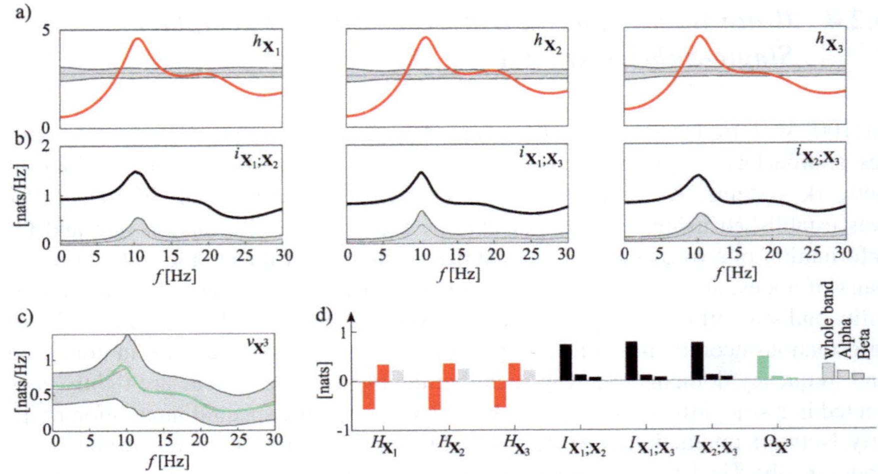

Fig. 6.8 **a** Spectral ER profiles. **b** Spectral MIR profiles. **c** Spectral OIR of order 3. The surrogate/bootstrapping distributions are depicted as shaded grey area, median (grey solid line) and percentiles (black solid lines, computed with 5% significance level). **d** ER, MIR and OIR values integrated in the whole band (left bars), the alpha (α, middle bars) and the beta (β, right bars) frequency bands. The figure is adapted from [100]

then filtered (band-pass, 2–45 Hz; notch, 59–61 Hz) and finally epoched to extract 20 trials of 4 s each. All trials were then reduced to zero mean and unit variance. We selected 6 EEG electrodes located over the contralateral and ipsilateral motor areas, and grouped them in 3 blocks, i.e. $\mathbf{X}_1 = [C_3, C_1]$, $\mathbf{X}_2 = [C_2, C_4]$ and $\mathbf{X}_3 = [F_z, C_z]$. For each trial, a VAR model (2.6) was identified through OLS, fixing the model order to 10. Then, the estimated VAR parameters were used to compute the spectral ER, MIR and OIR profiles, as detailed at the beginning of Sect. 4.3.2. Finally, time domain counterparts for the α and β brain rhythms, as well as over the whole frequency range were obtained by integrating the interaction measures over the relevant frequency ranges (i.e., $\alpha = [7 - 15]$ Hz, $\beta = [18 - 26]$ Hz, $f \in [0 - f_s/2]$ Hz, respectively). Surrogate and bootstrap data analyses were applied as detailed in Appendix A to assess the statistical significance of the computed measures, with $N_s = 100$ iterations and $\alpha = 0.05$ significance level.

Results and Discussion. Figure 6.8 reports the grand-average over trials of the frequency profiles of ER (panel *a*), MIR (panel *b*), and OIR (panel *c*), computed during the motor execution task and depicted over the frequency range $f \in [1 - 30]$ Hz. The spectral ERs (panel *a*), compared with the 97.5th percentile of the surrogate distributions (top black solid lines), show statistically significant oscillations 10 Hz (α band) (see THE DYNAMICS OF THE SINGLE NODE). The oscillations in β band were found to be non significant since they are all below the 2.5th percentile of the surrogate distributions. Integrated values of ER demonstrate the presence of oscillations in α and β bands as well (panel *d*), in accordance with the physiology of the motor execution as demonstrated in several works on this topic [106, 131, 153]. The spectral MIRs and their integrated values are shown in panels *b* and *d*, respectively. The MIR shows statistical significance across all potential pairs of processes within both α and β bands (see COUPLED DYNAMICS BETWEEN PAIRS OF NODES). Although the difference is not so apparent when compared to other possible pairs, the highest MIR value emerges when examining the interaction between $\mathbf{X}_1$ and $\mathbf{X}_2$ ($i_{\mathbf{X}_1;\mathbf{X}_2}$). This underscores the presence of a robust dynamical coupling between the two brain hemispheres during the execution of a motor task [154]. The spectral OIR and the corresponding integrated values in the time domain, α and β bands are reported in panels *c* and *d*. The study of the interaction of order 3 in both time and frequency domain reveals a statistically significant redundant contribution in the network with a prominent peak in the α band (see AN APPROACH TO THE STATISTICAL ASSESSMENT OF HIGH- ORDER INTERACTIONS). This result can be related with the prevalence of redundancy in EEG dynamics during motor task execution, previously highlighted in other studies [106, 130]. Remarkably, the dominance of redundancy may be ascribed to the effects of volume conduction that blur the information identified at the level of scalp EEG sensors [93].

6.2.5 High-Order Behaviours Uncover the Hierarchical Organization of Interactions in the Motor Network of the Human Brain

Understanding brain dynamics during motor tasks is a significant challenge in neuroscience, often limited to studying pairwise interactions, which however may overlook high-order interactions. An exemplary application to EEG signals acquired in healthy subjects performing a motor task execution is shown in Sect. 6.2.1 [101], where, despite prevalence of redundancy for HOIs was found for signals recorded from channels located on central, contralateral and ipsilateral scalp areas, we also showed that specific combinations of these signals give rise to synergistic HOIs reflecting the emergence of interaction mechanisms not retrievable from a pairwise analysis.

A recent study [155] employs the new framework proposed in [100] to analyze single-node, pairwise and high-order interactions within the motor network using EEG data from 10 healthy subjects performing motor tasks. As far as we know, only a limited number of studies has explored HOIs using EEG signals (see, e.g., [101, 106, 130, 156]) and none of these have thoroughly investigated the occurrence of HOIs within the MN itself or how motor task execution impacts the synergy/redundancy balance. Indeed, e.g., in [106], the study of HOIs within the MN was conducted at the scalp sensor level and results may be influenced by volume conduction effects, while [130] focused on stroke patients and analyzed the modulation of HOIs among the MN, default mode network and executive control network due to rehabilitation treatments, without extensively analyzing how HOIs may arise from emergent behaviours occurring within the MN (see Sect. 6.2.3). It is known that motor tasks inherently involve the coordination of multiple brain regions, leading to complex interactions between neurons and networks that give rise to emergent behaviors, which are often difficult to detect. A more detailed investigation of HOIs within the MN could shed light on the intricate mechanisms underlying motor task execution, including the roles of both contralateral and ipsilateral areas, as well as the role of interhemispheric connections, which remains debated in the literature [154]. We tried to accomplish this demanding task.

EEG Dataset Description and Pre-Processing. Ten healthy young individuals (males, 9 right-handed, age range: 30.2±3.9 years) were enrolled in this study. Each session comprised two experimental conditions: *(i)* a resting state (REST) and *(ii)* a handgrip (HG) task. During the resting state session, EEG data were collected over a 3-minute period of closed eyes. For the handgrip task, participants performed a series of handgrip squeezes, with closed eyes, in a structured manner over a 5-minute period, consisting of five 60-second blocks alternating between 30 s of REST and 30 s of HG, squeezing a soft ball using their dominant arm with a frequency of 1 Hz.

EEG acquisition was performed using a cap with 60 Ag/AgCl monopolar electrodes placed on the scalp according to the International 10/20 system. Impedances were maintained below 5 kΩ. The online reference electrode was positioned between Cz and Cpz. The signals were sampled at a frequency of 1 kHz using the Synamps 2/RT EEG system (Neuroscan). The experimental sessions took place in the Robotic Laboratory, operated by the *Consiglio Nazionale delle Ricerche at the Presidio di Riabilitazione dell'Ospedale Valduce Villa Beretta, Costa Masnaga (LC), Italy*. Written informed consent was obtained from each subject before inclusion in the study. The study was reviewed and approved by the local Ethics Committee at *A. Manzoni Hospital, Lecco*, and was conducted in compliance with the *Declaration of Helsinki*.

The preprocessing of EEG signals summarized in the first block of Fig. 6.9 was performed through EEGLab toolbox [134] by exploiting MatLab (The Mathworks, Inc.). Scalp signals were first band-pass filtered ([1−45] Hz) and then down-sampled to reduce the sampling frequency 256 Hz. If a channel had a correlation lower than 0.8 to its robust estimate (based on other channels) for the 60% of the recording or had more line noise relative to its signal than three times standard deviation from the channel population mean, it was discarded. Then, ICA was applied by exploiting the RUNICA Infomax algorithm [136]. ICA components related with physiological artifacts (e.g., cardiac activity, ocular blinks, saccads, muscular activity, line noise, etc.) were manually removed and the originally removed channels were reconstructed by the interpolation of the neighbor signals. The resulting dataset was then re-referenced to a common average value.

Source Localization and Time Series Extraction. The brain current source density distributions were extracted using the eLORETA approach, which implements a discrete, linear, weighted minimum norm inverse solution [138]. Brain source activ-

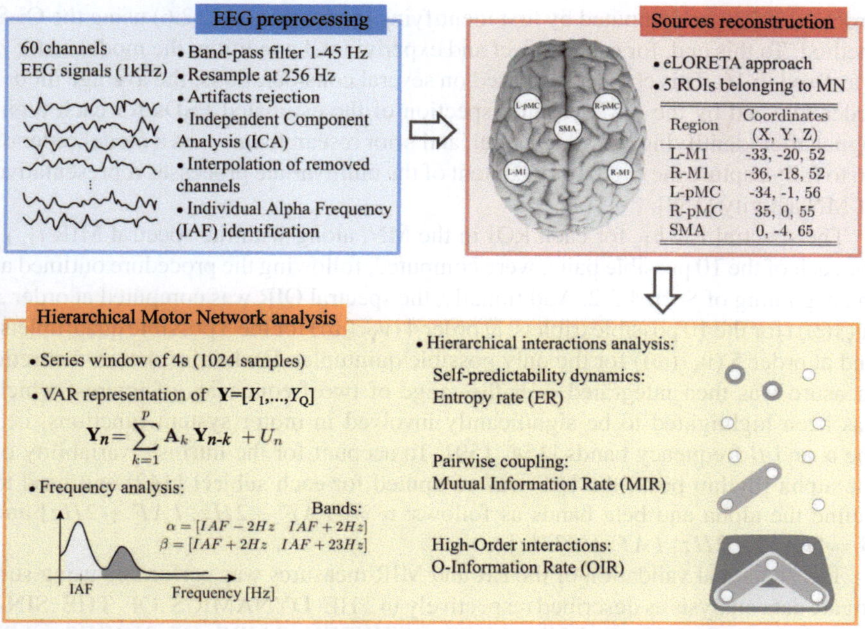

Fig. 6.9 Schematic representation of the EEG analysis pipeline: EEG preprocessing, sources reconstruction, and hierarchical analysis of the Motor Network dynamics. The figure is adapted from [155]

ity was reconstructed only in the cortical grey matter (6239 isotropic voxels with 5 mm spatial resolution) in the MNI152 space [157]. Five ROIs within the MN were selected based on previous literature, including the primary motor cortex (M1) and premotor cortex (pMC) in both the left (L) and right (R) hemispheres, as well as the supplementary motor area (SMA) [130, 131]. The centroid coordinates in the MNI152 space, along with the number of voxels, are reported in the second block of Fig. 6.9. All voxels within an 8 mm radius from a seed were considered part of the specific ROI, with the constraint of non-overlapping regions [140]. In cases of overlap, voxels were assigned to the nearest centroid region. Five time series were obtained by averaging the magnitude of source activity across all voxels within each ROI. To ensure homogeneous representation across the population, for left-handed subjects, the time series for the left and right ROIs were inverted to reflect ipsilateral (i) and contralateral (c) areas relative to the hand performing the task. Additionally, the ROI signals during the task period were segmented into shorter 4-s windows, with the initial windows for the REST and HG periods, each containing 1024 samples to overcome the presence of possible non-stationarities in the EEG signals.

Data and Statistical Analysis. As graphically resumed in the third block of Fig. 6.9, the time and spectral interaction measures of ER, MIR and OIR from order $N = 1$ to order $N = 5$ were computed by first identifying the VAR model (2.6) using the OLS method. To this end, for each subject and experimental condition, the model order p was fixed to 16. This choice was based on several considerations: the average model order selected by the AIC, visual inspection of the estimated PSDs for each brain signal after identifying the VAR model, and prior research that used a model order of 16 to fully capture the oscillatory content of the multivariate processes representative of MN activity [158].

The spectral ER h_{Y_i} for each ROI in the MN, along with the spectral MIR $i_{Y_i;Y_j}$ for each of the 10 possible pairs, were computed, following the procedure outlined at the beginning of Sect. 4.3.2. Additionally, the spectral OIR was computed at order 3 ($\nu_{\mathbf{Y}^3}(\omega)$) for the 10 possible triplets, at order 4 ($\nu_{\mathbf{Y}^4}(\omega)$) for the 5 possible quadruplets, and at order 5 ($\nu_{\mathbf{Y}^5}(\omega)$) for the only possible quintuplet. Each information-theoretic measure was then integrated over the range of two frequencies of interest which has been highlighted to be significantly involved in motor system functions, i.e., the α and β frequency bands [154, 159]. To account for the intrinsic variability of the alpha rhythm peak, the IAF was computed for each subject [142] and used to define the alpha and beta bands as follows: $\alpha = [IAF - 2Hz, IAF + 2Hz]$ and $\beta = [IAF + 2Hz, IAF + 23Hz]$.

The statistical validation of the ER and MIR measures was performed using surrogate data analysis as described respectively in THE DYNAMICS OF THE SINGLE NODE, COUPLED DYNAMICS BETWEEN PAIRS OF NODES, while the block bootstrap data generation procedure was exploited to statistically validate OIR measures (AN APPROACH TO THE STATISTICAL ASSESSMENT OF HIGH- ORDER INTERACTIONS). Regardless of the information measure considered, for both experimental conditions and each subject, the number of surrogates N_s was set to 100 for each of the five time series within the MN, and the significance level was set to 0.05. Lastly, to test whether the motor task execution modulated the information measures at a given order, a statistical comparison between the distributions obtained across subjects during the REST and HG phases was performed. Since the normality of distributions was not verified and given the small number of subjects, we used the *paired non-parametric Wilcoxon signed-rank test*, with a significance level of 5%.

Results. Figure 6.10 shows the distributions of the ER values evaluated in the selected regions within the MN at REST and during the HG task. All five ROIs exhibit a statistically significant decrease in the complexity of brain dynamics in both the α (panel *a*) and β (panel *b*) frequency bands, with this trend observed in 80–100% of subjects. Moreover, the statistical significance of the ER values is greater in the α band (100% regardless of the experimental condition) compared to the β band, where a higher percentage of significance is observed during HG. Although not statistically tested, the complexity of brain regions within the MN generally appears lower in the β band.

The results of the pairwise analysis are summarized in Fig. 6.11 depicting the distributions over the whole group of subjects of the MIR values for each possible pair of ROIs at REST and during HG. Regardless the frequency band analyzed, the

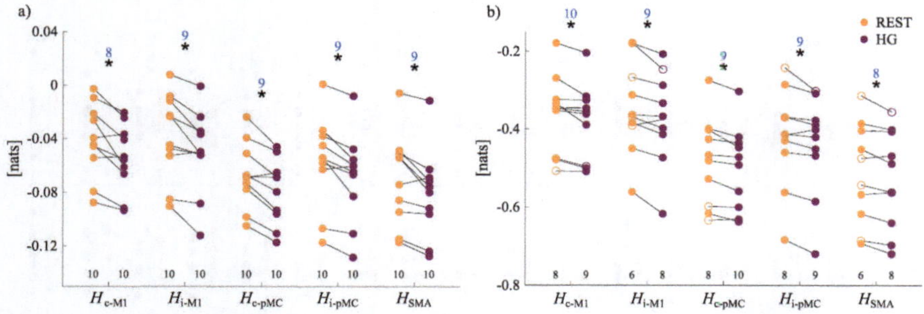

Fig. 6.10 Distributions of individual values of the ER measure evaluated for each ROI within the MN during REST (orange) and HG (purple) conditions in the **a** α and **b** β frequency bands (filled circles: significant ER values; numbers below: total number of significant values). *Wilcoxon signed-rank test* ($p < 0.05$): REST versus HG; blue integers above: number of subjects showing the significant reduction in HG. The figure is adapted from [155]

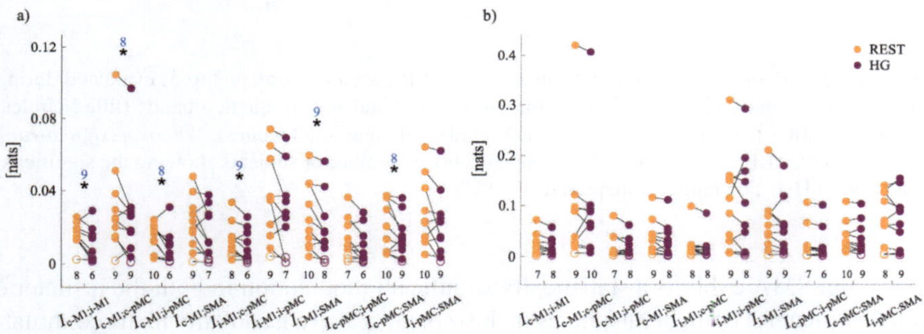

Fig. 6.11 Distributions of individual values of the MIR measure evaluated for each pair of ROIs within the MN during REST (orange) and HG (purple) conditions in the **a** α and **b** β frequency bands (filled circles: significant MIR values; numbers below: total number of significant values). *Wilcoxon signed-rank test* ($p < 0.05$): REST versus HG; blue integers above: number of subjects showing the significant reduction in HG. The figure is adapted from [155]

pairwise interactions, measured through the MIR, tend to decrease during the execution of the motor task, though the modulation is statistically significant only in the α band for almost all the couples with the exception of $I_{c-M1;SMA}$, $I_{i-M1;i-pMC}$, $I_{c-pMC;i-pMC}$ and $I_{i-pMC;SMA}$ (panel *a*). Moreover, surrogate data analysis reveals the presence of statistically significant pairwise interactions in 80–90% of the subjects. This also occurs in the β band even if none of dynamical interactions show statistically significant modulation with the experimental condition (panel *b*).

Figure 6.12 displays the trends of HOIs measured through the OIR, integrated over the α (panel *a*) and β (panel *b*) frequency bands, for each subject in the experimental group performing the motor task. Overall, there is an increment of the OIR quantity with the order of interaction, regardless of the experimental condition and the frequency range analyzed. Only a small subset of triplets shows a statistically significant decrement of the OIR with the motor task. Specifically, the triplet involving M1,

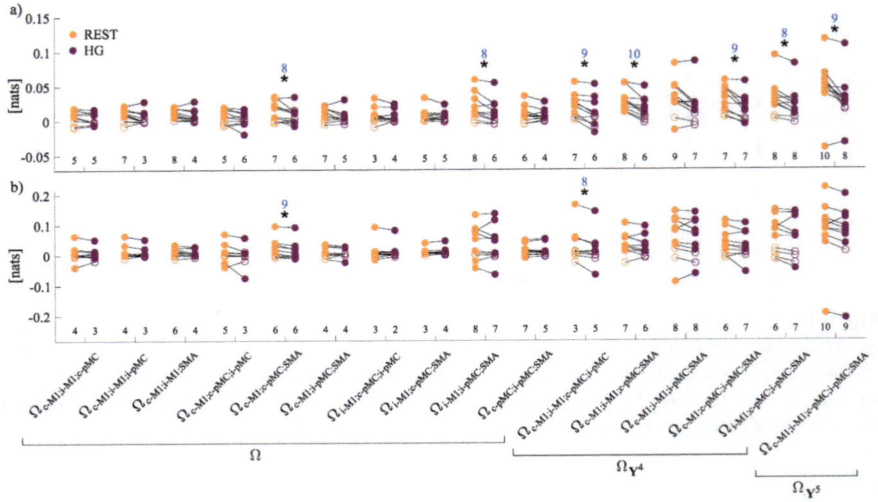

Fig. 6.12 Distributions of individual values of the OIR measures (orders 3 to 5) evaluated during REST (orange) and HG (purple) conditions in the **a** α and **b** β frequency bands (filled circles: significant OIR values; numbers below: total number of significant values). *Wilcoxon signed-rank test* ($p < 0.05$): REST versus HG; blue integers above: number of subjects showing the significant reduction in HG. The figure is adapted from [155]

pMC, and SMA exhibits a statistically significant modulation in both the ipsilateral and contralateral brain hemispheres in the α band (panel a), and only in the contralateral hemisphere in the β band (panel b). In the α band, all quadruples exhibit a significant decrease in the OIR measure during HG, except for $\Omega_{c-M1;i-M1;i-pMC;SMA}$. In contrast, in the β band, only the quadruplet $c - M1; i - M1; c - pMC; i - pMC$ shows a statistically significant reduction in OIR associated with motor task execution. While this trend is evident in 80% of subjects, surrogate analysis reveals that only a small proportion (30% at REST and 50% during HG) have OIR values above the threshold of significance (panel b). Finally, the analysis of the OIR at order five, i.e., $\Omega_{c-M1;i-M1;c-pMC;i-pMC;SMA}$, reveals a statistically significant presence of HOIs at REST (100% of subjects), which slightly decreases during HG (90% of subjects). The modulation of this OIR measure with HG is statistically significant only in the α band, supporting the results obtained for lower orders where HOIs were clearly identified.

Overall, the modulation of information-theoretic measures at any order of interaction is most pronounced in the α frequency band, where the strongest modulation occurs in the contralateral hemisphere, despite the involvement of both hemispheres. Moreover, the analysis of MIR and HOIs at orders 3 and 4 reveals a reduction in interhemispheric information flow mediated by the SMA. In contrast, the analysis of the β frequency band shows almost exclusive involvement of the contralateral hemisphere at order 3, as evidenced by the triplets $c - M1; c - pMC; SMA$. At order 4, modulation of HOIs is observed with involvement limited to M1 and pMC in both hemispheres.

Discussion and Conclusion. This study presents the first comprehensive characterization, of node-specific, pairwise, and high-order dynamical interactions in the human brain during a motor task performed by ten healthy volunteers. The interactions within the contra- and ipsilateral motor networks were hierarchically categorized across five brain regions based on reconstructed source activity. The entropy rate, mutual information rate, and O-information rate were employed to reveal complex brain dynamics in the motor network and assess the impact of motor task execution on: *(i)* the predictability of internal dynamics within each brain region of the MN; *(ii)* the modifications in dynamically shared information between pairs of brain signals; and *(iii)* the balance between synergy and redundancy across groups of brain regions.

The investigation of brain dynamics complexity has been quite limited, with only a few studies analyzing brain signals at the sensor level over the past two decades [160–163]. Our analysis revealed a reduction in the complexity of the individual dynamics within both the contralateral and ipsilateral hemispheres of the motor network during movement execution, with greater persistence in the alpha frequency band. This reduction can be related with previous studies obtained at the level of scalp sensors and with both linear and non-linear estimators which have shown how motor task execution is related to a decreased complexity of brain signals [160] and how the emergence of a dominant rhythm can be responsible for an increase of predictability in brain dynamics [22]. The stronger evidence for this significant trend in the alpha band can be attributed to the physiological desynchronization of μ rhythms in both hemispheres during motor execution tasks, which leads to a reduction in PSD [164]. Since the entropy rate can be regarded as a logarithmic equivalent of PSD, our findings on the brain's self-predictability in motor areas can be directly linked to this phenomenon, which is observable through simple Fourier analysis of the brain signals.

Pertaining the interactions of order two, we exploited the concept of MIR which provides valuable insights on the information flow brain regions. Given the extensive research on pairwise connectivity in the absence of tasks, often referred to as *intrinsic connectivity* [154], we first discuss the anatomical presence of these connections at rest. Specifically, by examining a comprehensive database of invasive connectivity studies in non-human primates [165], we identified connections in the macaque brain: between the SMA and both the ipsilateral and contralateral M1 [166], between the SMA and both the ipsilateral [167] and contralateral pMC [168], and between the pMC and both ipsilateral and contralateral M1 [166]. Our findings, derived from a data-driven approach, revealed statistically significant dynamic coupling between all cortical area combinations in nearly 100% of subjects, indicating the existence of similar intrinsic connections in the human brain. Moreover, our analysis revealed a statistically significant decrease in the strength of pairwise interactions in both contralateral and ipsilateral hemispheres within the α frequency band during motor task performance. From a physiological point of view, this decrease may be related to the event-related potentials associated with movement preparation and execution [169, 170].

Previous studies performed in the domain of the brain sources through the use of Granger-based connectivity estimators suggested a modulation of the coupling between pMC, M1 and SMA in contra- and ipsilateral hemispheres in the α frequency band during motor execution [158]. Indeed, in [11, 171] the authors exploited Dynamic Causal Modeling highlighting a strong connectivity between SMA and M1 even if during a motor imagery task, while the use of conditional Granger causality emphasized forward and backward effective connectivity between SMA and other brain regions [172, 173]. Our results of generalized decoupling in α frequency band are in line with the aforementioned studies and may be also related with the strong suppressive influence of SMA exerted on M1 and on other brain regions during motor imagery task [107].

At the best of our knowledge, the only work which uses the MIR measure on EEG signals for analyzing motor execution tasks is [106]. This study, which focused on scalp-level analysis, reported a statistically significant reduction in the β frequency band for the information that is dynamically shared between group of sensors. Specifically, significant reductions in MIR were observed between electrodes associated with M1 and pMC, as well as between M1, pMC, and SMA. Although our study also found a statistically significant reduction in MIR within the α frequency band, it is important to note that these results are not directly comparable to those of [106]. Indeed, the latter study did not account for volume conduction effects or the IAF, which may influence the results.

Regarding interhemispheric connections, several studies have documented anatomical links between M1-M1 [166, 174, 175] and pMC-pMC [168, 176] in monkeys, which can be correlated with similar connections in the human brain at rest [154]. Our findings revealed statistically significant dynamic coupling in 80% of subjects for M1-M1 and 70% for pMC-pMC, indicating that comparable intrinsic connections are present in the human brain, as observed in primates. The analysis of MIR showed a statistically significant decrease in coupling between ipsilateral M1 and contralateral M1 during the unilateral HG task, with a less pronounced modulation of the pMC-pMC coupling. This observation is consistent with the well-documented phenomenon of interhemispheric coupling showing that unilateral motor tasks reduce connectivity between primary motor cortices across hemispheres, while coupling between ipsilateral and contralateral premotor cortices remains relatively stable [154, 177].

From the analysis of higher-order interactions, we found an increment of the redundancy with the order of the analyzed multiplet which is coherent with previous studies in literature analyzing high-order interdependences over brain signals with different approaches [28, 106, 178]. Moreover, the predominance of redundancy in the MN, observed in both contralateral and ipsilateral hemispheres, can be related to a recent study [60], which characterized neuronal profiles of synergy and redundancy across different brain networks at rest by analyzing fMRI data from healthy participants. Using a framework that does not consider dynamic interactions among brain regions, the authors observed a predominance of redundancy within the MN.

Despite being based on fMRI time series, their findings are consistent with our results. However, the physiological interpretation of this finding remains debated in the literature, particularly because the results stem from different types of brain signals, tasks and experimental groups. The level of redundancy is modulated by the motor task execution to an extent that depends on the analyzed multiplet and it is more evident in α frequency band, where we found almost all the statistically significant differences at any order of interaction. Specifically, when analyzing the interactions of order 3, we found a statistically significant decrement of the OIR only for the triplets $\Omega_{L-M1;L-pMC;SMA}$ and $\Omega_{R-M1;R-pMC;SMA}$, which are representative of the activity of the contra- and ispilateral MN. This clearly demonstrates the presence of emergent behaviors in the brain that require higher-order measures to be unraveled, as also demonstrated in [106]. Moreover, this result can be partially related to previous studies suggesting a strong interaction and involvement of M1, SMA, and pMC in the planning and execution of hand motor tasks [11, 107, 154, 159], even though these were obtained using a dyadic representation of the interaction within the MN. Our findings regarding the involvement of the ipsilateral MN are consistent with previous research highlighting the crucial role of ipsilateral hemisphere in the planning and execution of limb movements [179].

The modulation of OIR during motor task execution at orders 4 and 5 was statistically significant in almost all quadruplets and quintuplets analyzed, highlighting a reduction of the redundancy or an increase of synergy in the interactions characterizing the motor network. This may be physiologically linked to the interhemispheric coupling of neural activity. Prior studies have shown that during unimanual motor tasks, both M1 regions exert mutual inhibitory influences on each other [180, 181], potentially involving inhibitory GABAergic interneurons, although this mechanism remains a topic of ongoing debate [182].

Further Remarks and Limitations. Our work represents a seminal study aimed at gathering initial insights and testing the hypothesis that the motor network features higher-order mechanisms underpinning motor function. Nonetheless, the small sample size represents a key limitation of this study. In this work, the computation of all the information-theoretic measures reported relies on solving a linear regression problem, whose estimation accuracy is significantly affected by the number of available data samples [160]. For our purpose, we had 1024 samples (4 s) available for each of the 5 brain regions, and we could not fit more than the 5 ROIs with a model order equal to 16 to avoid that the linear problem became ill-posed [106, 148]. Although the literature is still debated, some studies suggest that brain signals exhibit significant complex fluctuations indicative of nonlinear processes [183]. Nonlinear estimators are therefore recommended to fully capture the intrinsic nature of these signals [184]. However, many nonlinear estimation methods require long data segments for reliable application and often assume signal stationarity, which may not be feasible with extended EEG recordings. Additionally, model-free approaches that can reliably estimate higher-order interactions are still lacking, making it challenging to fully characterize dynamic interactions between brain regions. As a final remark, the five ROIs were chosen based on previous works [130, 131] where the centroid coordinates in the MNI space have been used to ensure that the selected brain areas were similar in volume. Then, it is important to note that modifying the selection of ROIs or the definition of conduction volume could impact the obtained results. However, given the limited spatial resolution of EEG (approximately 1–2 cm), small variations in ROI size around the same centroid coordinates should have a relatively minor effect on the outcomes.

Overall, our results demonstrated that motor task execution: *(i)* reduces the complexity of brain dynamics, making them more predictable due to the physiological desynchronization of μ rhythms; *(ii)* diminishes the strength of the pairwise dynamics within both ipsilateral and contralateral motor networks, as well as in interhemispheric interactions, as a result of the suppressive influence of the supplementary motor area on several brain regions and transcallosal inhibition; and *(iii)* modulates the OIR value as a result of hierarchical interactions between anatomically connected brain regions, reflecting a decrease in redundancy (or an increase in synergy) at different orders.

6.3 Summary

In the field of Network Neuroscience, neural activity is investigated using static/dynamic measures of brain functional connectivity in different experimental conditions and patho-physiological states. We opened this chapter with a detailed discussion of the main differences between fMRI and EEG connectivity analyses, which have led us to differentiate the type of investigation performed on the available data. Indeed, while in the case of fMRI we have followed existing approaches which disregard temporal correlations between the variables and thus perform a zero-lag analysis mostly based on undirected functional connectivity measures [3], EEG analysis is known to require more sophisticated techniques taking into account the causal relationships among the investigated processes [36]. Nevertheless, shifting from traditional time-averaged measures to capturing the finer temporal fMRI fluctuations of functional connectivity could be essential for understanding brain functions in diverse states, as well as individual differences and internal state changes thereby enabling fMRI assessments at the single-subject level (see, e.g., [185, 186]). However, *fMRI-based dynamic functional connectivity analysis* presents significant challenges, requiring an effort to avoid false effects, an understanding of the hemodynamic nature of fMRI signals, and awareness of non-stationary artifacts within the data [185]. Further, we have shown how pairwise approaches to the study of brain connectivity, although highly effective, cannot fully capture the interplay among the multiple units of a complex neural system [28, 39] (see, e.g., the theoretical example of Sect. 3.3.1). Recognizing and modeling *high-order functional structures* has become a crucial and evolving area of Network Neuroscience [39, 40], where HOIs have been suggested as fundamental components of brain interactions (see, e.g., [41, 42]).

Overall, the application of OI-based high-order information measures to resting-state BOLD signals has demonstrated that high-order synergies represent a kind of shadow structure emerging from resting-state brain activity and missed by bivariate functional connectivity approaches, which indeed do not provide an overall map of the statistical structure of the network. Furthermore and more interestingly, the utilization of OI gradients evidenced peculiar connectivity networks contributing to redundancy and synergy in the large-scale organization of fMRI networks, and confirmed that gradients of increasing order tend to highlight less redundant/more synergistic interactions thus acquiring a key role in OI-based high-order network analyses. Remarkably, we showed that the healthy brain probed by fMRI features *(i)* a default mode network probably associated to the least redundancy, and *(ii)* a somatosensor network which instead could play a major redundant role; *(iii)* moreover, we detected the emergence of a synergistic circuit in the resting brain made of visual, somatomotor and ventral attention which has not been observed in previous studies.

On the other hand, applications of dynamic hierarchical information measures [28, 100] to EEG signals documented an overall prevalence of redundancy for high-order brain interactions, as well as the emergence of synergistic circuits not retrievable from a pairwise analysis.

Dominance of redundancy for HOIs emerging from central, contra- and ipsilateral scalp areas, with prominent peaks in the α band, was found during motor task execution, probably ascribed to the effects of volume conduction that blur the information identified at the level of scalp EEG sensors [93]. Based on reconstructed source activity, the interactions within the contra- and ipsi-lateral motor networks were then hierarchically categorized across five brain regions. Results demonstrated that motor task execution modulates the high-order interactions between anatomically connected brain regions, reflecting a decrease in redundancy (or an increase in synergy) at different orders. The latter finding is of remarkable interest and reflects the emergence of high-order interaction mechanisms not retrievable from EEG signals acquired on the scalp. A predominance of redundant interactions among the default mode, the executive control and the motor networks was also found both before and after the rehabilitation following subacute stroke events. After rehabilitation, the role of the default mode as a common source of causal interactions from the executive control and the motor networks allowed to detect an increased shift toward less redundant and more synergistic behaviors.

References

1. Martijn P Van Den Heuvel and Hilleke E Hulshoff Pol. "Exploring the brain network: a review on resting-state fMRI functional connectivity". In: *European neuropsychopharmacology* 20.8 (2010), pp. 519–534.
2. Olaf Sporns. "Structure and function of complex brain networks". In: *Dialogues in clinical neuroscience* (2022).
3. Karl J Friston. "Functional and effective connectivity in neuroimaging: a synthesis". In: *Human brain mapping* 2.1-2 (1994), pp. 56–78.

4. RC. Craddock, RL. Tungaraza, and MP. Milham. "Connectomics and new approaches for analyzing human brain functional connectivity". In: *Gigascience* 4 (2015), p. 13. https://doi.org/10.1186/s13742-015-0045-x.
5. O. Sporns and DS. Bassett. "Editorial: New Trends in Connectomics". In: *Network Neuroscience* 2.2 (2018), pp. 125–127. https://doi.org/10.1162/netn_e_00052.
6. Huifang E Wang et al. "A systematic framework for functional connectivity measures". In: *Frontiers in neuroscience* 8 (2014), p. 405.
7. Bin He et al. "Electrophysiological Brain Connectivity: Theory and Implementation". In: *IEEE Transactions on Biomedical Engineering* 66.7 (2019). Conference Name: IEEE Transactions on Biomedical Engineering, pp. 2115–2137. ISSN: 1558-2531. https://doi.org/10.1109/TBME.2019.2913928.
8. André M. Bastos and Jan Mathijs Schoffelen. "A tutorial review of functional connectivity analysis methods and their interpretational pitfalls". In: *Frontiers in Systems Neuroscience* 9.JAN2016 (2016). Publisher: Frontiers Research Foundation. ISSN: 16625137. https://doi.org/10.3389/fnsys.2015.00175.
9. Jun Cao et al. "Brain functional and effective connectivity based on electroencephalography recordings: A review". In: *Human Brain Mapping* 43.2 (2022), pp. 860–879.
10. AR McIntosh and F Gonzalez-Lima. "Network interactions among limbic cortices, basal forebrain, and cerebellum differentiate a tone conditioned as a Pavlovian excitor or inhibitor: fluorodeoxyglucose mapping and covariance structural modeling". In: *Journal of neurophysiology* 72.4 (1994), pp. 1717–1733.
11. Karl J Friston, Lee Harrison, and Will Penny. "Dynamic causal modelling". In: *Neuroimage* 19.4 (2003), pp. 1273–1302.
12. Olivier David et al. "Dynamic causal modeling of evoked responses in EEG and MEG". In: *NeuroImage* 30.4 (2006), pp. 1255–1272.
13. John Geweke. "Measurement of linear dependence and feedback between multiple time series". In: *Journal of the American statistical association* 77.378 (1982), pp. 304–313.
14. Thomas Schreiber. "Measuring information transfer". In: *Physical review letters* 85.2 (2000), p. 461.
15. Y Saito and H Harashima. *Tracking of information within multichannel EEG record causal analysis in eeg. Yamaguchi N, Fujisawa K (eds) Recent advances in* {EEG} *and* {EMG} *data processing*. 1981.
16. Luiz Antonio Baccalá et al. "Studying the interaction between brain structures via directed coherence and a causality test". In: *Applied Signal Processing* (1998).
17. Luiz A Baccalá and Koichi Sameshima. "Partial directed coherence: a new concept in neural structure determination". In: *Biological cybernetics* 84.6 (2001), pp. 463–474.
18. Koichi Sameshima and Luiz Antonio Baccalá. "Using partial directed coherence to describe neuronal ensemble interactions". In: *Journal of neuroscience methods* 94.1 (1999), pp. 93–103.
19. M. J. Kaminski and K. J. Blinowska. "A new method of the description of the information flow in the brain structures". en. In: *Biol. Cybern.* 65.3 (1991), pp. 203–210. ISSN: 1432-0770. https://doi.org/10.1007/BF00198091. (visited on 10/12/2022).
20. Lionel Barnett, Adam B Barrett, and Anil K Seth. "Granger causality and transfer entropy are equivalent for Gaussian variables". In: *Physical review letters* 103.23 (2009), p. 238701.
21. Adam B Barrett, Lionel Barnett, and Anil K Seth. "Multivariate Granger causality and generalized variance". In: *Physical Review E* 81.4 (2010), p. 041907.
22. Luca Faes, Silvia Erla, and Giandomenico Nollo. "Measuring connectivity in linear multivariate processes: definitions, interpretation, and practical analysis". In: *Computational and mathematical methods in medicine* 2012 (2012).
23. Izrail Moiseevich Gelfand and AM IAglom. *Calculation of the amount of information about a random function contained in another such function*. American Mathematical Society Providence, 1959.
24. Tyrone E Duncan. "On the calculation of mutual information". In: *SIAM Journal on Applied Mathematics* 19.1 (1970), pp. 215–220.

25. Daniel Chicharro. "On the spectral formulation of Granger causality". In: *Biological cybernetics* 105 (2011), pp. 331–347.
26. Fernando E Rosas et al. "Quantifying high-order interdependencies via multivariate extensions of the mutual information". In: *Physical Review E* 100.3 (2019), p. 032305.
27. Sebastiano Stramaglia et al. "Quantifying dynamical high-order interdependencies from the o-information: an application to neural spiking dynamics". In: *Frontiers in Physiology* 11 (2021), p. 595736.
28. Luca Faes et al. "A new framework for the time-and frequency-domain assessment of high-order interactions in networks of random processes". In: *IEEE Transactions on Signal Processing* 70 (2022), pp. 5766–5777.
29. Tomas Scagliarini et al. "Gradients of O-information: Low-order descriptors of high-order dependencies". In: *Physical Review Research* 5.1 (2023), p. 013025.
30. Bharat B Biswal, Joel Van Kylen, and James S Hyde. "Simultaneous assessment of flow and BOLD signals in resting-state functional connectivity maps". In: *NMR in Biomedicine* 10.4-5 (1997), pp. 165–170.
31. John C Gore et al. "Principles and practice of functional MRI of the human brain". In: *The Journal of clinical investigation* 112.1 (2003), pp. 4–9.
32. David M Cole, Stephen M Smith, and Christian F Beckmann. "Advances and pitfalls in the analysis and interpretation of resting-state FMRI data". In: *Frontiers in systems neuroscience* 4 (2010), p. 1459.
33. Gianvincenzo Sparacia et al. "Resting-State Functional Magnetic Resonance Imaging for Surgical Neuro-Oncology Planning: Towards a Standardization in Clinical Settings". In: *Brain Sciences* 11.12 (2021), p. 1613.
34. Baxter P Rogers et al. "Assessing functional connectivity in the human brain by fMRI". In: *Magnetic resonance imaging* 25.10 (2007), pp. 1347–1357.
35. Paolo M Rossini et al. "Methods for analysis of brain connectivity: An IFCN-sponsored review". In: *Clinical Neurophysiology* 130.10 (2019), pp. 1833–1858.
36. Jennifer Rizkallah et al. "Exploring the Correlation Between M/EEG Source–Space and fMRI Networks at Rest". en. In: *Brain Topogr* 33.2 (2020), pp. 151–160. ISSN: 1573-6792. https://doi.org/10.1007/s10548-020-00753-w. (visited on 10/12/2022).
37. Fernando Lopes da Silva. "EEG and MEG: relevance to neuroscience". In: *Neuron* 80.5 (2013), pp. 1112–1128.
38. Steven J Luck. *An introduction to the event-related potential technique*. MIT press, 2014.
39. Federico Battiston et al. "Networks beyond pairwise interactions: Structure and dynamics". In: *Physics reports* 874 (2020), pp. 1–92.
40. Danielle S Bassett and Olaf Sporns. "Network neuroscience". In: *Nature neuroscience* 20.3 (2017), pp. 353–364.
41. Giulio Tononi, Olaf Sporns, and Gerald M Edelman. "A measure for brain complexity: relating functional segregation and integration in the nervous system." In: *Proceedings of the National Academy of Sciences* 91.11 (1994), pp. 5033–5037.
42. Andrea I Luppi et al. "What it is like to be a bit: an integrated information decomposition account of emergent mental phenomena". In: *Neuroscience of Consciousness* 2021.2 (2021), niab027.
43. Joseph T Lizier et al. *Information decomposition of target effects from multi-source interactions: Perspectives on previous, current and future work*. 2018.
44. Luca Faes et al. "Information decomposition in multivariate systems: definitions, implementation and application to cardiovascular networks". In: *Entropy* 19.1 (2016), p. 5.
45. Sebastiano Stramaglia et al. "Expanding the transfer entropy to identify information circuits in complex systems". In: *Physical Review E* 86.6 (2012), p. 066211.
46. Alberto Porta et al. "Quantifying net synergy/redundancy of spontaneous variability regulation via predictability and transfer entropy decomposition frameworks". In: *IEEE Transactions on Biomedical Engineering* 64.11 (2017), pp. 2628–2638.
47. Kaidi Shao, Nikos K Logothetis, and Michel Besserve. "Information Theoretic Measures of Causal Influences during Transient Neural Events". In: *arXiv preprint* arXiv:2209.07508 (2022).

48. Ruben Sanchez-Romero and Michael W Cole. "Combining multiple functional connectivity methods to improve causal inferences". In: *Journal of cognitive neuroscience* 33.2 (2021), pp. 180–194.
49. Moritz Günther, Jan W Kantelhardt, and Ronny P Bartsch. "The reconstruction of causal networks in physiology". In: *Frontiers in Network Physiology* 2 (2022).
50. Jeffrey S Anderson et al. "Reproducibility of single-subject functional connectivity measurements". In: *American journal of neuroradiology* 32.3 (2011), pp. 548–555.
51. Laura Sparacino et al. "Statistical Approaches to Identify Pairwise and High-Order Brain Functional Connectivity Signatures on a Single-Subject Basis". In: *Life* 13.10 (2023), p. 2075.
52. Dieter Haeussinger et al. "Hepatic encephalopathy". In: *Nature Reviews Disease Primers* 8.1 (2022), p. 43.
53. J de Ville de Goyet. "Extrahilar mesenterico-left portal shunt to relieve extrahepatic portal hypertension after partial liver transplant". In: *Transplantation* 53 (1992), pp. 231–232.
54. Vincenzina Lo Re et al. "Cognitive outcomes in patients treated with neuromuscular electrical stimulation after coronary artery bypass grafting". In: *Frontiers in Neurology* 14 (2023), p. 1209905.
55. Bharat Biswal et al. "Functional connectivity in the motor cortex of resting human brain using echo-planar MRI". In: *Magnetic resonance in medicine* 34.4 (1995), pp. 537–541.
56. Gianvincenzo Sparacia et al. "Resting-state functional connectome in patients with brain tumors before and after surgical resection". In: *World Neurosurgery* 141 (2020), e182–e194.
57. Stephen M Smith et al. "Resting-state fMRI in the human connectome project". In: *Neuroimage* 80 (2013), pp. 144–168.
58. Thomas F Varley et al. "Multivariate information theory uncovers synergistic subsystems of the human cerebral cortex". In: *Communications biology* 6.1 (2023), p. 451.
59. Conor Finn and Joseph T Lizier. "Generalised measures of multivariate information content". In: *Entropy* 22.2 (2020), p. 216.
60. Andrea I Luppi et al. "A synergistic core for human brain evolution and cognition". In: *Nature Neuroscience* 25.6 (2022), pp. 771–782.
61. Yuri Antonacci et al. "Measuring the Balance Between Synergy and Redundancy in Network Systems by Using Information Theory". In: *Mediterranean Conference on Medical and Biological Engineering and Computing*. Springer. 2023, pp. 145–154.
62. Tomas Scagliarini et al. "Gradients of O-information highlight synergy and redundancy in physiological applications". In: *Frontiers in Network Physiology* 3 (2024), p. 1335808.
63. Robin AA Ince. "Measuring multivariate redundant information with pointwise common change in surprisal". In: *Entropy* 19.7 (2017), p. 318.
64. David C Van Essen et al. "The Human Connectome Project: a data acquisition perspective". In: *Neuroimage* 62.4 (2012), pp. 2222–2231.
65. BT Thomas Yeo et al. "The organization of the human cerebral cortex estimated by intrinsic functional connectivity". In: *Journal of neurophysiology* (2011).
66. Pedro AM Mediano et al. "Towards an extended taxonomy of information dynamics via integrated information decomposition". In: *arXiv preprint* arXiv:2109.13186 (2021).
67. AM Barrett, Olga Boukrina, and Soha Saleh. "Ventral attention and motor network connectivity is relevant to functional impairment in spatial neglect after right brain stroke". In: *Brain and cognition* 129 (2019), pp. 16–24.
68. Xiao Zhu et al. "Higher functional connectivity of ventral attention and visual network to maintain cognitive performance in white matter hyperintensity". In: *Aging and Disease* 14.4 (2023), p. 1472.
69. Karl J Friston, Christopher D Frith, et al. "Schizophrenia: a disconnection syndrome". In: *Clin Neurosci* 3.2 (1995), pp. 89–97.
70. Christos A Frantzidis et al. "Functional disorganization of small-world brain networks in mild Alzheimer's Disease and amnestic Mild Cognitive Impairment: an EEG study using Relative Wavelet Entropy (RWE)". In: *Frontiers in aging neuroscience* 6 (2014), p. 224.

71. Noa Fogelson et al. “Functional connectivity abnormalities during contextual processing in schizophrenia and in Parkinson’s disease”. In: *Brain and cognition* 82.3 (2013), pp. 243–253.
72. Cornelis J. Stam and Jaap C. Reijneveld. “Graph theoretical analysis of complex networks in the brain”. eng. In: *Nonlinear Biomedical Physics* 1.1 (2007), p. 3. ISSN: 1753-4631. https://doi.org/10.1186/1753-4631-1-3.
73. Mark D Holmes. “Dense array EEG: methodology and new hypothesis on epilepsy syndromes”. In: *Epilepsia* 49 (2008), pp. 3–14.
74. Killian Kleffner-Canucci et al. “A novel hydrogel electrolyte extender for rapid application of EEG sensors and extended recordings”. In: *Journal of neuroscience methods* 206.1 (2012), pp. 83–87.
75. Mikhail A Lebedev and Miguel AL Nicolelis. “Brain–machine interfaces: past, present and future”. In: *TRENDS in Neurosciences* 29.9 (2006), pp. 536–546.
76. Stefan Debener et al. “Single-trial EEG–fMRI reveals the dynamics of cognitive function”. In: *Trends in cognitive sciences* 10.12 (2006), pp. 558–563.
77. Jonathan R Wolpaw and E Winter Wolpaw. “Brain-computer interfaces: something new under the sun”. In: *Brain-computer interfaces: principles and practice* 14 (2012).
78. Sven Bestmann and Eva Feredoes. “Combined neurostimulation and neuroimaging in cognitive neuroscience: past, present, and future”. In: *Annals of the New York Academy of Sciences* 1296.1 (2013), pp. 11–30.
79. Camarin E. Rolle et al. “Functional connectivity using high density EEG shows competitive reliability and agreement across test/retest sessions”. en. In: *Journal of Neuroscience Methods* 367 (2022), p. 109424. ISSN: 0165-0270. https://doi.org/10.1016/j.jneumeth.2021.109424. https://www.sciencedirect.com/science/article/pii/S0165027021003599 (visited on 11/25/2022).
80. Michel Le Van Quyen et al. “Large-Scale Microelectrode Recordings of High-Frequency Gamma Oscillations in Human Cortex during Sleep”. In: *Journal of Neuroscience* 30.23 (2010), pp. 7770–7782. ISSN: 0270-6474, 1529-2401. https://doi.org/10.1523/JNEUROSCI.5049-09.2010.
81. J. Jacobs et al. “High-Frequency Oscillations (HFOs) in Clinical Epilepsy”. In: *Progress in Neurobiology* 98.3 (2012), pp. 302–315. ISSN: 03010082. https://doi.org/10.1016/j.pneurobio.2012.03.001.
82. Andres D. Grosmark and György Buzsáki. “Diversity in Neural Firing Dynamics Supports Both Rigid and Learned Hippocampal Sequences”. In: *Science* 351.6280 (2016), pp. 1440–1443. DOI: https://doi.org/10.1126/science.aad1935. eprint: https://www.science.org/doi/pdf/10.1126/science.aad1935.
83. Stephen V. Gliske et al. “Effect of Sampling Rate and Filter Settings on High Frequency Oscillation Detections”. In: *Clinical Neurophysiology: Official Journal of the International Federation of Clinical Neurophysiology* 127.9 (2016), pp. 3042–3050. ISSN: 1872-8952. https://doi.org/10.1016/j.clinph.2016.06.029.
84. Juan Bolea et al. “Influence of Heart Rate in Non-linear HRV Indices as a Sampling Rate Effect Evaluated on Supine and Standing”. In: *Frontiers in Physiology* 7 (2016), p. 501. ISSN: 1664-042X. https://doi.org/10.3389/fphys.2016.00501.
85. H. Jing and M. Takigawa. “Low Sampling Rate Induces High Correlation Dimension on Electroencephalograms from Healthy Subjects”. In: *Psychiatry and Clinical Neurosciences* 54.4 (2000), pp. 407–412. ISSN: 1323-1316. https://doi.org/10.1046/j.1440-1819.2000.00729.x.
86. D.B. Geselowitz. “The zero of potential”. In: *IEEE Engineering in Medicine and Biology Magazine* 17.1 (1998), pp. 128–136. https://doi.org/10.1109/51.646230.
87. Christoph M Michel et al. “EEG source imaging”. In: *Clinical neurophysiology* 115.10 (2004), pp. 2195–2222.
88. Christoph M Michel and Denis Brunet. “EEG source imaging: a practical review of the analysis steps”. In: *Frontiers in neurology* 10 (2019), p. 325.
89. Christos Koutlis, Vasilios K Kimiskidis, and Dimitris Kugiumtzis. “Comparison of Causality Network Estimation in the Sensor and Source Space: Simulation and Application on EEG”. In: *Frontiers in Network Physiology* 1 (2021), p. 706487.

90. Paul L Nunez and Ramesh Srinivasan. *Electric fields of the brain: the neurophysics of EEG*. Oxford University Press, USA, 2006.
91. Stefan Haufe et al. "A critical assessment of connectivity measures for EEG data: a simulation study". In: *Neuroimage* 64 (2013), pp. 120–133.
92. Margarita Papadopoulou, Karl Friston, and Daniele Marinazzo. "Estimating directed connectivity from cortical recordings and reconstructed sources". In: *Brain topography* 32 (2019), pp. 741–752.
93. Frederik Van de Steen et al. "Critical comments on EEG sensor space dynamical connectivity analysis". In: *Brain topography* 32 (2019), pp. 643–654.
94. B. Cea-Cañas et al. "Connectivity strength of the EEG functional network in schizophrenia and bipolar disorder". In: *Prog Neuropsychopharmacol Biol Psychiatry* 98 (2020), p. 109801. https://doi.org/10.1016/j.pnpbp.2019.109801.
95. Luyao Jin et al. "Frequency nesting interactions in the subthalamic nucleus correlate with the step phases for Parkinson's disease". In: *Front. Physiol.* 13 (2022), p. 890753.
96. Y. Liang et al. "Altered Functional Connectivity after Epileptic Seizure Revealed by Scalp EEG". In: *Neural Plast.* 2020 (2020), p. 8851415. https://doi.org/10.1155/2020/8851415.
97. Floriana Pichiorri et al. "Brain–computer interface boosts motor imagery practice during stroke recovery". In: *Annals of neurology* 77.5 (2015), pp. 851–865. https://doi.org/10.1002/ana.24390.
98. G. Chiarion and L. Mesin. "Functional connectivity of eeg in encephalitis during slow biphasic complexes". In: *Electronics (Switzerland)* 10.23 (2021). ISSN: 20799292. https://doi.org/10.3390/electronics10232978.
99. Giovanni Chiarion et al. "Connectivity analysis in EEG data: a tutorial review of the state of the art and emerging trends". In: *Bioengineering* 10.3 (2023), p. 372.
100. Laura Sparacino et al. "Measuring hierarchically-organized interactions in dynamic networks through spectral entropy rates: theory, estimation, and illustrative application to physiological networks". In: *Neurocomputing* 630 (2025), p. 129675.
101. Laura Sparacino et al. "Quantifying High-Order Interactions in Complex Physiological Networks: A Frequency-Specific Approach". In: *International Conference on Complex Networks and Their Applications*. Springer. 2022, pp. 301–309.
102. Christian Gerloff et al. "Functional coupling and regional activation of human cortical motor areas during simple, internally paced and externally paced finger movements." In: *Brain: a journal of neurology* 121.8 (1998), pp. 1513–1531.
103. Gerwin Schalk et al. "BCI2000: a general-purpose brain-computer interface (BCI) system". In: *IEEE Transactions on biomedical engineering* 51.6 (2004), pp. 1034–1043.
104. Ary L Goldberger et al. "PhysioBank, PhysioToolkit, and PhysioNet: components of a new research resource for complex physiologic signals". In: *circulation* 101.23 (2000), e215–e220.
105. RE Passingham. "Supplementary motor cortex and self-intiated movement". In: *Neural programming* (1989), pp. 13–24.
106. Yuri Antonacci et al. "Measuring high-order interactions in rhythmic processes through multivariate spectral information decomposition". In: *IEEE Access* 9 (2021), pp. 149486–149505.
107. Gert Pfurtscheller and FH Lopes Da Silva. "Event-related EEG/MEG synchronization and desynchronization: basic principles". In: *Clinical neurophysiology* 110.11 (1999), pp. 1842–1857.
108. Ivan Kotiuchyi et al. "A framework to assess the information dynamics of source EEG activity and its application to epileptic brain networks". In: *Brain Sciences* 10.9 (2020), p. 657.
109. Gerwin Schalk and Eric C Leuthardt. "Brain-computer interfaces using electrocorticographic signals". In: *IEEE reviews in biomedical engineering* 4 (2011), pp. 140–154.
110. N Jeremy Hill et al. "Recording human electrocorticographic (ECoG) signals for neuroscientific research and real-time functional cortical mapping". In: *Journal of Visualized Experiments (JoVE)* 64 (2012), e3993.
111. Hyunmin Moon et al. "Electrocorticogram (ECoG): Engineering Approaches and Clinical Challenges for Translational Medicine". In: *Advanced Materials Technologies* (2024), p. 2301692.

112. Toru Yanagawa et al. "Large-scale information flow in conscious and unconscious states: an ECoG study in monkeys". In: *PloS one* 8.11 (2013), e80845.
113. Satu Palva and J Matias Palva. "New vistas for α-frequency band oscillations". In: *Trends in neurosciences* 30.4 (2007), pp. 150–158.
114. Sylvain Chauvette et al. "Properties of slow oscillation during slow-wave sleep and anesthesia in cats". In: *Journal of Neuroscience* 31.42 (2011), pp. 14998–15008.
115. Michael Murphy et al. "Source modeling sleep slow waves". In: *Proceedings of the National Academy of Sciences* 106.5 (2009), pp. 1608–1613.
116. Pascal Fries. "Neuronal gamma-band synchronization as a fundamental process in cortical computation". In: *Annual review of neuroscience* 32 (2009), pp. 209–224.
117. Bernard J Baars. "The conscious access hypothesis: origins and recent evidence". In: *Trends in cognitive sciences* 6.1 (2002), pp. 47–52.
118. Andrea I Luppi et al. "A synergistic workspace for human consciousness revealed by integrated information decomposition". In: *BioRxiv* (2020).
119. Pedro AM Mediano et al. "Beyond integrated information: A taxonomy of information dynamics phenomena". In: *arXiv preprint* arXiv:1909.02297 (2019).
120. Marilyn Gatica et al. "High-order interdependencies in the aging brain". In: *Brain connectivity* 11.9 (2021), pp. 734–744.
121. Rubén Herzog et al. "Genuine high-order interactions in brain networks and neurodegeneration". In: *Neurobiology of Disease* 175 (2022), p. 105918.
122. Joshua Sarfaty Siegel et al. "Disruptions of network connectivity predict impairment in multiple behavioral domains after stroke". In: *Proceedings of the National Academy of Sciences* 113.30 (2016), E4367–E4376.
123. Christian Grefkes and Gereon R Fink. "Reorganization of cerebral networks after stroke: new insights from neuroimaging with connectivity approaches". In: *Brain* 134.5 (2011), pp. 1264–1276.
124. Brenton Hordacre et al. "Resting state functional connectivity is associated with motor pathway integrity and upper-limb behavior in chronic stroke". In: *Neurorehabilitation and neural repair* 34.6 (2020), pp. 547–557.
125. Cathy Stinear. "Prediction of recovery of motor function after stroke". In: *The Lancet Neurology* 9.12 (2010), pp. 1228–1232.
126. Caihong Wang et al. "Altered functional organization within and between resting-state networks in chronic subcortical infarction". In: *Journal of Cerebral Blood Flow & Metabolism* 34.4 (2014), pp. 597–605.
127. Zhiyong Zhao et al. "Altered intra-and inter-network functional coupling of resting-state networks associated with motor dysfunction in stroke". In: *Human Brain Mapping* 39.8 (2018), pp. 3388–3397.
128. Changwei W Wu et al. "Synchrony between default-mode and sensorimotor networks facilitates motor function in stroke rehabilitation: a pilot fMRI study". In: *Frontiers in neuroscience* 14 (2020), p. 548.
129. Zhiyong Zhao et al. "Altered effective connectivity of the primary motor cortex in stroke: a resting-state fMRI study with Granger causality analysis". In: *PloS one* 11.11 (2016), e0166210.
130. I Pirovano et al. "Rehabilitation modulates high-order interactions among large-scale brain networks in subacute stroke". In: *IEEE Transactions on Neural Systems and Rehabilitation Engineering* 31 (2023), pp. 4549–4560.
131. Ileana Pirovano et al. "Resting state EEG directed functional connectivity unveils changes in motor network organization in subacute stroke patients after rehabilitation". In: *Frontiers in Physiology* 13 (2022), p. 862207.
132. AR Fugl-Meyer et al. "The post-stroke hemiplegic patient. 1. a method for evaluation of physical performance. Vol. 7". In: *Scandinavian journal of rehabilitation medicine. 1AD* (), p. 13.
133. Maureen K Holden et al. "Clinical gait assessment in the neurologically impaired: reliability and meaningfulness". In: *Physical therapy* 64.1 (1984), pp. 35–40.

134. Arnaud Delorme and Scott Makeig. "EEGLAB: an open source toolbox for analysis of single-trial EEG dynamics including independent component analysis". In: *Journal of neuroscience methods* 134.1 (2004), pp. 9–21.
135. Tim R Mullen et al. "Real-time neuroimaging and cognitive monitoring using wearable dry EEG". In: *IEEE transactions on biomedical engineering* 62.11 (2015), pp. 2553–2567.
136. Scott Makeig et al. "Independent component analysis of electroencephalographic data". In: *Advances in neural information processing systems* 8 (1995).
137. Luca Pion-Tonachini, Ken Kreutz-Delgado, and Scott Makeig. "ICLabel: An automated electroencephalographic independent component classifier, dataset, and website". In: *NeuroImage* 198 (2019), pp. 181–197.
138. Roberto D Pascual-Marqui. "Discrete, 3D distributed, linear imaging methods of electric neuronal activity. Part 1: exact, zero error localization". In: *arXiv preprint* arXiv:0710.3341 (2007).
139. Cory S Inman et al. "Altered resting-state effective connectivity of fronto-parietal motor control systems on the primary motor network following stroke". In: *Neuroimage* 59.1 (2012), pp. 227–237.
140. Jessica Samogin et al. "Frequency-dependent functional connectivity in resting state networks". In: *Human brain mapping* 41.18 (2020), pp. 5187–5198.
141. Marcus E Raichle. "The restless brain". In: *Brain connectivity* 1.1 (2011), pp. 3–12.
142. Wolfgang Klimesch. "EEG alpha and theta oscillations reflect cognitive and memory performance: a review and analysis". In: *Brain research reviews* 29.2-3 (1999), pp. 169–195.
143. Gaëlle Doucet et al. "Brain activity at rest: a multiscale hierarchical functional organization". In: *Journal of neurophysiology* 105.6 (2011), pp. 2753–2763.
144. Marcus E Raichle et al. "A default mode of brain function". In: *Proceedings of the national academy of sciences* 98.2 (2001), pp. 676–682.
145. Maurizio Corbetta and Gordon L Shulman. "Control of goal-directed and stimulus-driven attention in the brain". In: *Nature reviews neuroscience* 3.3 (2002), pp. 201–215.
146. Zaira Romeo et al. "Electrophysiological signatures of resting state networks predict cognitive deficits in stroke". In: *Cortex* 138 (2021), pp. 59–71.
147. Wen Geng et al. "Reduced functional network connectivity is associated with upper limb dysfunction in acute ischemic brainstem stroke". In: *Brain Imaging and Behavior* (2022), pp. 1–9.
148. Yuri Antonacci et al. "Information transfer in linear multivariate processes assessed through penalized regression techniques: validation and application to physiological networks". In: *Entropy* 22.7 (2020), p. 732.
149. Alois Schlögl. "A comparison of multivariate autoregressive estimators". In: *Signal processing* 86.9 (2006), pp. 2426–2429.
150. John F Geweke. "Measures of conditional linear dependence and feedback between time series". In: *Journal of the American Statistical Association* 79.388 (1984), pp. 907–915.
151. Yonghong Chen, Steven L Bressler, and Mingzhou Ding. "Frequency decomposition of conditional Granger causality and application to multivariate neural field potential data". In: *Journal of neuroscience methods* 150.2 (2006), pp. 228–237.
152. Vasily A Vakorin, Olga A Krakovska, and Anthony R McIntosh. "Confounding effects of indirect connections on causality estimation". In: *Journal of neuroscience methods* 184.1 (2009), pp. 152–160.
153. Filippo Cona et al. "Changes in EEG power spectral density and cortical connectivity in healthy and tetraplegic patients during a motor imagery task". In: *Computational intelligence and neuroscience* 2009 (2009).
154. Christian Grefkes et al. "Dynamic intra-and interhemispheric interactions during unilateral and bilateral hand movements assessed with fMRI and DCM". In: *Neuroimage* 41.4 (2008), pp. 1382–1394.
155. Yuri Antonacci et al. "Spectral Information Dynamics of Cortical Signals Uncover the Hierarchical Organization of the Human Brain's Motor Network". In: *IEEE Transactions on Biomedical Engineering* (2024).

156. Luca Faes et al. "Quantifying High-Order Interactions in Cardiovascular and Cerebrovascular Networks". In: *2022 12th Conference of the European Study Group on Cardiovascular Oscillations (ESGCO)*. IEEE. 2022, pp. 1–2.
157. John Mazziotta et al. "A probabilistic atlas and reference system for the human brain: International Consortium for Brain Mapping (ICBM)". In: *Philosophical Transactions of the Royal Society of London. Series B: Biological Sciences* 356.1412 (2001), pp. 1293–1322.
158. Laura Astolfi et al. "Estimation of the cortical connectivity by high-resolution EEG and structural equation modeling: simulations and application to finger tapping data". In: *IEEE Transactions on Biomedical Engineering* 52.5 (2005), pp. 757–768.
159. Gert Pfurtscheller, Christa Neuper, and Johannes Berger. "Source localization using eventrelated desynchronization (ERD) within the alpha band". In: *Brain topography* 6.4 (1994), pp. 269–275.
160. Yuri Antonacci et al. "Assessment of EEG Brain Dynamics in Time and Frequency Domains through Information-Theoretic Measures". In: *2024 IEEE 22nd Mediterranean Electrotechnical Conference (MELECON)*. IEEE. 2024, pp. 305–310.
161. Silvia Erla et al. "k-Nearest neighbour local linear prediction of scalp EEG activity during intermittent photic stimulation". In: *Medical engineering & physics* 33.4 (2011), pp. 504–512.
162. X-S Zhang, Rob J Roy, and Erik W Jensen. "EEG complexity as a measure of depth of anesthesia for patients". In: *IEEE transactions on biomedical engineering* 48.12 (2001), pp. 1424–1433.
163. Joydeep Bhattacharya. "Complexity analysis of spontaneous EEG". In: *Acta neurobiologiae experimentalis* 60.4 (2000), pp. 495–501.
164. Santiago Arroyo et al. "Functional significance of the mu rhythm of human cortex: an electrophysiologic study with subdural electrodes". In: *Electroencephalography and clinical Neurophysiology* 87.3 (1993), pp. 76–87.
165. Klaas Enno Stephan. "The history of CoCoMac". In: *Neuroimage* 80 (2013), pp. 46–52.
166. Eric M Rouiller et al. "Transcallosal connections of the distal forelimb representations of the primary and supplementary motor cortical areas in macaque monkeys". In: *Experimental brain research* 102 (1994), pp. 227–243.
167. Giuseppe Luppino et al. "Corticocortical connections of area F3 (SMA-proper) and area F6 (pre-SMA) in the macaque monkey". In: *Journal of Comparative Neurology* 338.1 (1993), pp. 114–140.
168. Driss Boussaoud et al. "Callosal connections of dorsal versus ventral premotor areas in the macaque monkey: a multiple retrograde tracing study". In: *BMC neuroscience* 6 (2005), pp. 1–18.
169. Riitta Salmelin et al. "Functional segregation of movement-related rhythmic activity in the human brain". In: *Neuroimage* 2.4 (1995), pp. 237–243.
170. Michael T Jurkiewicz et al. "Post-movement beta rebound is generated in motor cortex: evidence from neuromagnetic recordings". In: *Neuroimage* 32.3 (2006), pp. 1281–1289.
171. Christian H Kasess et al. "The suppressive influence of SMA on M1 in motor imagery revealed by fMRI and dynamic causal modeling". In: *Neuroimage* 40.2 (2008), pp. 828–837.
172. Qing Gao, Xujun Duan, and Huafu Chen. "Evaluation of effective connectivity of motor areas during motor imagery and execution using conditional Granger causality". In: *Neuroimage* 54.2 (2011), pp. 1280–1288.
173. Huafu Chen et al. "Evaluation of the effective connectivity of supplementary motor areas during motor imagery using Granger causality mapping". In: *Neuroimage* 47.4 (2009), pp. 1844–1853.
174. AB Jenny. "Commissural projections of the cortical hand motor area in monkeys". In: *Journal of Comparative Neurology* 188.1 (1979), pp. 137–145.
175. George R Leichnetz. "Afferent and efferent connections of the dorsolateral precentral gyrus (area 4, hand/arm region) in the macaque monkey, with comparisons to area 8". In: *Journal of Comparative Neurology* 254.4 (1986), pp. 460–492.
176. B Marconi et al. "Callosal connections of dorso-lateral premotor cortex". In: *European Journal of Neuroscience* 18.4 (2003), pp. 775–788.

177. Deborah J Serrien, Richard B Ivry, and Stephan P Swinnen. “Dynamics of hemispheric specialization and integration in the context of motor control”. In: *Nature Reviews Neuroscience* 7.2 (2006), pp. 160–166.
178. Simone Valenti et al. “Assessing High-Order Interdependencies Through Static O-Information Measures Computed on Resting State fMRI Intrinsic Component Networks”. In: *International Conference on Image Analysis and Processing*. Springer. 2022, pp. 386–397.
179. Hao Ding et al. “The role of ipsilateral motor network in upper limb movement”. In: *Frontiers in Physiology* 14 (2023), p. 1199338.
180. A Ferbert et al. “Interhemispheric inhibition of the human motor cortex.” In: *The Journal of physiology* 453.1 (1992), pp. 525–546.
181. Eric M Wassermann et al. “Effects of transcranial magnetic stimulation on ipsilateral muscles”. In: *Neurology* 41.11 (1991), pp. 1795–1795.
182. Zafiris J Daskalakis et al. “The mechanisms of interhemispheric inhibition in the human motor cortex”. In: *The Journal of physiology* 543.1 (2002), pp. 317–326.
183. Neep Hazarika, Ah Chung Tsoi, and Alex A Sergejew. “Nonlinear considerations in EEG signal classification”. In: *IEEE Transactions on signal Processing* 45.4 (1997), pp. 829–836.
184. Cornelis J Stam. “Nonlinear dynamical analysis of EEG and MEG: review of an emerging field”. In: *Clinical neurophysiology* 116.10 (2005), pp. 2266–2301.
185. Jingyuan E Chen, Mikail Rubinov, and Catie Chang. “Methods and considerations for dynamic analysis of functional MR imaging data”. In: *Neuroimaging Clinics* 27.4 (2017), pp. 547–560.
186. Leonardo Novelli and Adeel Razi. “A mathematical perspective on edge-centric brain functional connectivity”. In: *Nature Communications* 13.1 (2022), p. 2693.

Chapter 7
Conclusions and Future Directions

In this book, we have drawn a pathway which connects univariate to bivariate to high-order network analysis, as well as static and dynamic approaches to the study of complex physiological systems, in the currently developing fields of Network Physiology and Network Neuroscience.

7.1 A Picture of the Complex Interactions in Physiological and Brain Networks of Multiple Nodes

To sum up our considerations, give me the chance to tell you a story about what you have read up to now.

Imagine to be an external observer looking at a big picture on a white wall, where two nodes of two different colors are depicted. ***Why do they have diverse colors?***: they have a different level of *information content*. The information content associated to the two nodes mainly depends on their intrinsic time and spectral features, but may vary over time due to internal/external drivers of the node dynamics. The *entropy* and *entropy rate* quantify this property in the case of random variables and processes, respectively; their computation is essential to retrieve basic knowledge on the *activity* of the single node in a variety of experimental conditions and physiological states. In static analysis, *activity* simply refers to the state of the system mapped by the node under examination, whereas in dynamic analysis, the *activity* is quantified as the current state of the system (at the present time), the latter dependent on the past states, and thus it takes into account the temporal statistical structure of the process. An exemplary reminder is the beat-to-beat time series of arterial compliance, which we characterized at a single-node-level in Sect. 5.1 [1]: this exploratory investigation led us to draw some preliminary conclusions about its dynamic nature in response to the postural stress. In that case, the utilization of our new measure of linear self-predictability [1] was crucial in identifying both time and frequency-specific patterns of self-dependencies in the process. Still, we do not know if the observed behaviors,

L. Sparacino, *Spectral Information Dynamics in Network Neuroscience and Physiology*, Studies in Computational Intelligence 1235,
https://doi.org/10.1007/978-3-032-05416-6_7

mainly highlighting the importance of low frequency fluctuations when the process has to cope with the physiological perturbations due to the orthostatic challenge, are truly due to self-effects or rather led by unobserved drivers of compliance variability. For this reason, let us move a bit forward.

If you look at the picture, two directional links of different widths have appeared connecting the two nodes. ***Why do they have diverse widths?*** They are associated with diverse underlying physiological mechanisms which retain distinct intensity and significance. Here, you can investigate the *mutual information shared* between the nodes as the amount of information obtained about one node by observing the other, as well as *causality patterns* between the two nodes, thus identifying the emergence of a preferred direction of interaction leading the dynamics of the couple. Cardiorespiratory, cardiovascular and cerebrovascular closed-loop interactions have been widely investigated through classical dynamic measures of, e.g., *mutual information rate* and *Granger causality* (see, e.g., Sects. 5.4 [2], 5.5 [3] and 5.6 [4]), where the well-known physiological mechanisms of respiratory sinus arrhythmia, cardiac baroreflex and cerebral autoregulation have been explored in patho-physiological states and in response to different external stressors). Exemplary applications to cardiovascular, cardiorespiratory and cerebrovascular closed-loop systems evidenced well-known behaviors including, e.g., the increase of the causal interaction along the baroreflex in healthy subjects and the drop of cardiovascular interactions in patients prone to develop postural-related syncope at the low frequencies, the increase of the information transferred along the pressure-to-flow link with the postural stress, suggesting a progressive weakening of cerebral autoregulation in syncope subjects, as well as the tilt-induced decrease of the rate of information shared between cardiac and respiratory variables, reflecting a decrease of the strength of cardiorespiratory interactions due to sympathetic activation and vagal withdrawal. Nevertheless, the white wall you are looking at, where the two nodes and the two links are depicted with their different properties, is slightly becoming more crowded. Indeed, despite unobserved confounders of physiological closed-loop systems have always been neglected in bivariate analysis, they still play a key role in determining the observed emergent behaviors (see, e.g., the role of arterial carbon dioxide in guiding cerebral blood flow variability independently of blood pressure changes [4–6], or the respiratory-related fluctuations of arterial pressure transmitted to the heart via the baroreflex [7]). Moreover, in the context of static/dynamic analyses of fMRI/EEG brain connectivity mostly based on bivariate measures of (directional) coupling such as *mutual information* or *directed coherence*, respectively, it has been widely shown how, although pairwise approaches are highly effective, computationally convenient and of more intuitive understanding (see, e.g., the application to fMRI BOLD signals in Sect. 6.1.1), such measures cannot fully capture the interplay among the multiple units of a complex neural system, as also demonstrated in the theoretical example of Sect. 3.3.1.

Therefore, let the picture evolve to denser shades together with your awareness that a complete description of a complex physiological network should require the involvement of all the variables participating in information sharing circuits within the network itself. This means that new nodes appear on the wall, and that links become hyperlinks grouping from two to N distinct nodes. In the context of Network Physiology, analyses exploiting hierarchical interaction measures have highlighted the redundant nature of cardiovascular, cardiorespiratory and cerebrovascular interactions, and documented the relevance of separating low and high frequency contributions to elicit the driving role of respiration on cardiovascular and cerebrovascular variables; this allowed to confirm that HOIs can have different nature for different rhythms because synergistic and redundant behaviors generally alternate in different bands of the frequency spectrum. On the other hand, the application of OI-based high-order information measures to resting-state BOLD signals has demonstrated that high-order synergies represent a kind of shadow structure emerging from resting-state brain activity and missed by bivariate functional connectivity approaches [8]. The utilization of OI gradients evidenced peculiar connectivity networks contributing to redundancy and synergy in the large-scale organization of fMRI networks, and confirmed that gradients of increasing order tend to highlight less redundant/more synergistic interactions thus acquiring a key role in OI-based high-order network analyses [9]. Further, exemplary applications of dynamic hierarchical information measures to EEG signals (see, e.g., [10–12]) documented an overall prevalence of redundancy for high-order brain interactions, as well as the emergence of synergistic circuits not retrievable from a pairwise analysis.

7.2 Simultaneous Evaluation of Information-Theoretic Measures at Different Hierarchical Levels

At this point, you may think that the picture is too complicated to be fully understood, and that the task of a complete inclusion is nearly impossible, and you may be right. Indeed, a comprehensive characterization of the intricate dynamics taking place in different patho-physiological conditions and in response to a variety of internal/external stressors is quite burdensome due to practical issues related to, e.g., the lack of non-invasive assessment techniques of meaningful drivers of cardiovascular and cerebrovascular interactions (such as the activity exerted by the autonomy nervous system on cardiac, vascular and respiratory variables) or computational load, significantly increasing with the number of network nodes. Nonetheless, in the past years many research studies have gracefully moved forward towards the characterization of physiological sub-systems involving variables of clinical relevance in the field of Network Physiology, such as cardiac, vascular, respiratory and cerebral variables [13, 14]. Further, recognizing and modeling high-order functional structures, characterized by statistical interactions involving more than two network units, has become a crucial and evolving area of Network Neuroscience [15, 16], where high-order interactions have been suggested as fundamental components of complexity and functional integration in brain networks [17], and proposed to be linked to emergent mental phenomena and consciousness [18].

In this frame, our newly developed framework, based on a comprehensive assessment of single-node, pairwise and especially high-order effects in complex systems of multiple nodes [12] acquires a meaningful relevance, laying a smooth groundwork from which future studies on complex physiological and brain interactions involving more than two sub-systems may have the opportunity to draw new and insightful conclusions on the nature of the emergent interaction patterns. We remark that, in spite of the doubtless usefulness of such information measures for the separate assessment of self-predictability, causality, autonomy and high-order relationships among the data in networks of multiple nodes, the innovative potential of our framework mainly resides in the possibility to perform a *simultaneous* evaluation of these nontrivial relationships at different hierarchical levels involving different subgroups of nodes. Then, comparing the resulting patterns emerging at the multiple investigated levels would shed light on the complementarities of the information measures, as well as on the nature of the explored networks intended as structural pairwise links and/or redundant/synergistic informational circuits. For instance, we showed how dynamic high-order measures such as the O-information rate can correctly capture the synergistic and/or redundant characters of interaction pathways involving multiple processes with given structures (see, e.g., panels *f, g* of Figs. 4.8 and 4.9 in Sect. 4.3.3.1), as well as that pairwise approaches such as the mutual information rate may fail depending on the underlying structural links among the processes (panels *c, e* of Fig. 4.9 in Sect. 4.3.3.1).

Crucially, the expansion of the single-node (entropy rate), pairwise (mutual information rate) and high-order (O-information rate and its gradient) measures in the frequency domain allows to provide a spectral representation of the information processed in the analyzed network, as well as to focus on specific frequency bands related to oscillations with physiological meaning. This acquires a remarkable emphasis when dealing with cardiovascular, cerebrovascular, respiratory and neural signals, rich of oscillatory content located at different scales. The possibility to identify, e.g., frequency-specific redundant/synergistic informational circuits may allow to interpret the intricate physiological mechanisms underlying diverse behaviors during the resting state or task accomplishment, such as motor execution tasks (see, e.g., [10] and Sect. 6.2.5), in response to a given stressor (see, e.g., [10, 13, 19]), in the presence of pathological states affecting, e.g., brain regions, such as subacute ischemic stroke (see, e.g., [11] and Sect. 6.2.5).

7.3 Future Perspectives

The application of our hierarchical high-order network analysis to physiological signals, including cardiovascular, cardiorespiratory and cerebrovascular time series, as well as to neural networks probed by fMRI/EEG, offers a powerful framework for unraveling the complex interdependencies underlying human physiology and brain dynamics, paving the way for more comprehensive assessments of healthy and diseased conditions. Future directions will likely focus on enhancing computational methods to capture non-linear and multiscale interactions across different physiological systems and brain regions. We remark that the approach proposed in this book is highly robust but limited to linear interactions, whereas current model-free solutions are impractical as we move from univariate to bivariate and high-order analysis due to the issue of curse of dimensionality. Forthcoming works might seek a compromise between fully parametric models and fully model-free approaches (e.g., by investigating the role of local prediction models). Furthermore, it is worth noting that high-order measures evidencing the balance between redundancy and synergy in the analyzed network (such as the OI/OIR) do not put in evidence multiplets of variables which are both redundant and synergistic with equal strength, and often obscure the detection of HOIs. Expansions to dynamic processes of well-known approaches which separately evaluate redundancy and synergy, such as the partial information decomposition, are thus envisaged. In this book, we mentioned the significant role that our newly developed partial information rate decomposition framework applied to random processes has in accurately disentangling the network structure while considering the temporal statistical structure of the processes. Future applications to networks of physiological time series should focus on these aspects to better elicit unique rates of information shared and/or transferred among the nodes of the network. Last but not least, the incorporation of wearable technologies and continuous monitoring systems would enable real-time analysis and personalized health tracking; integrating these measures with multimodal data, such as combining fMRI with electrophysiological recordings or behavioral assessments, could offer deeper insights into brain function in a variety of patho-physiological states. Applying these techniques in clinical settings would have a key role for early detection of disease, patient stratification, and individualized treatment plans. Advances in machine learning are expected to play a major role in automating these analyses, ultimately improving our ability to predict and manage physiological dysfunction.

References

1. Laura Sparacino et al. "A method to assess linear self-predictability of physiologic processes in the frequency domain: application to beat-to-beat variability of arterial compliance". In: *Frontiers in Network Physiology* 4 (2024), p. 1346424.
2. Riccardo Pernice et al. "Assessment of Cardiorespiratory Interactions During Spontaneous and Controlled Breathing: Linear Parametric Analysis". In: *2022 12th Conference of the European Study Group on Cardiovascular Oscillations (ESGCO)*. IEEE. 2022, pp. 1–2.
3. Riccardo Pernice et al. "Spectral decomposition of cerebrovascular and cardiovascular interactions in patients prone to postural syncope and healthy controls". In: *Autonomic Neuroscience* (2022), p. 103021.
4. Laura Sparacino et al. "A method to assess granger causality, isolation and autonomy in the time and frequency domains: theory and application to cerebrovascular variability". In: *IEEE Transactions on Biomedical Engineering* (2023).
5. Simone Cencetti, Gabriele Bandinelli, and Alfonso Lagi. "Effect of PCO2 changes induced by head-upright tilt on transcranial Doppler recordings". In: *Stroke* 28.6 (1997), pp. 1195–1197.
6. RB Panerai et al. "Effect of CO2 on dynamic cerebral autoregulation measurement". In: *Physiological measurement* 20.3 (1999), p. 265.

7. Luciano Bernardi et al. “Arterial baroreceptors as determinants of 0.1 Hz and respiration-related changes in blood pressure and heart rate spectra”. In: *Frontiers of blood pressure and heart rate analysis*. IOS Press, 1997, pp. 241–252.
8. Laura Sparacino et al. “Statistical Approaches to Identify Pairwise and High-Order Brain Functional Connectivity Signatures on a Single-Subject Basis”. In: *Life* 13.10 (2023), p. 2075.
9. Tomas Scagliarini et al. “Gradients of O-information highlight synergy and redundancy in physiological applications”. In: *Frontiers in Network Physiology* 3 (2024), p. 1335808.
10. Laura Sparacino et al. “Quantifying High-Order Interactions in Complex Physiological Networks: A Frequency-Specific Approach”. In: *International Conference on Complex Networks and Their Applications*. Springer. 2022, pp. 301–309.
11. I Pirovano et al. “Rehabilitation modulates high-order interactions among large-scale brain networks in subacute stroke”. In: *IEEE Transactions on Neural Systems and Rehabilitation Engineering* 31 (2023), pp. 4549–4560.
12. Laura Sparacino et al. “Measuring hierarchically-organized interactions in dynamic networks through spectral entropy rates: theory, estimation, and illustrative application to physiological networks”. In: *Neurocomputing* 630 (2025), p. 129675.
13. Luca Faes et al. “A new framework for the time-and frequency-domain assessment of high-order interactions in networks of random processes”. In: *IEEE Transactions on Signal Processing* 70 (2022), pp. 5766–5777.
14. Plamen Ch Ivanov. “The new field of network physiology: building the human physiolome”. In: *Frontiers in Network Physiology* 1 (2021), p. 711778.
15. Danielle S Bassett and Olaf Sporns. “Network neuroscience”. In: *Nature neuroscience* 20.3 (2017), pp. 353–364.
16. Federico Battiston et al. “Networks beyond pairwise interactions: Structure and dynamics”. In: *Physics reports* 874 (2020), pp. 1–92.
17. Giulio Tononi, Olaf Sporns, and Gerald M Edelman. “A measure for brain complexity: relating functional segregation and integration in the nervous system.” In: *Proceedings of the National Academy of Sciences* 91.11 (1994), pp. 5033–5037.
18. Andrea I Luppi et al. “What it is like to be a bit: an integrated information decomposition account of emergent mental phenomena”. In: *Neuroscience of Consciousness* 2021.2 (2021), niab027.
19. Gorana Mijatovic et al. “Assessing High-Order Links in Cardiovascular and Respiratory Networks via Static and Dynamic Information Measures”. In: *IEEE Open Journal of Engineering in Medicine and Biology* (2024).

Appendix A

This chapter presents the use of surrogate (Surrogate Data Analysis) and bootstrap (Bootstrap Data Analysis) data analyses to statistically validate the time and frequency domain interaction measures defined in Chaps. 3 and 4 for static and dynamic networks, respectively, in the linear parametric framework. Specifically, while surrogate methods are generally exploited to validate single-node (The Dynamics of the Single Node) and pairwise activities in bivariate (Coupled Dynamics Between Pairs of Nodes, Autonomous Dynamics in Bivariate Processes) or multivariate (Link-Specific Analysis of HOIs) settings, bootstrap approaches are well suited to the statistical validation of high-order measures characterizing the interactions between the multiple nodes of a complex network (An Approach to the Statistical Assessment of High-Order Interactions). Moreover, we also employed bootstrap data to assess the statistical significance of pole-specific measures of LSP as discussed in Spectral Measures of Self-predictability of Single Network Nodes.

Validation is performed at the level of individual realizations of the observed variables ($\{V_1, \ldots, V_M\}$ and $\{Y_1, \ldots, Y_Q\}$ for static and dynamic systems, respectively) obtained in the form of the sets of time series $\mathbf{v}_m = \{v_m(1), \ldots, v_m(L)\}$, $\mathbf{y}_q = \{y_q(1), \ldots, y_q(L)\}$, where $m = 1, \ldots, M$; $q = 1, \ldots, Q$ and L is the length of the time series.

Surrogate Data Analysis

It is common practice to statistically validate the estimated metrics of univariate and bivariate interactions involving one or two observed network nodes, which corresponds to assess whether the investigated (blocks of) time series own significant self-dependencies or are significantly coupled, i.e., whether the estimated value of autonomy or pairwise connectivity is significantly nonzero. Originally, the method

L. Sparacino, *Spectral Information Dynamics in Network Neuroscience and Physiology*, Studies in Computational Intelligence 1235,
https://doi.org/10.1007/978-3-032-05416-6

of surrogate data was proposed to investigate the existence of non-linear dynamics in time series [1, 2] but later it was exploited to test the significance of coupling measures in EEG recordings [3]. Generally speaking, due to practical estimation problems, nonzero values of the estimated connectivity index can occur even in absence of a real coupling between the two considered series. To face this issue, the statistical significance of a given coupling measure has been typically assessed by estimating its distribution and comparing it to a given arbitrary threshold. However, a rigorous and more powerful method would consist in defining a threshold level on the basis of statistical criteria derived from the sampling (theoretical or empirical) distribution of the used estimator. Theoretical approaches were largely used to assess the statistical significance of the coherence [4] and directed coherence [5] estimators [6], but they present some limitations which cannot be neglected in real applications [7]. Therefore, the empirical distribution of the considered index for the estimation of a threshold level has been used in place of theoretical approaches [8], commonly obtained by exploiting the method of surrogate data. According to this approach, the coupling index is computed over a set of surrogate time series, which are derived from the original series by a procedure mimicking their properties, but destroying their coupling. The confidence interval (CI) of the empirical distribution is then computed under the null hypothesis of full uncoupling between the time series; the $100(1-\alpha)$th percentile of the distribution (which represents the threshold value) is then compared with the observed value, and the null hypothesis is accepted or rejected at the α significance level depending on the position of the observed value with respect to the threshold [9]. Indeed, if the index assessed over the original series is above the threshold, the null hypothesis is rejected with type I error probability below α. The same concepts can be exploited to assess the statistical significance of measures of self-dependencies, characterizing the information content of a given time series.

In the past decades, different algorithms have been proposed to generate surrogate time series sharing some given properties with the original but being uncoupled, to validate measures of coupling, causality and self-dependencies. In the following, we will discuss these approaches in contexts where the observed network is composed of one or two nodes, which allow the statistical validation of entropy rate and time domain self-predictability measures (The Dynamics of the Single Node), as well as mutual information rates, transfer entropies (Coupled Dynamics Between Pairs of Nodes), isolation and autonomy measures (Autonomous Dynamics in Bivariate Processes). Moreover, the case of link-specific analysis of HOIs will be discussed (Link-Specific Analysis of HOIs).

The Dynamics of the Single Node

Let us assume to observe only one network unit. Measures of single-node activity, e.g., the ER (4.1) or the LSP (4.3) for dynamic processes, can be statistically validated through a simple procedure described below.

Randomly Shuffled (RS) Surrogates. They are realizations of i.i.d. stochastic processes with the same mean, variance and probability distribution as the original series, generated by randomly permuting in temporal order the samples of the original series [10]. This procedure destroys the autocorrelation function.

Statistical validation of the time domain ER [11] and LSP [12] measures is performed by generating *RS* surrogates, according to the null hypotheses of fully unpredictable process without temporal statistical structure, and absence of self-dependencies within the investigated process, respectively. The procedure is repeated N_s times to obtain the surrogate series $\mathbf{y}_q^s$ ($q \in 1, \ldots, Q$; $s = 1, \ldots, N_s$). The time domain ER and LSP can be then estimated on each surrogate, or block of surrogates if the processes in $\mathbf{Y}$ can be grouped into M blocks $\{\mathbf{X}_1, \ldots, \mathbf{X}_M\}$, yielding the surrogate distributions from which the significance thresholds are derived taking the $100(\alpha)$th and $100(1-\alpha)$th percentiles, respectively, where α is the prescribed significance level. The original time domain ER and LSP values are deemed as statistically significant if they stand below and above the corresponding thresholds, respectively. As regards the spectral counterparts, the spectral ER profiles (4.5) are estimated on each surrogate and integrated to get surrogate ER values in specific frequency bands, yielding the surrogate distributions from which the significance thresholds are derived taking the $100(\frac{\alpha}{2})$th and $100(1-\frac{\alpha}{2})$th percentiles [11]. The original frequency domain ER value, i.e., the value obtained from integration of the spectral ER profile within a given frequency band, is deemed as statistically (non) significant if it stands (below) above the (lower) higher threshold.

Validating the spectral LSP (4.9) is more challenging. As far as we know, no techniques exist which are able to focus on the statistical validation of a single oscillation with a given frequency without altering other spectral patterns of the investigated process. For this reason, in [12] we proposed an empirical approach based on bootstrap data analysis [13] which will be detailed in Bootstrap Data Analysis.

A theoretical example showing the use of *RS* surrogates to statistically validate time domain LSP measures can be found in Sect. 4.1.3, while applications of the proposed method to physiological variables are shown in Sect. 5.1, where the LSP is tested, and Sects. 6.2.4 and 6.2.5, where time domain and spectral ER measures are validated.

Coupled Dynamics Between Pairs of Nodes

When two nodes are observed, more sophisticated approaches are needed to validate the computed measures of pairwise interactions between the nodes.

Iterative Amplitude Adjusted FT (iAAFT) surrogates. This method represents an advancement over the *FT* method [1], which exploits *phase randomized* surrogates, i.e., realizations of linear stochastic processes with the same power spectra as the original series, obtained by a phase randomization procedure applied independently to each series. Similarly, *iAAFT* surrogates [2] are realizations of linear stochastic processes with the same autocorrelations and probability distributions as the original series, while the power spectra are the best approximation of the original ones according to the number of iterates. The procedure generates surrogate time series by computing the FT of the original series, substituting the Fourier phases with random numbers uniformly distributed between 0 and 2π, and then performing the inverse FT. To address the main limitation of the FT method, consisting in the distortion of the amplitude distribution when such distribution is not Gaussian, an iterative procedure is implemented, which alternately ensures that the surrogate series maintains both the same power spectrum and amplitude distribution as the original series.

iAAFT surrogate time series which preserve the individual linear correlation properties of two series but destroy any correlation between them have been exploited to statistically validate MI measures for static systems (see, e.g., [14], as well as MIR and TE measures for dynamic systems (see, e.g., [15]). In the latter case, the null hypothesis of full uncoupling is often used in directionality analysis as it is compatible with the absence of causal relation [16–22]. The procedure is repeated N_s times to obtain the set of surrogate series $\mathbf{v}_m^s$ ($m = 1, \ldots, M$; $s = 1, \ldots, N_s$, in the case of static analysis), or $\mathbf{y}_q^s$ ($q = 1, \ldots, Q$; $s = 1, \ldots, N_s$, in the case of dynamic analysis). The MI measures can be then estimated on each surrogate pair $\{\mathbf{v}_i^s, \mathbf{v}_j^s\}$ ($i, j = 1, \ldots, M, i \neq j$) thus yielding the surrogate distributions $I^s(\mathbf{v}_j; \mathbf{v}_i)$; similarly, the MIR and TE measures can be estimated on each surrogate pair $\{\mathbf{y}_i^s, \mathbf{y}_j^s\}$ ($i, j = 1, \ldots, Q, i \neq j$), or on pairs of blocks of surrogates if the processes in $\mathbf{Y}$ can be grouped into M blocks $\{\mathbf{X}_1, \ldots, \mathbf{X}_M\}$, yielding the surrogate distributions $I^s_{\mathbf{y}_j;\mathbf{y}_i}$, and $T^s_{\mathbf{y}_j \to \mathbf{y}_i}$, $T^s_{\mathbf{y}_j \to \mathbf{y}_i}$. From the the surrogate distributions, the significance thresholds are derived taking the $100(1-\alpha)$th percentiles. Finally, the original MI, MIR and TE values are deemed as statistically significant if they stand above the corresponding thresholds. Remarkably, the same procedure applies to both time and frequency domain MIR and TE values, the latter obtained by integrating the spectral MIR and TE profiles within specific frequency bands of interest. Moreover, we note that *iAAFT* surrogates are used for TD and GC measures as well, which are equivalent to MIR and TE measures up to a factor 2 under the hypothesis of Gaussianity [23, 24].

It is remarkable to highlight that surrogates preserving the power spectrum of the original series (*FT*, *iAAFT*) are recommended to avoid false coupling detections in the presence of oscillations occurring at nearby frequencies but due to different mechanisms, as may frequently happen with brain oscillations [9].

iAAFT surrogates have been exploited in our works (see, e.g., [15, 25, 26]) to assess the statistical significance of pairwise interaction measures (e.g., the MIR). Exemplary applications are presented in Sects. 5.4, 5.11, and 6.2.3.

Autonomous Dynamics in Bivariate Processes

In the context of a dynamic bivariate analysis where $\mathbf{Y} = [Y_1, Y_2]$ is the observed process, to test the statistical significance of the time and spectral domain GI (4.35, 4.36) and GA (4.21, 4.44) measures, a method using explicit model equations extracted from the data can be implemented, as we proposed in [27].

The method generates surrogates of the observed time series Y_1 and Y_2 according to the null hypothesis of absence of causal coupling from Y_1 to Y_2 to test the GI (H_1), or absence of internal dynamics within the process Y_2 to test the GA (H_2). Specifically, each driver series, say Y_1, is fitted with the ARX model (2.4a), while the target series, say Y_2, is fitted with the AR model (2.2) to test H_1 and with the X model (2.9) to test H_2. Then, in each case, pairs of surrogate time series are generated feeding the models with noise realizations obtained shuffling randomly the samples of the estimated residuals. N_s pairs of surrogate time series are obtained iterating this procedure, and the time domain and spectral measures of GI and GA are computed at each iteration. The significance of the measures, computed either in the time domain or integrating the spectral function over the spectral bands of interest, is assessed comparing the values obtained on the original time series with the CIs of the surrogate distribution computed with significance α. Specifically, the GI and GA are deemed as statistically significant if their value is respectively below the αth percentile of the GI distribution over surrogates generated under H_1, and above the $(100 - \frac{\alpha}{2})$th or below the $\frac{\alpha}{2}$th of the GA distribution over surrogates generated under H_2. A representative application to cerebrovascular variability is illustrated in Sect. 5.6.

Link-Specific Analysis of HOIs

Link-specific analysis of HOIs requires the assessment of the statistical significance of the iS and ciS measures, which has relevance for the identification of limit values of the B-index and for the corresponding inference of the network structure. As done in [26], such assessment can be performed using the method of surrogate data. Under the null hypothesis of absence of iS or absence of ciS, the values of iS and ciS computed from the original data are compared with the distribution of iS/ciS computed from the surrogate sets using a test based on percentiles, run with significance α; then, the null hypothesis is rejected (accepted) and the original iS/ciS measure is deemed as statistically significant (non-significant) if the original values of iS or ciS are larger (smaller) than the $(1 - \alpha)$th percentile of the corresponding surrogate distribution. Accordingly, when the estimates of iS and/or ciS are detected as non significant the B-index is set to 1, −1, or to NaN according to Fig. 3.4.

In the analysis of *static systems mapped by random variables*, the procedure described above is implemented using the MI or the cMI as discriminating statistic for the quantification of iS or ciS. In this case, realizations of the multiple observed random variables are available in the form of numeric sequences, and surrogate data are generated by independently shuffling in random order the sequences relevant to each variable, so as to make the surrogate variables independent while preserving their marginal distributions (*RS* surrogates). On the other hand, in the analysis of *dynamic systems mapped by random processes* whose realizations are multivariate time series, the MIR or the cMIR were used as discriminating statistic and surrogate time series were generated using the *iAAFT* procedure which preserves the individual properties of each time series while destroying the interactions among them.

Theoretical examples showing the use of *RS* and *iAAFT* surrogates to statistically validate static and dynamic link-specific measures of HOIs (i.e., the B-index and the B-index Rate, respectively) can be found in Sects. 3.3.2 and 4.3.3.4, respectively. An exemplary application to physiological variables is presented in Sect. 5.11.

Bootstrap Data Analysis

The bootstrap method [28] has been used to identify CIs for high-order interaction measures such as the OIR and the OIR gradient (and their correspondent static versions, i.e., the OI and OI gradient) and thus assess their statistical significance in a network of multiple interacting nodes [11, 14] (An Approach to the Statistical Assessment of High-Order Interactions). Moreover, we also exploited this method to assess the statistical significance of pole-specific measures of linear self-predictability, as in [12] (Spectral Measures of Self-predictability of Single Network Nodes).

In both cases, the *block bootstrap* data generation procedure [13], consisting in a *block resampling* of a given time series, can be followed to generate, starting from the time series $\mathbf{v}_m$ (static) or $\mathbf{y}_q$ (dynamic), N_b *bootstrap pseudo-series* $\mathbf{v}_m^b = \{v_m^b(1), \ldots, v_m^b(L)\}$ or $\mathbf{y}_q^b = \{y_q^b(1), \ldots, y_q^b(L)\}$ $(b = 1, \ldots, N_b)$, which maintain all the individual properties of the original time series, i.e., mean, variance and probability distribution. Block resampling of the observed time series is not the only approach to bootstrapping non-i.i.d. data. Almost immediately following Efron's paper [28], the *residual-based bootstrap* for linear regression was introduced and studied [29–32]. The residual-based bootstrap amounts to a i.i.d. bootstrap of the estimated residuals. If the assumed AR model holds true, then the above residual-based bootstrap works well for the sample mean and other more complicated statistics [13].

Spectral Measures of Self-predictability of Single Network Nodes

A method for the statistical validation of spectral LSP measures was proposed in [12], which exploits the residual-based bootstrap. Specifically, the bootstrap pseudo-series $\mathbf{y}_q^b = \{y_q^b(1), \ldots, y_q^b(L)\}$ $(b = 1, \ldots, N_b)$ are generated by feeding the AR model (2.2) identified on the original time series $\mathbf{y}_q$ with *bootstrap pseudo-residuals*. The procedure creates the bootstrap pseudo-residuals $\mathbf{u}_{y_q}^b = \{u_{y_q}^b(1), \ldots, u_{y_q}^b(L)\}$ by joining together $k = \frac{L}{b_s}$ non-overlapping blocks chosen randomly from the set $\{B_1, \ldots, B_k\}$, where b_s is the size of each block, $B_m = \{\mathbf{u}_{y_q}(l), \ldots, \mathbf{u}_{y_q}(l + b_s - 1)\}$

and l is chosen randomly from the set $\{1, \ldots, L - b_s + 1\}$. After generation of the bootstrap time series $\mathbf{y}_q^b$ from the original AR model coefficients $a_{y,k}^{(ar)}$, $k = 1, \ldots, p$, and the bootstrap pseudo-residuals $\mathbf{u}_{y_q}^b$, the time domain (4.3) and spectral (4.9) LSP profiles are recomputed from the new, full-size bootstrap series $\mathbf{y}_q^b$ to get the estimates $S_{\mathbf{y}_q}^b$ and $s_{\mathbf{y}_q}^{(\kappa)^b}(\bar{f})$, respectively, with $\kappa = 1, \ldots, K$; the spectral profiles are then integrated in the desired band F to get the estimates $s_{\mathbf{y}_q}^{(F)^b}$. The procedure is iterated for $b = 1, \ldots, N_b$ to construct bootstrap distributions. The significance threshold $s_{\mathbf{y}_q}^{(F)^\alpha}$ is derived taking the αth percentile of the distribution. In order to assess the existence of significant self-predictability in F, we exploit the fact that the spectral profile $s_{\mathbf{y}_q}^{(F)^b}(\bar{f})$ oscillates around the value $S_{\mathbf{y}_q}$, which does not vary with $\bar{f}$. Then, the core of the procedure lies in evaluating the degree of emergence of the peak of $s_{\mathbf{y}_q}^{(F)^b}(\bar{f})$ in F with respect to the mean value assumed by the same spectral profile if the oscillation is not present, i.e., $S_{\mathbf{y}_q}$. To this purpose, the original pole-specific spectral LSP value integrated in F, $s_{\mathbf{y}_q}^{(F)}$, is deemed as statistically significant if $s_{\mathbf{y}_q}^{(F)^\alpha} > \frac{1}{\Delta F}\int_F S_{\mathbf{y}_q}(\bar{f})d f$, where ΔF is the bandwidth and $S_{\mathbf{y}_q}(\bar{f})$ is the spectral profile of the original time domain LSP, equal to $S_{\mathbf{y}_q}\ \forall f$.

A theoretical example showing the use of residual-based bootstrap to statistically validate spectral LSP measures can be found in Sect. 4.1.3, while an application of the proposed method to beat-to-beat time series of arterial compliance is reported in Sect. 5.1.

An Approach to the Statistical Assessment of High-Order Interactions

A novel method based on bootstrap data generation for the statistical validation of HOI measures was first explored in [14] to assess the significance of OI measures computed on fMRI data, and lately exploited in [11] to check statistical significance of OIR measures computed on both physiological and neural networks. Specifically, the procedure creates the bootstrap pseudo-series $\mathbf{v}_m^b$ ($\mathbf{y}_q^b$) by joining together $k = \frac{L}{b_s}$ non-overlapping blocks chosen randomly from the set $\{B_1, \ldots, B_k\}$, where b_s is the size of each block, $B_m = \{\mathbf{v}_m(l), \ldots, \mathbf{v}_m(l + b_s - 1)\}$ ($B_m = \{\mathbf{y}_q(l), \ldots, \mathbf{y}_q(l + b_s - 1)\}$) and l is chosen randomly from the set $\{1, \ldots, L - b_s + 1\}$. Then, the OI (OIR) is recomputed at each order N from the new, full-size bootstrap series $\mathbf{v}_{i_1}^b, \ldots, \mathbf{v}_{i_N}^b$ ($\mathbf{y}_{i_1}^b, \ldots, \mathbf{y}_{i_N}^b$), with $i_1, \ldots, i_N \in \{1, \ldots, M\}$, $N \leq M$, to get the estimate $\Omega^b(\mathbf{v}^N)$ ($\Omega^b(\mathbf{y}^N)$). The procedure is iterated for $b = 1, \ldots, N_b$ to construct bootstrap distributions.

Confidence intervals of the bootstrap distributions can be exploited to check the statistical significance of the absolute OI(R) values and the OI(R) increments. Specifically, when a given OI bootstrap distribution comprises the zero threshold at the α significance level, i.e., if the zero value is below the $100(1 - \frac{\alpha}{2})$th and above the

$100(\frac{\alpha}{2})$th percentile of that distribution, the corresponding OI(R) measure is deemed as not statistically significant. Moreover, the proposed bootstrap method allows to check whether the OI(R) gradient (3.12, 4.52) due to the addition of a putative target $\mathbf{v}_i$ ($\mathbf{y}_i$) to a given lower-order group of variables $\mathbf{v}_{-i}^N$ ($\mathbf{y}_{-i}^N$) of order $N-1$, with $N = 4, \ldots, M$, is significant or not, i.e., if the OI value computed for $\mathbf{v}^N$ ($\mathbf{y}^N$) significantly differs from the same measure computed for $\mathbf{v}_{-i}^N$ ($\mathbf{y}_{-i}^N$). To do that for each order, the bootstrap distributions of the OI computed for all the multiplets of two consecutive orders N and $N-1$ can be exploited. Fixing the multiplet at order N, all the roots of that multiplet at the preceding order $N-1$, i.e., lower-order multiplets whose elements are all contained in the high-order multiplet, are identified. Then, for each root, the lower-order and the high-order bootstrap distributions are compared by means of the *parametric Student t-test for unpaired data*. Finally, the corresponding OI increment is deemed as significant when the difference between the two distributions is significant at the α significance level according to the statistical test. Remarkably, the same procedures applies to both time and frequency domain OIR values, the latter obtained by integrating the spectral OIR profiles within specific frequency bands.

Applications of the proposed methodology to physiological data can be found in Sect. 5.8, 6.1.1, 6.1.2 and 6.2.5.

Statistical Significance of the Difference Between Experimental Conditions

When the pairwise and high-order interaction measures (e.g., the MI/MIR, and OI/OIR) are computed on a *single-subject basis in two different experimental conditions*, the bootstrap distributions can be employed to assess the significance of the difference between the two conditions through a statistical test, as we did in [14]. To this end, the bootstrap data generation procedure is executed in both the analyzed experimental conditions, and the *parametric Student t-test for unpaired data* is then employed to assess the statistical significance of the difference between pairs of bootstrap distributions for a given measure.

Applications to physiological data can be found in Sects. 5.2 and 6.1.1, respectively.

References

1. James Theiler et al. "Testing for nonlinearity in time series: the method of surrogate data". In: *Physica D: Nonlinear Phenomena* 58.1–4 (1992), pp. 77–94.
2. Thomas Schreiber and Andreas Schmitz. "Improved surrogate data for nonlinearity tests". In: *Physical review letters* 77.4 (1996), p. 635.

3. Jean-Philippe Lachaux et al. "Measuring phase synchrony in brain signals". en. In: *Human Brain Mapping* 8.4 (1999), pp. 194–208. ISSN: 1097-0193.
4. Lambert H Koopmans. *The spectral analysis of time series*. Elsevier, 1995.
5. Michael Eichler. "On the evaluation of information flow in multivariate systems by the directed transfer function". In: *Biological cybernetics* 94 (2006), pp. 469–482.
6. Ilenia Toppi et al. "Testing the significance of connectivity networks: comparison of different assessing procedures". In: *IEEE Transactions on Biomedical Engineering* 63.12 (2016), pp. 2461–2473.
7. Alberto Porta and Luca Faes. "Wiener–Granger causality in network physiology with applications to cardiovascular control and neuroscience". In: *Proceedings of the IEEE* 104.2 (2015), pp. 282–309.
8. RE Challis and RI Kitney. "Biomedical signal processing (in four parts) Part 3 The power spectrum and coherence function". In: *Medical and Biological Engineering and Computing* 29 (1991), pp. 225–241.
9. Luca Faes et al. "Surrogate data analysis for assessing the significance of the coherence function". In: *IEEE transactions on biomedical engineering* 51.7 (2004), pp. 1156–1166.
10. Milan Palus. "Detecting phase synchronization in noisy systems". In: *Physics Letters A* 235.4 (1997), pp. 341–351.
11. Laura Sparacino et al. "Measuring hierarchically-organized interactions in dynamic networks through spectral entropy rates: theory, estimation, and illustrative application to physiological networks". In: *Neurocomputing* 630 (2025), p. 129675.
12. Laura Sparacino et al. "A method to assess linear self-predictability of physiologic processes in the frequency domain: application to beat-to-beat variability of arterial compliance". In: *Frontiers in Network Physiology* 4 (2024), p. 1346424.
13. Dimitris N Politis. "The impact of bootstrap methods on time series analysis". In: *Statistical Science* (2003), pp. 219–230.
14. Laura Sparacino et al. "Statistical Approaches to Identify Pairwise and High-Order Brain Functional Connectivity Signatures on a Single-Subject Basis". In: *Life* 13.10 (2023), p. 2075.
15. I Piovano et al. "Rehabilitation modulates high-order interactions among large-scale brain networks in subacute stroke". In: *IEEE Transactions on Neural Systems and Rehabilitation Engineering* 31 (2023), pp. 4549–4560.
16. Maciej Kaminski et al. "Evaluating causal relations in neural systems: Granger causality, directed transfer function and statistical assessment of significance". es. In: *Biol Cybern* 85.2 (2001), pp. 145–157. ISSN: 1432-0770. DOI: 10.1007/s004220000236.
17. Mario Chavez, Jacques Martinserie, and Michel Le Van Quyen. "Statistical assessments of nonlinear causality: application to epileptic EEG signals". In: *Journal of Neuroscience methods* 124.2 (2003), pp. 113–128.

18. Joseph R Lizier et al. "Multivariate information theoretic measures reveal directed information structure and task relevant changes in fMRI connectivity". In: *Journal of computational neuroscience* 30 (2011), pp. 85–107.
19. Martin Vernerika and Milan Palus. "Inferring the directionality of coupling with conditional mutual information". In: *Physical Review E* 77.2 (2008), p. 026224.
20. Bojan Murazza et al. "Interactions between cardiac, respiratory and EEG-δ oscillations in rats during anaesthesia". In: *The Journal of Physiology* 580.1 (2007), pp. 215–228.
21. Luca Faes, Giandomenico Nollo, and RI H Cross. "Assessment of Granger causality by nonlinear model identification: application to short-term cardiovascular variability". In: *Annals of biomedical engineering* 36 (2008), pp. 381–395.
22. Luca Faes, Alberto Porta, and Giandomenico Nollo. "Mutual nonlinear prediction as a tool to evaluate coupling strength and directionality in bivariate time series: comparison among different strategies based on k nearest neighbors". In: *Physical Review E* 78.2 (2008), p. 026201.
23. Adam B Barrett, Lionel Barnett, and Anil K Seth. "Multivariate Granger causality and generalized variance". In: *Physical Review E* 81.4 (2010), p. 041907.
24. Daniel Chicharro. "On the spectral formulation of Granger causality". In: *Biological cybernetics* 105 (2011), pp. 331–347.
25. Riccardo Pernice et al. "Assessment of Cardiorespiratory Interactions During Spontaneous and Controlled Breathing: Linear Parametric Analysis". In: *2022 12th Conference of the European Study Group on Cardiovascular Oscillations (ESGCO)*. IEEE. 2022, pp. 1–2.
26. Gorana Mijajlovic et al. "Assessing High-Order Links in Cardiovascular and Respiratory Networks via Static and Dynamic Information Measures". In: *IEEE Open Journal of Engineering in Medicine and Biology* (2024).
27. Laura Sparacino et al. "A method to assess granger causality, isolation and autonomy in the time and frequency domains: theory and application to cerebrovascular variability". In: *IEEE Transactions on Biomedical Engineering* (2023).
28. B Efron. "Bootstrap methods: another look at the jackknife". In: *The Annals of Statistics* 7 (1979), pp. 1–26.
29. David A Freedman. "Bootstrapping regression models". In: *The annals of statistics* 9.6 (1981), pp. 1218–1228.
30. David Freedman. "On bootstrapping two-stage least-squares estimates in stationary linear models". In: *The Annals of Statistics* 12.3 (1984), pp. 827–842.
31. Bradley Efron and Robert Tibshirani. "Bootstrap methods for standard errors, confidence intervals, and other measures of statistical accuracy". In: *Statistical science* (1986), pp. 54–75.
32. Bradley Efron and Robert J Tibshirani. *An introduction to the bootstrap*. 1993.

Appendix B

Extended Causal Modelling

The assessment of causality, intended as the influence that a driver process exerts on a given target, requires a proper handling of *instantaneous (zero-lag) effects*, i.e., effects from one series to another occurring within the same time lag [1]. In practical physiological time series modelling, instantaneous causality shows up whenever the time resolution of the measurements is lower than the time scale of the lagged causal influences occurring among the analysed processes. This non-delayed effect can arise due to non-physiological factors (e.g., unobserved confounders) or fast (i.e., within-beat) physiologically meaningful interactions [1, 2]. The importance of considering instantaneous effects in the analysis of physiological interactions, e.g., cardiovascular interactions, where zero-lag interdependencies are expected to contribute significantly to the baroreflex mechanism [2], and of cardiorespiratory interactions, where the information transfer from respiration to heart rate variability is expected to be negligible in the LF band [3], was previously documented [2, 4].

The classical AR models used to study causality do not incorporate instantaneous effects, which thus remain unexplained as correlations between the model residuals [5]. Several approaches have been proposed to account for instantaneous correlations in frequency domain measures, either using extended AR models that incorporate zero-lag effects after determining their direction [2, 6] or keeping them as undirected but including them in extended spectral causality measures [1, 7].

Herein, we show how these instantaneous effects can be accounted for by exploiting *extended AR models* incorporating them after determining their direction, generally known a priori thanks to, e.g., previous physiological knowledge [2, 6]. We start noting that the strictly causal VAR model introduced in Sect. 2.3.3 cannot account for instantaneous correlations. In fact, effects between the variables $Y_{i,n}$ and $Y_{j,n}$ $(i, j = 1, \dots, Q; i \neq j)$ are not described by any coefficient in (2.6), and are indeed translated into a correlation between the innovations, resulting in a non-diagonal covariance matrix $\Sigma_{\mathbf{U_Y}}^{(var)}$. Nevertheless, the bivariate process $\mathbf{Y}$ can also be described

L. Sparacino, *Spectral Information Dynamics in Network Neuroscience and Physiology*, Studies in Computational Intelligence 1235,
https://doi.org/10.1007/978-3-032-05416-6

including instantaneous effects into the interactions allowed by the model [2, 6], leading to the compact representation

$$\mathbf{Y}_n = \sum_{k=0}^{p} \mathbf{B}_{\mathbf{Y},k}^{(var)} \mathbf{Y}_{n-k} + \mathbf{W}_{\mathbf{Y},n}^{(var)}, \tag{7.1}$$

where the set of coefficients now includes those modeling instantaneous effects collected in the matrix $\mathbf{B}_{\mathbf{Y},0}^{(var)}$ and $\mathbf{W}_{\mathbf{Y},n}^{(var)}$ contains the Q innovation processes of the extended model. Since instantaneous correlations are now modeled through $\mathbf{B}_{\mathbf{Y},0}^{(var)}$, the innovations processes have a diagonal covariance matrix $\Lambda_{\mathbf{W_Y}}^{(var)} = \left[\mathbf{W}_{\mathbf{Y},n}^{(var)} \mathbf{W}_{\mathbf{Y},n}^{(var)\mathsf{T}}\right]$.

The procedure to identify the model (7.1) is as follows. Knowing $\mathbf{B}_{\mathbf{Y},0}^{(var)}$, it is possible to compute a so-called *mixing matrix* $\mathbf{L} = \left[\mathbf{I} - \mathbf{B}_{\mathbf{Y},0}^{(var)}\right]^{-1}$ and to derive the parameters of the extended model (7.1) from those of the strictly causal model (2.6) as $\mathbf{B}_{\mathbf{Y},k}^{(var)} = \mathbf{L}^{-1}\mathbf{A}_{\mathbf{Y},k}^{(var)}$, $\Lambda_{\mathbf{W_Y}}^{(var)} = \mathbf{L}^{-1}\Sigma_{\mathbf{U_Y}}^{(var)}(\mathbf{L}^{-1})^{\mathsf{T}}$ [2, 6]. The matrix of instantaneous effects is obtained solving the instantaneous model $\mathbf{U}_{\mathbf{Y},n}^{(var)} = \mathbf{L}\mathbf{W}_{\mathbf{Y},n}^{(var)}$ to get the mixing matrix $\mathbf{L}$; this task is typically performed imposing a causal order for the instantaneous effects and then following a permutation procedure that involves application of the Cholesky decomposition [2].

Here it is important to stress that this approach requires that instantaneous effects are imposed along a given direction (e.g., from $Y_{i,n}$ to $Y_{j,n}$, allowing $b_{Y_j Y_i} \neq 0$) and are neglected along the opposite direction (i.e., from $Y_{j,n}$ to $Y_{i,n}$, forcing $b_{Y_i Y_j} = 0$). This condition for identifiability of the extended model implies that the direction of zero-lag effects can be set in a plausible way. Conversely, the extended model (7.1) cannot be identified unequivocally, and the strictly causal model (2.6) should be used to model the process.

Remarkably, the extended representation including instantaneous effects allows to study the causal pairwise and high-order dependencies between the analyzed processes. The information-theoretic and predictability measures defined in Chap. 4, whose computation relies on the theoretical concepts of vector AR modelling detailed in Chap. 2, can be straightforwardly computed in the case of extended models by using (7.1) in place of (2.6). Some applicative examples are shown in Sects. 5.3 and 5.5.

References

1. D Nuzzi et al. "Extending the spectral decomposition of Granger causality to include instantaneous influences: application to the control mechanisms of heart rate variability". In: *Philosophical Transactions of the Royal Society A* 379.2212 (2021), p. 20200263.
2. Luca Faes et al. "A framework for assessing frequency domain causality in physiological time series with instantaneous effects". In: *Philosophical Transac-*

tions of the Royal Society A: Mathematical, Physical and Engineering Sciences 371.1997 (2013), p. 20110618.

3. Luca Faes, Giandomenico Nollo, and Alberto Porta. “Non-uniform multivariate embedding to assess the information transfer in cardiovascular and cardiorespiratory variability series”. In: *Computers in biology and medicine* 42.3 (2012), pp. 290–297.
4. L Schiatti et al. “Extended Granger causality: a new tool to identify the structure of physiological networks”. In: *Physiological measurement* 36.4 (2015), p. 827.
5. Luca Faes, Silvia Erla, and Giandomenico Nollo. “Measuring connectivity in linear multivariate processes: definitions, interpretation, and practical analysis”. In: *Computational and mathematical methods in medicine* 2012 (2012).
6. Riccardo Pernice et al. “Spectral decomposition of cerebrovascular and cardiovascular interactions in patients prone to postural syncope and healthy controls”. In: *Autonomic Neuroscience* (2022), p. 109021.
7. Luiz A Baccala and Koichi Sameshima. “Frequency domain repercussions of instantaneous Granger causality”. In: *Entropy* 23.8 (2021), p. 1037.

MIX
Papier aus verantwortungsvollen Quellen
Paper from responsible sources
FSC® C105338

If you have any concerns about our products,
you can contact us on
ProductSafety@springernature.com

In case Publisher is established outside the EU,
the EU authorized representative is:
Springer Nature Customer Service Center GmbH
Europaplatz 3, 69115 Heidelberg, Germany

Printed by Libri Plureos GmbH
in Hamburg, Germany